Essential Circuit Analysis using LTspice®

Farzin Asadi

Essential Circuit Analysis using LTspice®

 Springer

Farzin Asadi
Department of Electrical and Electronics Engineering
Maltepe University
Istanbul, Turkey

ISBN 978-3-031-09855-0   ISBN 978-3-031-09853-6   (eBook)
https://doi.org/10.1007/978-3-031-09853-6

This Springer imprint is published by the registered company Springer Nature Switzerland AG
The registered company address is: Gewerbestrasse 11, 6330 Cham, Switzerland

In loving memory of my father Moloud Asadi and my mother Khorshid Tahmasebi, always on my mind, forever in my heart.

Preface

A computer simulation is an attempt to model a real-life or hypothetical situation on a computer so that it can be studied to see how the system works. By changing variables in the simulation, predictions may be made about the behavior of the system. So, computer simulation is a tool to virtually investigate the behavior of the system under study.

Computer simulation has many applications in science, engineering, education, and even in entertainment. For instance, pilots use computer simulations to practice what they learned without any danger and loss of life.

A circuit simulator is a computer program which permits us to see the circuit behavior, that is, circuit voltages and currents, without making it. Use of circuit simulator is a cheap, efficient, and safe way to study the behavior of circuits. A circuit simulator even saves your time and energy. It permits you to test your ideas before you go wasting all that time building it with a breadboard or hardware just to find out it doesn't really work.

This book shows how a circuit can be analyzed in LTspice® environment. LTspice is a high-performance SPICE simulation software, schematic capture, and waveform viewer with enhancements and models for easing the simulation of analog circuits. LTspice was originally created by Linear Technology, which is now a part of Analog Devices. LTspice is the most widely distributed and used SPICE software in the industry.

Students of engineering (for instance, electrical, biomedical, mechatronics, and robotics, to name a few), engineers who work in industry, and anyone who wants to learn the art of circuit simulation with LTspice can benefit from this book.

This book contains 77 sample simulations. A brief summary of book chapters is given bellow:

Chapter 1 introduces the LTspice and shows how it can be used to analyze electric circuits. Students who take/took electric circuits I/II course can use this chapter as a reference to learn how to solve an electric circuit problem with the aid of computer. This chapter has 22 sample simulations.

Chapter 2 focuses on the simulation of electronic circuits (i.e., circuits which contain diode, transistor, ICs, etc.) with LTspice. Students who take/took electronic I/II course can use this chapter as a reference to learn how to analyze an electronic circuit with the aid of computer. This chapter has 31 sample simulations.

Chapter 3 focuses on the simulation of digital circuits with LTspice. Students who take/took digital design course can use this chapter as a reference to learn how to simulate a digital circuit using computer. This chapter has five sample simulations.

Chapter 4 focuses on the simulation of power electronics circuits with LTspice. Students who take/took power electronics/industrial electronics course can use this chapter as a reference to learn how to simulate a power electronic circuit with the aid of computer. This chapter has 19 sample simulations.

I hope this book will be useful to the readers, and I welcome comments on the book.

Istanbul, Turkey   Farzin Asadi

Contents

1 Simulation of Electric Circuits with LTspice®. 1
  1.1 Introduction 1
  1.2 Installation of LTspice 1
  1.3 Example 1: Simple Resistive Voltage Divider 7
  1.4 Example 2: .param Command 48
  1.5 Example 3: Potentiometer 50
  1.6 Example 4: Obtaining the DC Operating Point
       of the Circuit 52
  1.7 Example 5: DC Transfer Analysis 55
  1.8 Example 6: .ic Command. 60
  1.9 Example 7: RL Circuit Analysis 67
  1.10 Example 8: Switch Block. 69
  1.11 Example 9: Measurement of Phase Difference 78
  1.12 Example 10: Calculation of RMS and Average Values 82
  1.13 Example 11: .meas Command 89
  1.14 Example 12: Observing the Waveform 102
  1.15 Example 13: Calculation of Power Factor 112
  1.16 Example 14: Thevenin Equivalent Circuit. 116
  1.17 Example 15: Current-Dependent Voltage Source 123
  1.18 Example 16: Voltage-Dependent Current Source 126
  1.19 Example 17: Three-Phase Circuits. 129
  1.20 Example 18: .step Command. 137
  1.21 Example 19: Coupled Inductors. 145
  1.22 Example 20: Step Response of Circuits. 154
  1.23 Example 21: Impulse Response of the Circuit. 163
  1.24 Example 22: Exporting the Simulation Result
       into MATLAB 166
  1.25 Exercises 173
  References. 175

2 Simulation of Electronic Circuits with LTspice® 177
  2.1 Introduction 177
  2.2 Example 1: Transformer 177
  2.3 Example 2: Center Tap Transformer 188
  2.4 Example 3: Impedance Seen from Transformer 192
  2.5 Example 4: Input Impedance of Electric Circuits 195
  2.6 Example 5: Transfer Function of Linear Circuits 205

2.7   Example 6: DC Sweep Analysis 211
2.8   Example 7: I–V Characteristics of Zener Diode 215
2.9   Example 8: Common Emitter Amplifier 219
2.10   Example 9: FFT Analysis............................. 231
2.11   Example 10: .Four Command 239
2.12   Example 11: THD of Common Emitter Amplifier.......... 245
2.13   Example 12: Frequency Response of Common Emitter
       Amplifier 250
2.14   Example 13: Input Impedance of Common Emitter
       Amplifier 258
2.15   Example 14: Output Impedance of Common Emitter
       Amplifier 263
2.16   Example 15: Modeling a Custom Transistor
       with .model Command................................ 268
2.17   Example 16: Temperature Sweep 278
2.18   Example 17: Effect of Temperature on the Forward
       Voltage Drop of Diode............................... 284
2.19   Example 18: Noninverting op amp Amplifier 285
2.20   Example 19: Input Impedance of Noninverting
       op amp Amplifier.................................... 288
2.21   Example 20: Output Impedance of Noninverting
       op amp Amplifier.................................... 292
2.22   Example 21: Stability of op amp Amplifiers 294
2.23   Example 22: Addition of LM 741 op amp to LTspice 299
2.24   Example 23: Measurement of Common Mode Rejection
       Ratio (CMRR) of an op amp Difference Amplifier 317
2.25   Example 24: Measurement of CMRR for a Differential
       Pair Amplifier...................................... 323
2.26   Example 25: Differential Mode Input Impedance
       of Differential Pair.................................. 334
2.27   Example 26: Colpitts Oscillator........................ 340
2.28   Example 27: Optocoupler 344
2.29   Example 28: Astable Oscillator with NE 555 347
2.30   Example 29: Low Pass Filter.......................... 352
2.31   Example 30: High Pass Filter 357
2.32   Example 31: Band Pass Filter 358
2.33   Exercises 361
       References... 364

3   Simulation of Digital Circuits with LTspice® 365
3.1   Introduction 365
3.2   Example 1: Simulation of Logic Circuits 365
3.3   Example 2: Schmitt-Triggered Buffer Block.............. 374
3.4   Example 3: Flip Flop Blocks 377
3.5   Example 4: Counter Block............................ 382
3.6   Example 5: Two-Bit Binary Counter 386
3.7   Exercises 389
       References... 390

4   Simulation of Power Electronics Circuits with LTspice® 391
    4.1   Introduction 391
    4.2   Example 1: Buck Converter (I) 391
    4.3   Example 2: Buck Convert (II) 399
    4.4   Example 3: Making New Blocks 409
    4.5   Example 4: Operating Mode of DC–DC Converter 425
    4.6   Example 5: Efficiency of the Converter 427
    4.7   Example 6: Simulation of Circuits Containing LT IC's 431
    4.8   Example 7: Voltage Regulator Circuit 440
    4.9   Example 8: Measurement of Voltage Regulation 444
    4.10 Example 9: Dimmer Circuit 446
    4.11 Example 10: Single-Phase Half Wave Controlled Rectifier . . . 453
    4.12 Example 11: Single-Phase Full Wave Controlled
          Rectifier (I) 467
    4.13 Example 12: Single-Phase Full Wave Controlled
          Rectifier (II) 470
    4.14 Example 13: Three-Phase Controlled Rectifier 472
    4.15 Example 14: Harmonic Analysis of Rectifiers 490
    4.16 Example 15: Measurement of Power Factor
          for Rectifier Circuits 497
    4.17 Example 16: Single-Phase PWM Inverter 504
    4.18 Example 17: Three-Phase PWM Inverter 513
    4.19 Example 18: Harmonic Content of Three-Phase Inverter 524
    4.20 Example 19: Total Harmonic Distortion (THD)
          of Three-Phase Inverter 532
    4.21 Exercises 540
    References 543

Index 545

Simulation of Electric Circuits with LTspice®

1

1.1   Introduction

In this chapter, you will learn how to analyze electric circuits in LTspice. The theory behind the studied circuits can be found in any standard circuit theory text book [1–4]. It is a good idea to do some hand calculations for the circuits that are given and compare them with LTspice results.

1.2   Installation of LTspice

Installation of LTspice is quite easy. In order to install LTspice, search for "ltspice download" in Google (Fig. 1.1) and open the first result (Fig. 1.2).

Fig. 1.1 Searching for LTspice download page

Fig. 1.2 LTspice webpage

   Scroll down the screen until you see the download link (Fig. 1.3). Download the suitable version based on your operating system. After downloading the file, install it.

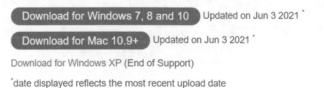

Fig. 1.3 Download link

The LTspice environment is shown in Fig. 1.4.

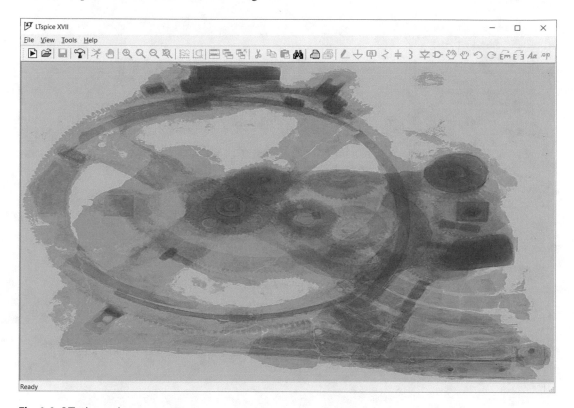

Fig. 1.4 LTspice environment

If you click the Help> Help Topics (Fig. 1.5), LTspice help appears on the screen (Fig. 1.6).

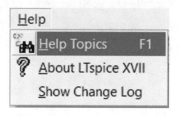

Fig. 1.5 Help menu

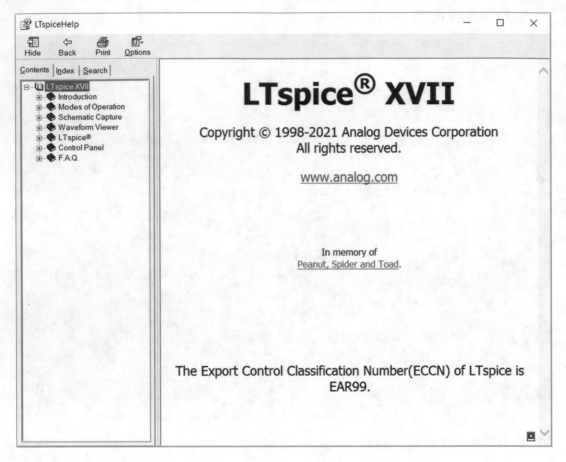

Fig. 1.6 LTspice Help window

The Dot Commands section of the help is a good reference to learn LTspice commands (Fig. 1.7).

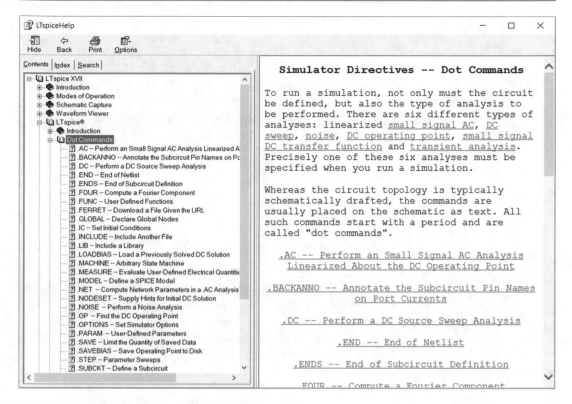

Fig. 1.7 Dot Commands section of LTspice Help

LTspice has an "examples" folder which contains many inspiring simulations. In order to open the "examples" folder, click the open icon (Fig. 1.8) and then go to the path that you installed LTspice (Fig. 1.9).

Fig. 1.8 Open icon

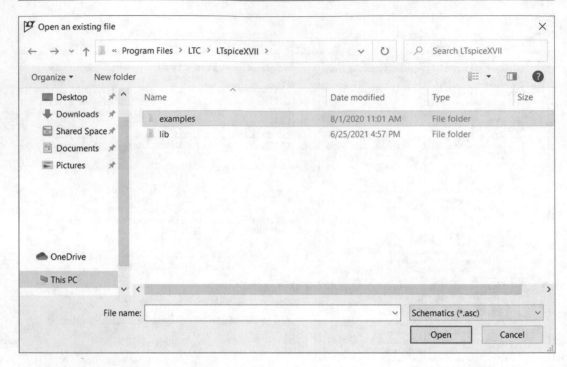

Fig. 1.9 Examples folder

Open the "examples" folder (Fig. 1.10).

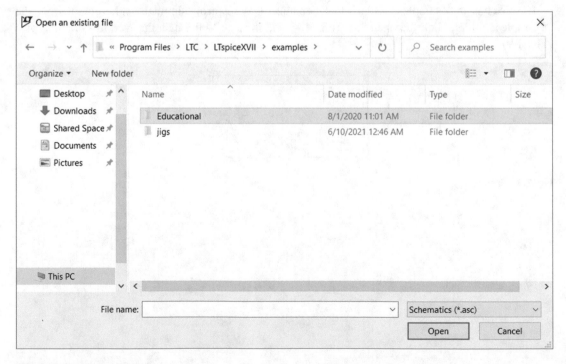

Fig. 1.10 Educational folder

Open the "Educational" folder (Fig. 1.11). Now you have access to the ready to use sample simulation files.

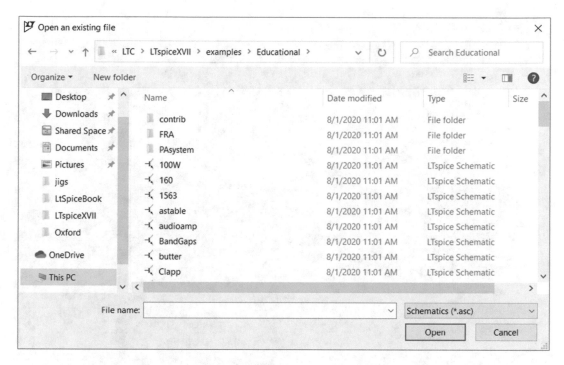

Fig. 1.11 Available sample simulation files

1.3   **Example 1: Simple Resistive Voltage Divider**

In this example, we want to simulate the behavior of following simple voltage divider circuit (Fig. 1.12) for [0, 100 ms] time interval.

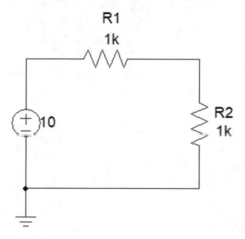

Fig. 1.12 Simple voltage divider circuit

Run the LTspice (Fig. 1.13).

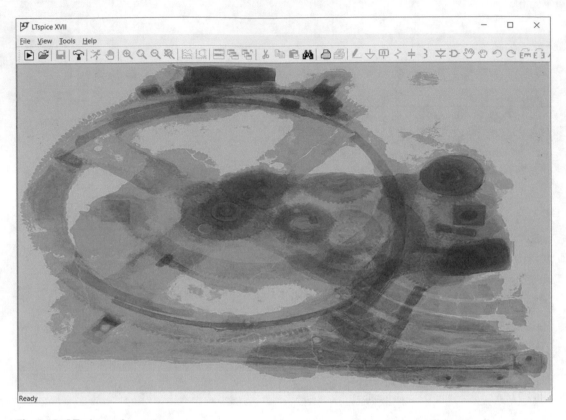

Fig. 1.13 LTspice environment

Click the New Schematic icon (Fig. 1.14). This opens a new schematic for you (Fig. 1.15).

Fig. 1.14 New Schematic icon

Fig. 1.15   New schematic file is opened

Click the resistor icon (Fig. 1.16). After clicking the resistor icon, the mouse pointer changes to a resistor. If you press the Ctrl+R, the resistor will be rotated. Press the Ctrl+R once and then click on the schematic to add a resistor to it (Fig. 1.17).

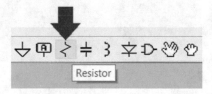

Fig. 1.16   Resistor icon

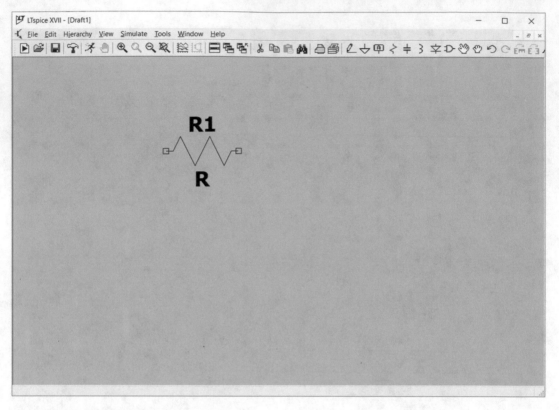

Fig. 1.17 Addition of a resistor to the schematic

Add another resistor to the schematic (Fig. 1.18), and after that, press the Esc key of your keyboard.

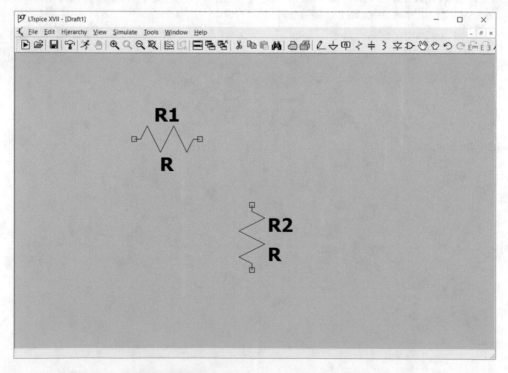

Fig. 1.18 Addition of second resistor to the schematic

Clicking the Resistor icon (Fig. 1.16) is not the only way to add resistors to the schematic. You can press the R key of your keyboard as well. The shortcut keys of LTspice are shown in Fig. 1.19. It is a good idea to memorize these shortcuts since they help you to draw the schematic easier and faster.

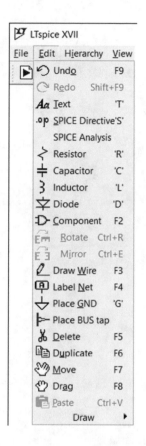

Fig. 1.19 Edit menu

Click on the Component icon (Fig. 1.20).

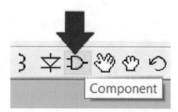

Fig. 1.20 Component icon

After clicking the Component icon, the Select Component Symbol window (Fig. 1.21) appears. Enter "voltage" in the component search box to find the voltage source block. The current source block can be found by searching for "current," as well (Fig. 1.22).

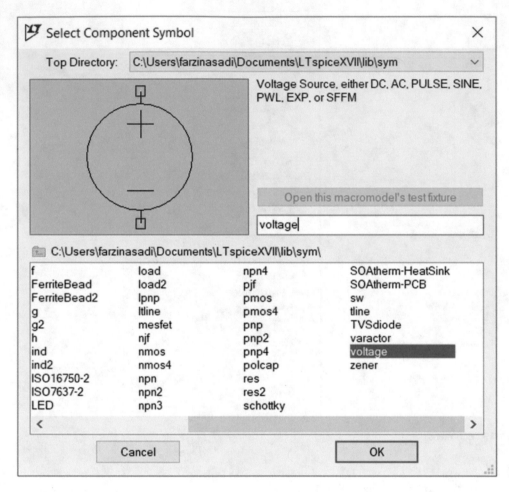

Fig. 1.21 Voltage source block

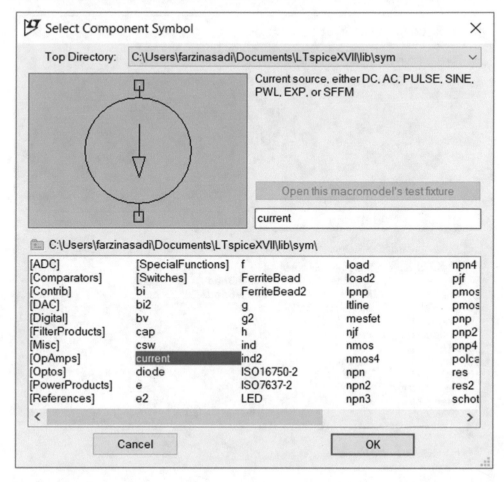

Fig. 1.22 Current source block

If you double click on the [ADC] (Fig. 1.23), the components inside it appear on the screen (Fig. 1.24). You can use the .. or Up One Level icon (Fig. 1.25) to return to the first page (Fig. 1.26).

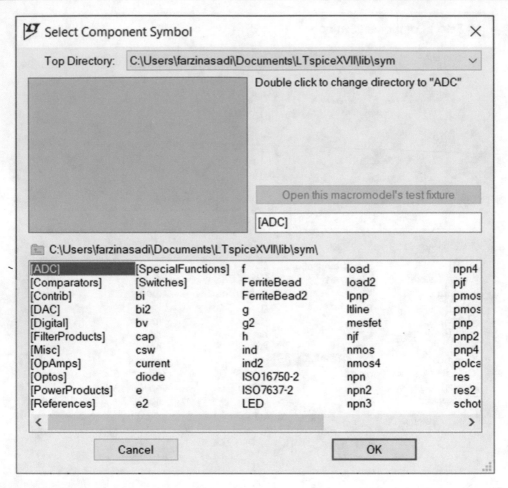

Fig. 1.23 ADC section

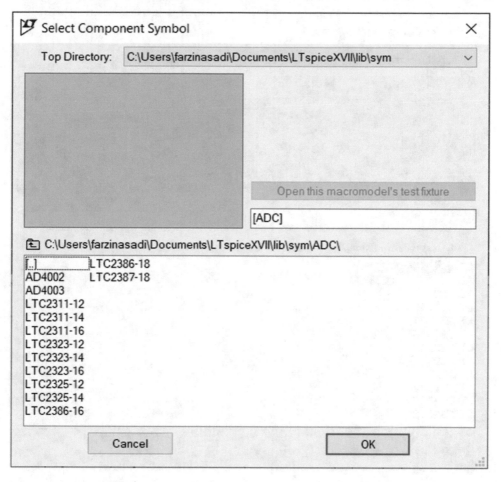

Fig. 1.24 Components inside the ADC section

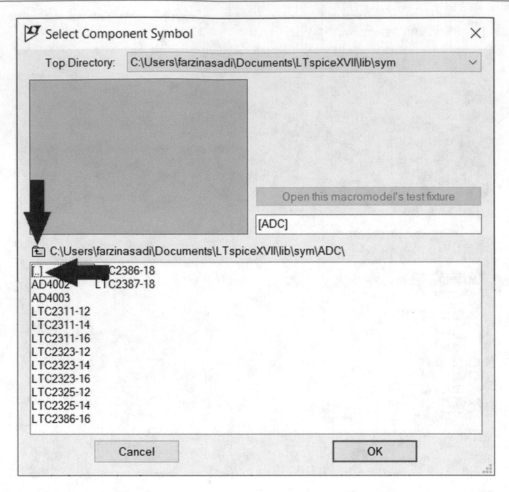

Fig. 1.25 Returning to the first page

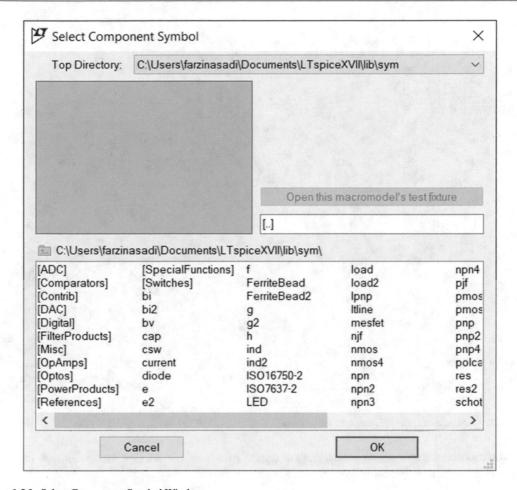

Fig. 1.26   Select Component Symbol Window

Add a voltage source block to the schematic (Fig. 1.27).

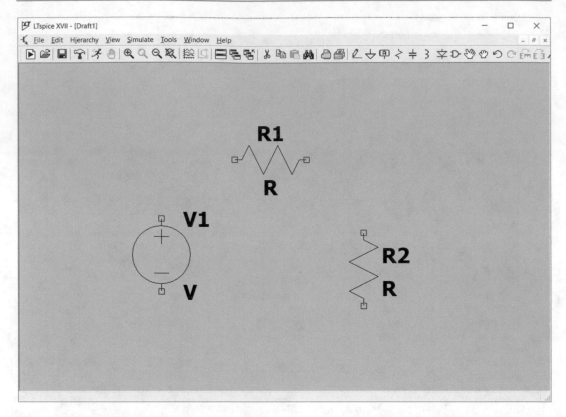

Fig. 1.27   Addition of a voltage source to the schematic

You can delete a component with the aid of scissor icon (Fig. 1.28). Click the scissor icon (or press F5) and then click on the component that needs to be removed.

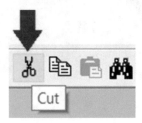

Fig. 1.28   Cut icon

You can duplicate a component with the aid of Copy icon (Fig. 1.29). Click the Copy icon (or press F6) and then click on the component that you want to make a copy of it. After clicking on the component, a copy of that component is attached to the mouse pointer. If you click on the schematic, the copied component will be attached to the schematic.

Fig. 1.29   Copy icon

Use the Wire icon (Fig. 1.30) to connect the components together (Fig. 1.31).

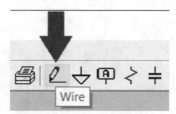

Fig. 1.30 Wire icon

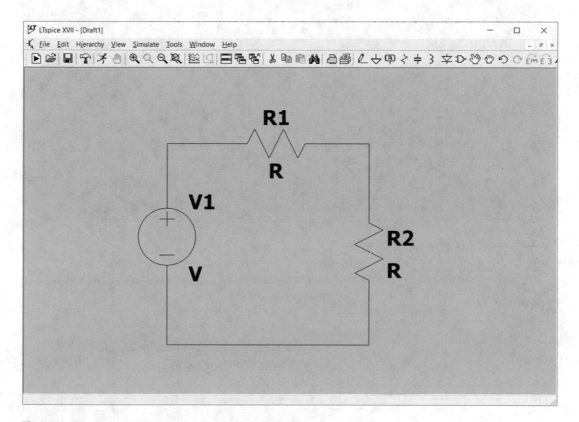

Fig. 1.31 Connecting the components together

Use the ground icon (Fig. 1.32) to add ground to your circuit (Fig. 1.33). If you try to run a schematic without ground, the error message shown in Fig. 1.34 appears.

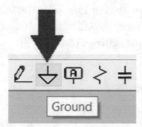

Fig. 1.32 Ground icon

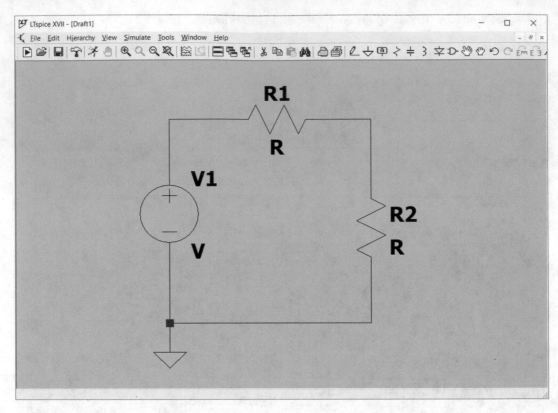

Fig. 1.33 Addition of ground to the schematic

Fig. 1.34 Error
message for a circuit
without ground

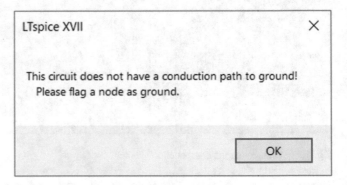

Right click on the voltage source V1 and enter 10 to the DC value[V] box (Fig. 1.35) and click the
OK button. If you click the Advanced button in Fig. 1.35, the window shown in Fig. 1.36 appears and
permits you to produce more complicated waveforms. In this example, we need a simple DC voltage
source, so there is no need to change the settings of Fig. 1.36.

Fig. 1.35 Voltage source settings

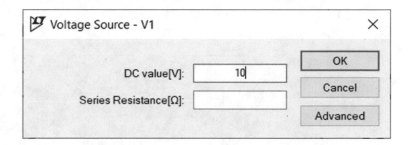

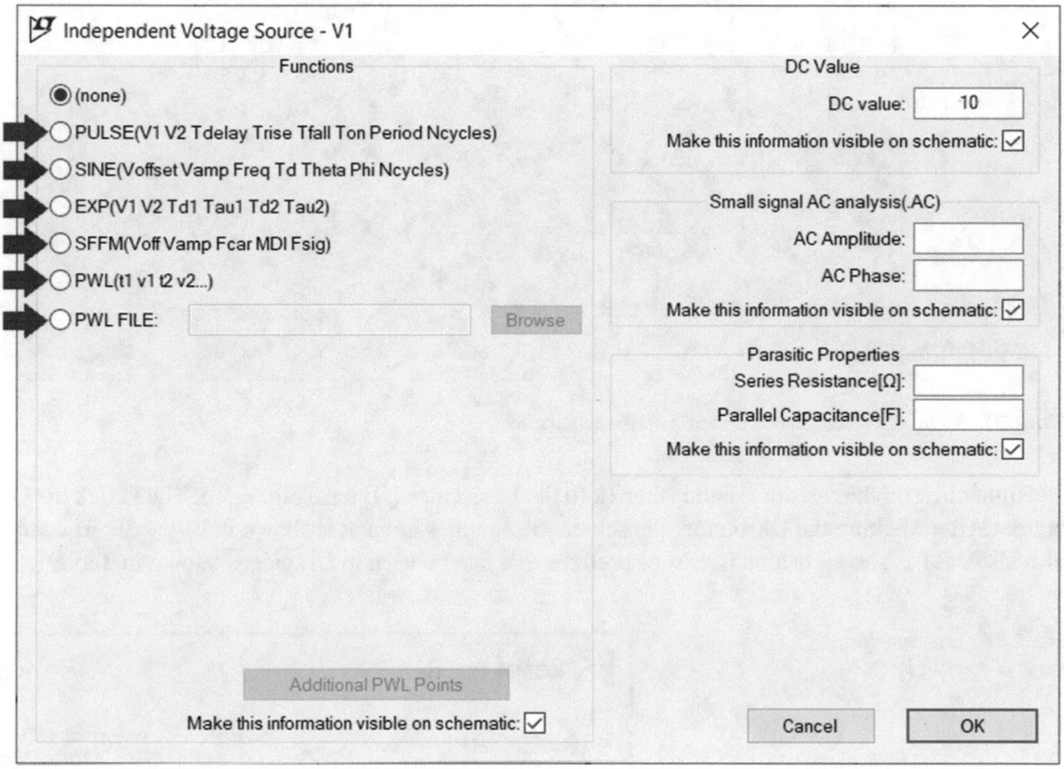

Fig. 1.36 Voltage source settings

After clicking the OK button of Fig. 1.36, the schematic changes to what is shown in Fig. 1.37. The value of DC voltage source is shown behind it.

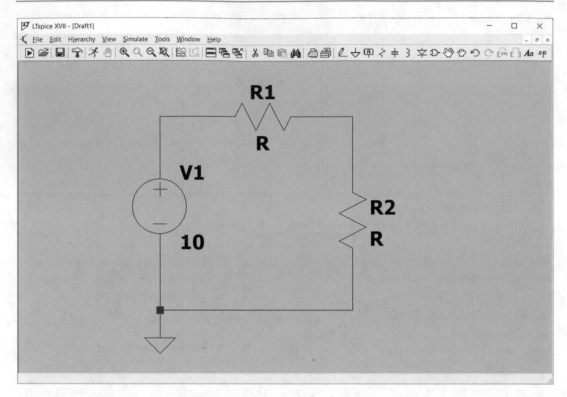

Fig. 1.37 Value of voltage source is shown on the schematic

Right click on the resistor R1 and enter 1k to the Resistance [Ω] box (Fig. 1.38). Then click the OK button. After clicking the OK button, the schematic changes to what is shown in Fig. 1.39. The value of resistor R1 is shown behind it. List of prefixes that can be used in LTspice is shown in Table 1.1.

Fig. 1.38 Entering the value of resistor R1

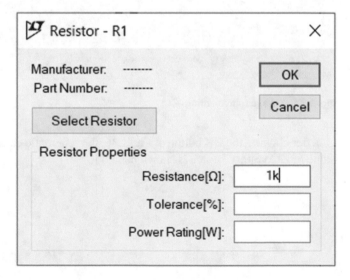

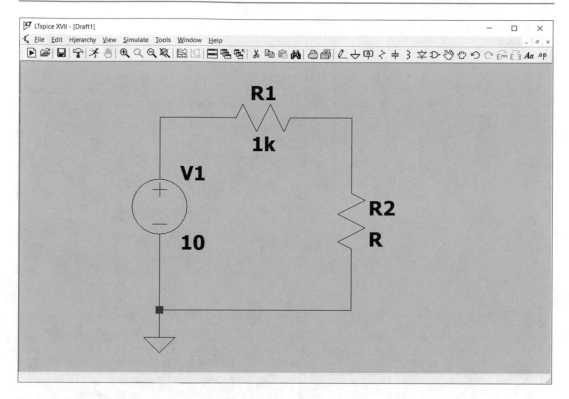

Fig. 1.39   Value of resistor R1 is shown on the schematic

Table 1.1   Prefixes that can be used in LTspice

Unit (prefix)	Unit	Multiple
T	Tera	10^{12}
G	Giga	10^{9}
Meg	Mega	10^{6}
k	Kilo	10^{3}
m	Milli	10^{-3}
u	Micro	10^{-6}
n	Nano	10^{-9}
p	Pico	10^{-12}
f	Femto	10^{-15}

Right click on the resistor R2 and enter 1k to the Resistance [Ω] box (Fig. 1.40). Then click the OK button. After clicking the OK button, the schematic changes to what is shown in Fig. 1.41. The value of resistor R2 is shown behind it.

Fig. 1.40 Entering the
value of resistor R2

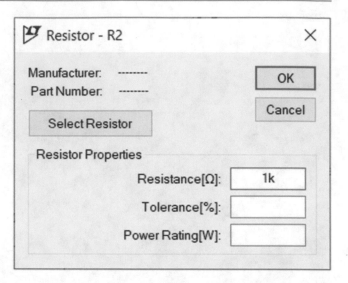

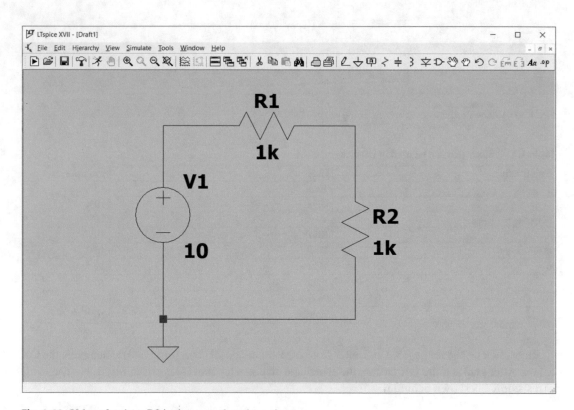

Fig. 1.41 Value of resistor R2 is shown on the schematic

You can change the components labels by right clicking on them and entering the new name. For instance, if you right click on the "R1" label in Fig. 1.41, the window shown in Fig. 1.42 appears and permits you to enter a new name to the resistor.

Fig. 1.42 Entering a new name

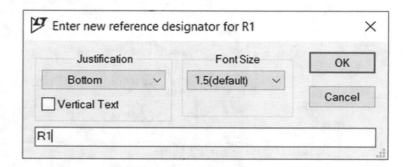

You can use the Label Net icon (Fig. 1.43) to give the desired names to the circuit nodes. After clicking the Label Net icon, the window shown in Fig. 1.44 appears and permits you to enter the desired name.

Fig. 1.43 Label Net icon

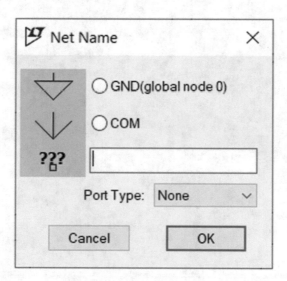

Fig. 1.44 Net Name window

Let's give the name "out" to the node which is connected to the upper terminal of resistor R2. To do this, click the Label Net icon and enter "out" to Net Name window and click the OK button (Fig. 1.45). After clicking the OK button, the entered name (out) is attached to the mouse pointer and a small square is under it. Place the square on the upper terminal of resistor R2 (Fig. 1.46) and click. After clicking, the upper terminal of R2 is renamed to "out" (Fig. 1.47).

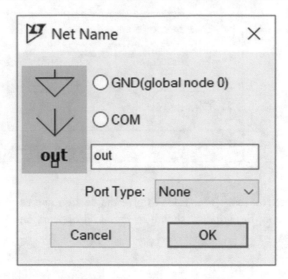

Fig. 1.45 Entering "out" to Net Name window

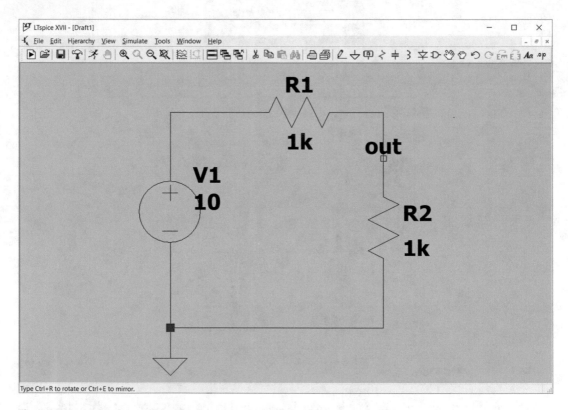

Fig. 1.46 Assigning the name "out" to upper terminal of R2 (=right terminal of R1)

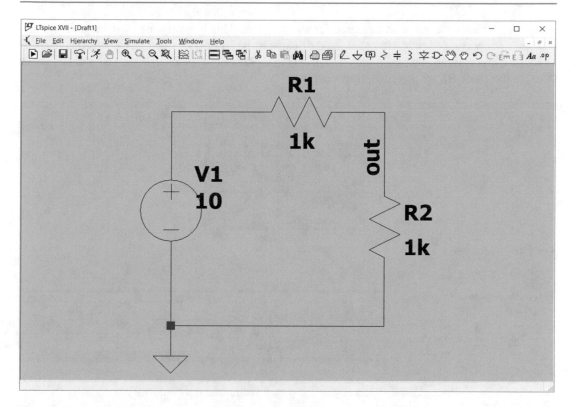

Fig. 1.47 Upper terminal of R2 (=right terminal of R1) is renamed to out

Rename the node connected to the positive terminal of voltage source to "in" (Fig. 1.48).

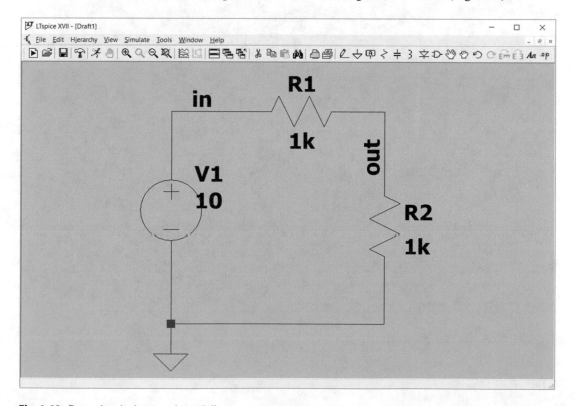

Fig. 1.48 Renaming the input node to "in"

You can use the Search icon to search for a component. This is a very useful in large schematics that have many components. After clicking the Search icon (Fig. 1.49), a text box is added to the right of toolbar, and you can enter the search term in it (Fig. 1.50). For instance, in Figs. 1.51 and 1.52, node "out" and resistor R2 are searched, respectively.

Fig. 1.49   Search icon

Fig. 1.50   Appeared
search box

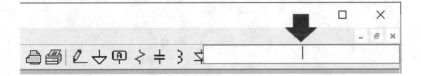

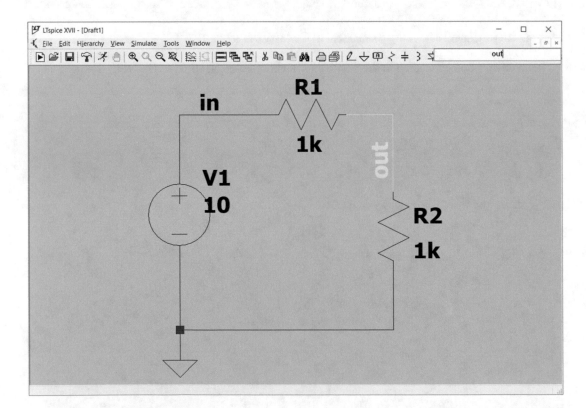

Fig. 1.51   Node "out" is highlighted

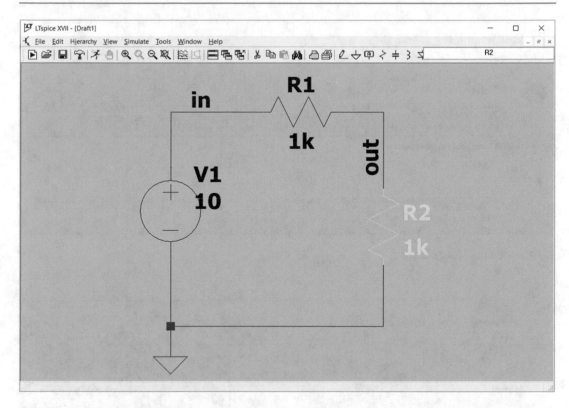

Fig. 1.52 Resistor R2 is highlighted

You can add text to the schematic by clicking the Text icon (Fig. 1.53). After clicking the Text icon, the window shown in Fig. 1.54 appears and permits you to enter the desired text. After entering the desired text, click the OK button and then click on the schematic to add the text to it (Fig. 1.55).

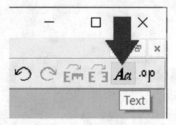

Fig. 1.53 Text icon

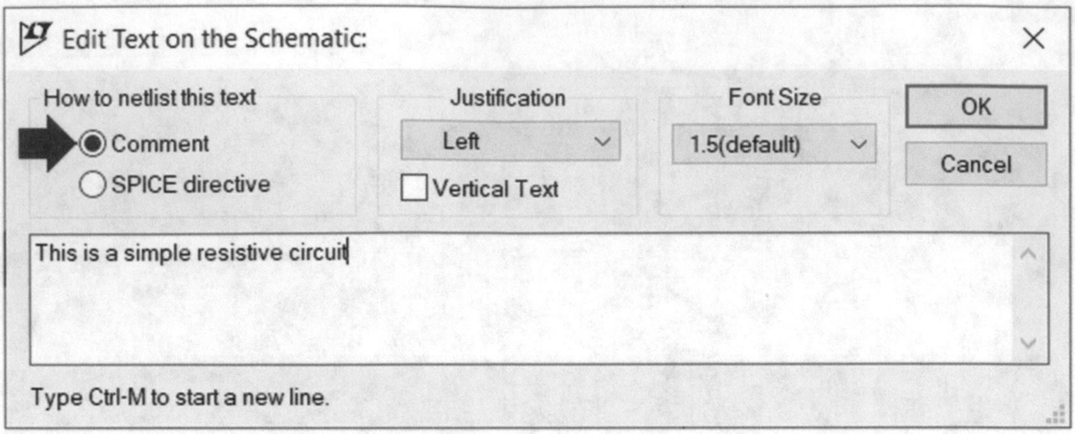

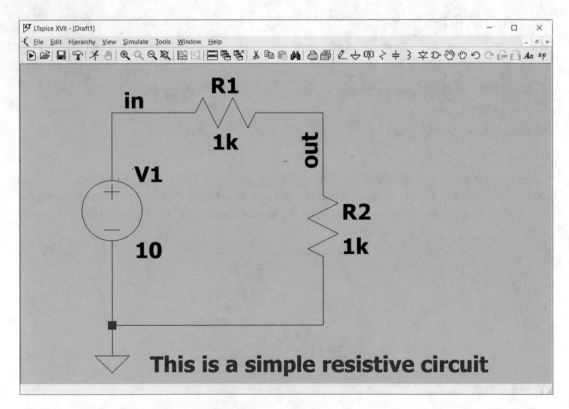

Fig. 1.54 Comment radio button

Fig. 1.55 Addition of a comment to the schematic

You can zoom in/out with the aid of mouse scroll button or the icons shown in Fig. 1.56. If you press the spacebar key of your keyboard, the best settings are selected to show the drawn schematic.

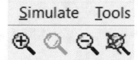

Fig. 1.56 Zoom icons

When you open a new schematic in LTspice, the default name Drafn (n shows a number) is assigned to it. If you want to save the file with your desired name, you need to use the File> Save As (Fig. 1.57). After Save As the file with desired name, you can click the Save icon (Fig. 1.58) to save the changes you apply to the schematic file.

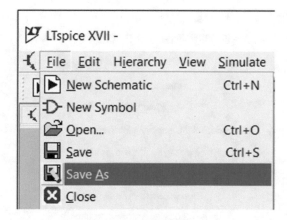

Fig. 1.57 File> Save As

Fig. 1.58 Save icon

The Run icon (Fig. 1.59) can be used to determine the type of simulation and run the simulation. In this example, we want to study the behavior of circuit for [0, 100 ms] time interval. In order to do this, click the Run icon, then enter 100m to the Stop time box (Fig. 1.60), and click the OK button. After clicking the OK button, the ".trans 100m" text is added to the schematic and a black window will be added to the LTspice environment (Fig. 1.61). The selected circuit waveforms will be shown in this black window. The ".trans 100m" line tells the LTspice to calculate the circuit voltages and currents for [0, 100 ms] interval.

LTspice XVII - [Draft1
File Edit Hierarch View
Run

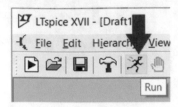

Fig. 1.59 Run icon

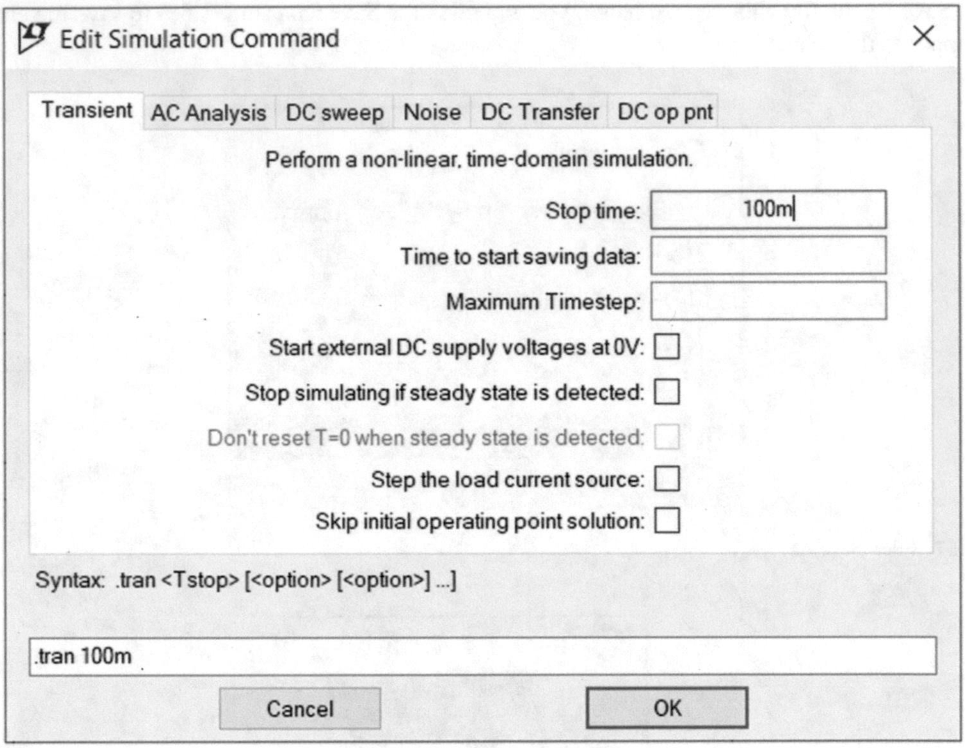

Edit Simulation Command
Transient
AC Analysis
DC sweep
Noise
DC Transfer
DC op pnt
Perform a non-linear, time-domain simulation.
Stop time:
100m
Time to start saving data:
Maximum Timestep:
Start external DC supply voltages at 0V:
Stop simulating if steady state is detected:
Don't reset T=0 when steady state is detected:
Step the load current source:
Skip initial operating point solution:
Syntax: .tran <Tstop> [<option> [<option>] ...]
.tran 100m
Cancel
OK

Fig. 1.60 Edit Simulation Command window

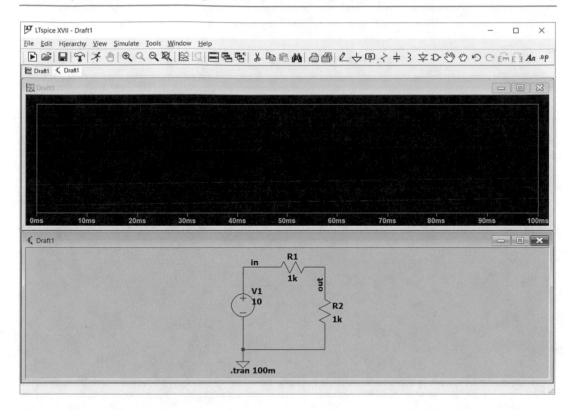

Fig. 1.61 Transient simulation is done

Put your mouse cursor on the node "out" and click it. After clicking the node, its voltage (with respect to ground) is shown (Fig. 1.62). According to Fig. 1.62, the voltage of node "out" is 5 V which is the correct value.

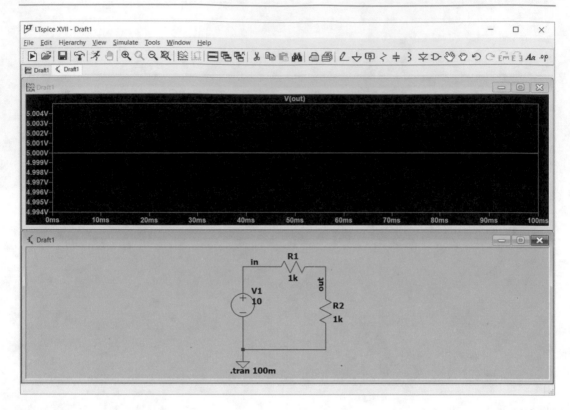

Fig. 1.62 Voltage of node "out"

If you right click on the time axis, the window shown in Fig. 1.63 appears. You can enter the range that you want to be shown in this window. For instance, if you want to see the [0, 10 ms] interval, you need to enter 0 and 10m to the Left and Right boxes, respectively (Fig. 1.64). After clicking the OK button, the [0,10 ms] time interval is shown (Fig. 1.65). If you right click on the vertical axis, the Vertical Axis window (Fig. 1.66) appears and permits you to enter the desired range. The upper bound of desired range must be entered to the Top box, and lower bound of desired range must be entered to the Bottom box.

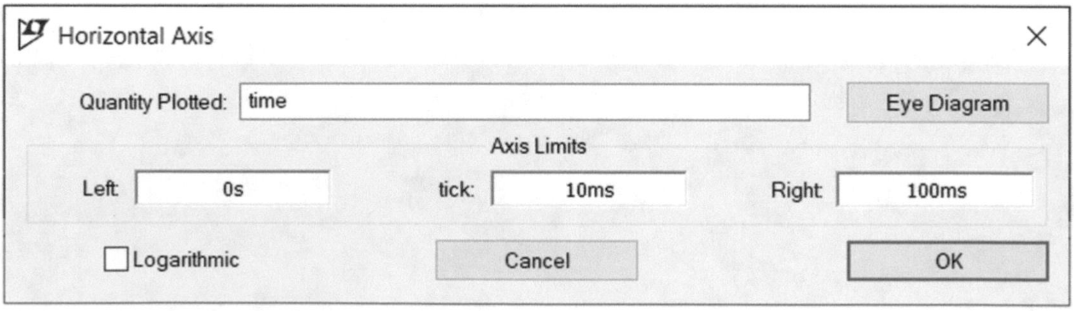

Fig. 1.63 Horizontal Axis window

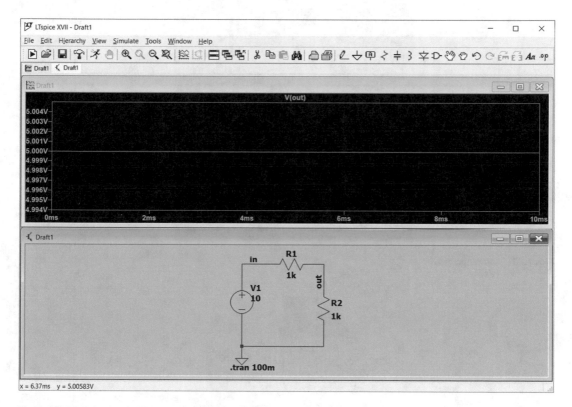

Fig. 1.64 Entering the new values to Horizontal Axis window

Fig. 1.65 Time interval of [0, 10 ms] is shown on the output window

Fig. 1.66 Vertical Axis window

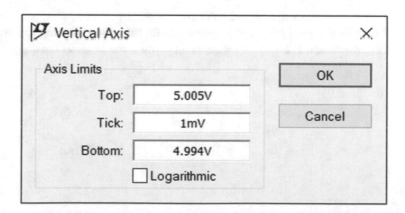

LTspice can select the best range for vertical axis automatically. In order to do this, right click on the graph and click the Autorange Y-axis (Fig. 1.67). You can click the Autorange icon (Fig. 1.68) as well.

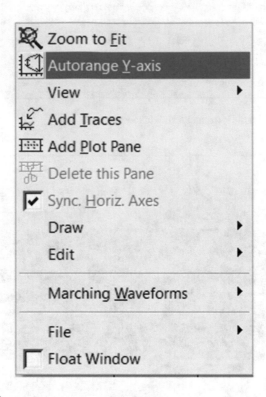

Fig. 1.67 Autorange Y-axis

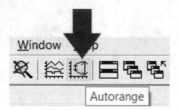

Fig. 1.68 Autorange icon

You can select the desired color for the graphs (traces), axis, back ground, etc. Click the Tools> Color Preferences (Fig. 1.69) in order to do this. After clicking the Tools> Color Preferences, the window shown in Fig. 1.70 appears and permits you to select the desired color for each item.

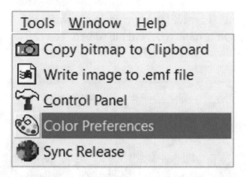

Fig. 1.69 Tools> Color Preferences

Fig. 1.70 Color Palette
Editor

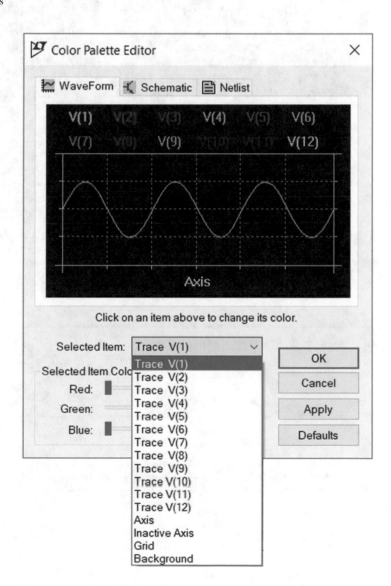

If you right click on the V(out) (Fig. 1.71), the window shown in Fig. 1.72 appears. If you click the Delete this Trace button, the graph is removed from the screen. You can change the color of graph with the aid of Default Color drop down list (Fig. 1.73).

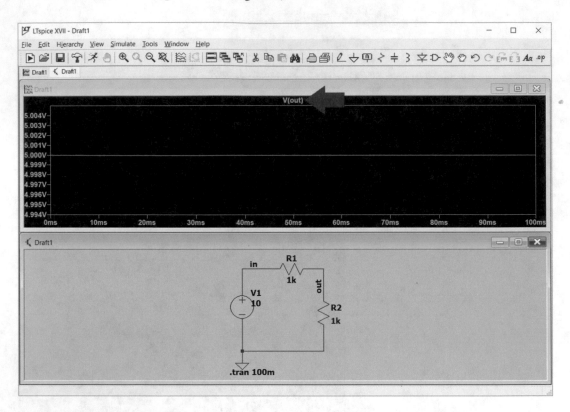

Fig. 1.71 Right clicking on V(out) opens the window shown in Fig. 1.72

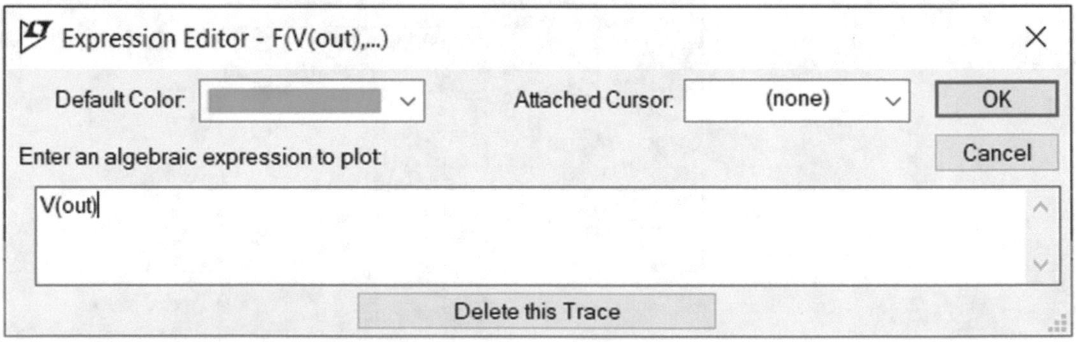

Fig. 1.72 Expression Editor window

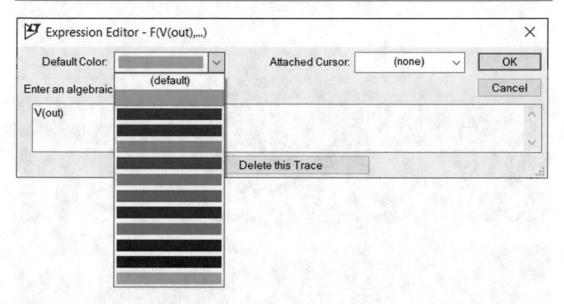

Fig. 1.73 Available colors

You can right click on the graph and check the Grid box (Fig. 1.74) to add grid to the graph (Fig. 1.75).

Fig. 1.74 Grid is checked

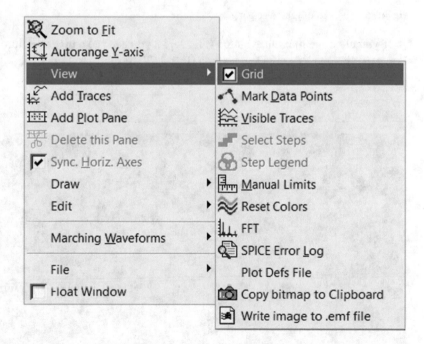

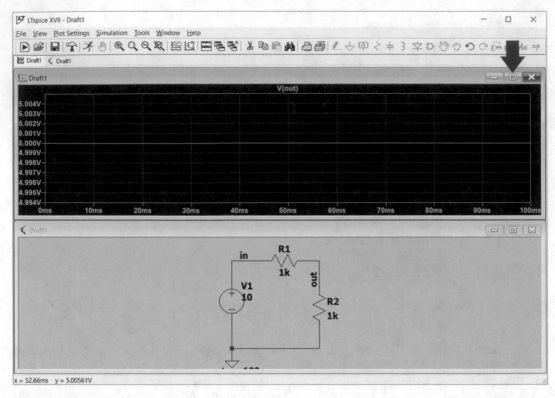

Fig. 1.75 Addition of grid to output graph

If you click the maximize button (Fig. 1.75), you can see a bigger picture of the graph (Fig. 1.76). You can click the Restore Down button (Fig. 1.77) to return to Fig. 1.75.

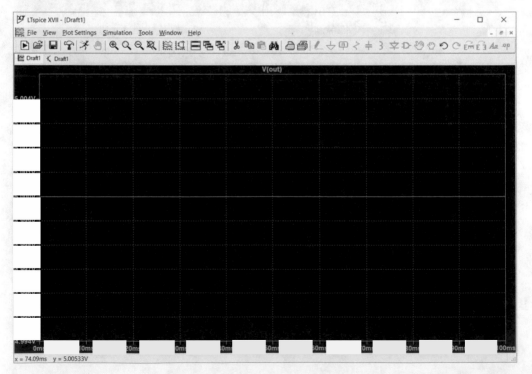

Fig. 1.76 Output graph is maximized

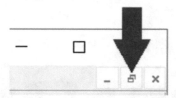

Fig. 1.77 Restore down button

You can copy the drawn graph into the clipboard and paste it in another software. In order to do this, right click on the graph and click the Copy bitmap to Clipboard (Fig. 1.78). This capability is very useful when you want to prepare a report/presentation and you want to show the circuit waveforms.

Fig. 1.78 Copy bitmap to Clipboard

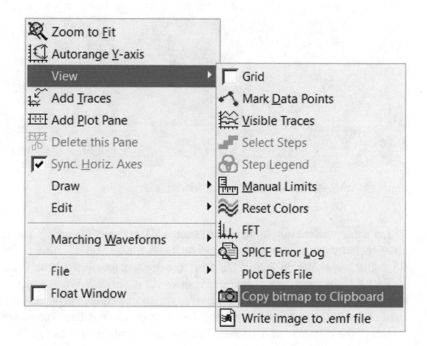

Let's measure voltage of resistor R1 (voltage difference between node "in" and "out," i.e., $V_{in} - V_{out}$). In order to do this, put the mouse cursor on the node "in" and hold down the mouse left button, this shows a red probe on the node "in." Don't release the mouse left button and drag the red probe toward node "out." When you reach the destination node (node "out"), the probe color becomes black. Now release the mouse left button. After releasing the mouse left button, the voltage difference between node "in" and node "out" appears on the screen (Fig. 1.79).

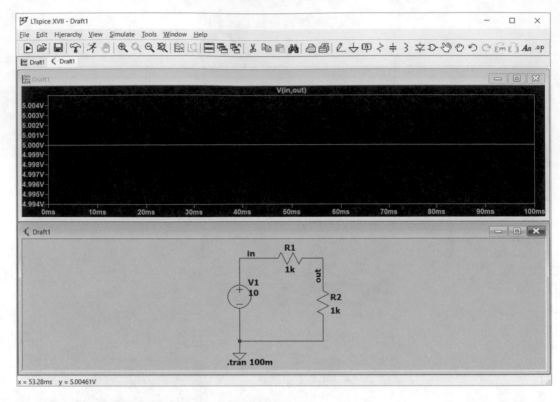

Fig. 1.79 Voltage difference between node "in" and "out"

Let's measure the current of voltage source V1, resistor R1 and R2. You can measure the current of an element by putting the mouse cursor on it and clicking on it. Current of voltage source V1, resistor R1 and R2 is shown in Figs. 1.80, 1.81, and 1.82, respectively. Note that current of source V1 and resistor R1 is negative, and current of resistor R2 is positive. When you put your mouse cursor on an element and click it, LTspice measures the current that enters the positive terminal of component. The current that enters the positive terminal is assumed to be positive. The positive terminal of the voltage source is the terminal with + label. The positive terminal of a diode is the anode terminal. Positive terminal of resistors, inductors, and capacitors has no special symbol. However, there are some rules that help you determine the positive terminal. If you add a resistor (or inductor or capacitor) to the schematic without any rotation, the positive terminal is the upper one (Fig. 1.83). If you press the Ctrl+R once before placing the component, the positive terminal is the right terminal. If you press the Ctrl+E after Ctrl+R, the positive terminal comes to left (Fig. 1.83).

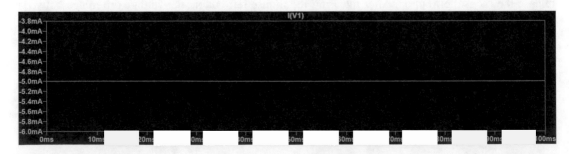

Fig. 1.80 Current drawn from voltage source

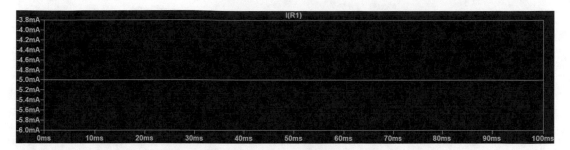

Fig. 1.81   Current through resistor R1

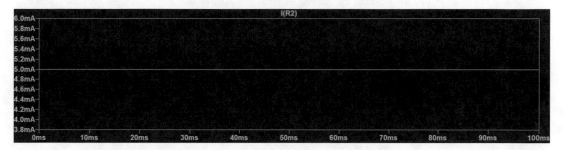

Fig. 1.82   Current through resistor R2

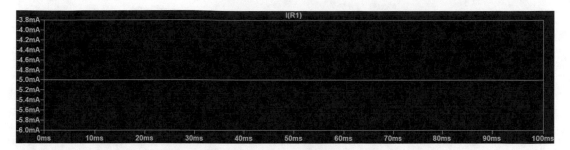

Fig. 1.83   (**a**) Default position of + and − terminals (**b**) Ctrl+R is pressed once (**c**) Ctrl+R is pressed once and after that Ctrl+E is pressed

The positive terminals of the circuit components are shown in Fig. 1.84. Resistor R1 is rotated by pressing the Ctrl+R, so the positive terminal is the right one. The circuit current comes out from the positive terminal of V1 and R1, so the current of these components is negative. The circuit current enters the positive terminal of resistor R2, so the current of resistor R2 is positive.

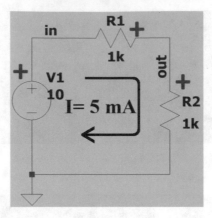

Fig. 1.84 Positive terminals of components

If you click on the I(R1) in Fig. 1.81 and change the I(R1) to –I(R1) (Fig. 1.85), the graph becomes positive (Fig. 1.86).

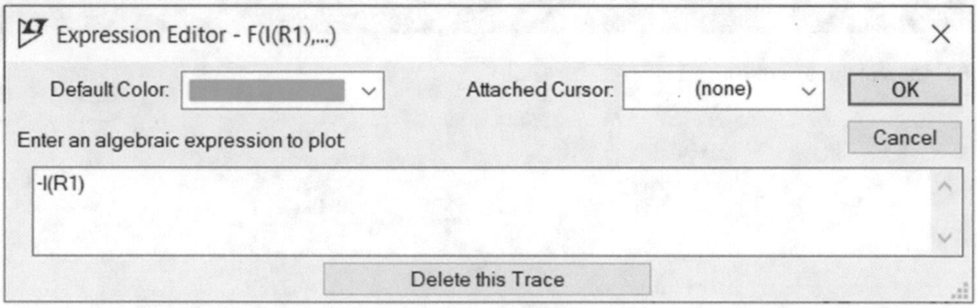

Fig. 1.85 –I(R1) is entered to the text box

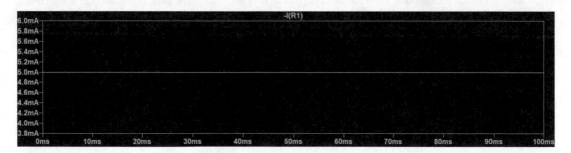

Fig. 1.86 Graph of –I(R1)

You can see the component power by holding down the Alt key and clicking on the component. For instance, the power of resistor R2 is shown in Fig. 1.87. The power is positive since the resistor consumes power. If you hold down the Alt key and click on the voltage source V1, you will see the graph shown in Fig. 1.88. The graph shown in Fig. 1.88 is negative since the power source supplies power to the circuit.

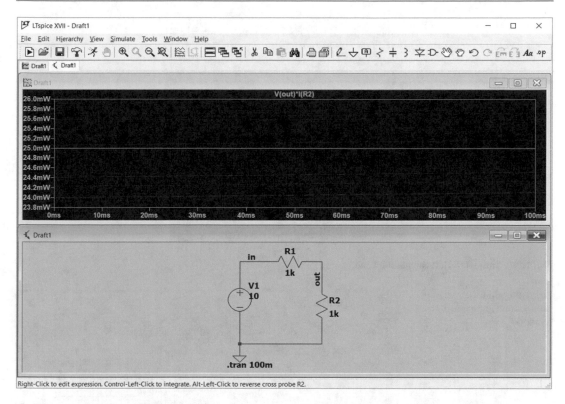

Fig. 1.87   Graph of V(out)*I(R2)

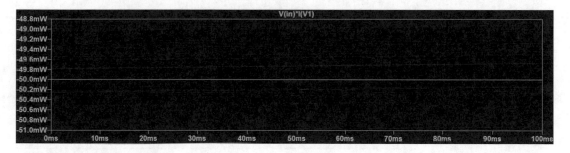

Fig. 1.88   Graph of V(in)*I(V1)

You can edit the simulation parameters by clicking the Simulate> Edit Simulation Cmd (Fig. 1.89). After clicking the Edit Simulation Cmd, the window shown in Fig. 1.90 appears and permits you to apply the desired changes to the simulation parameters. Enter 80m to the Time to start saving data box (Fig. 1.91) and click the OK button. After clicking the OK button, the schematic changes to what is shown in Fig. 1.92. If you run the simulation, the result shown in Fig. 1.93 appears. The points 0 and 20 ms of the graph shown in Fig. 1.93 show the behavior of the circuit at t = 80 ms and t = 100 ms, respectively. The behavior of the circuit for [0, 80 ms] interval is not shown in this graph.

Fig. 1.89   Simulate> Edit Simulation Cmd

Fig. 1.90   Edit
Simulation Command
window

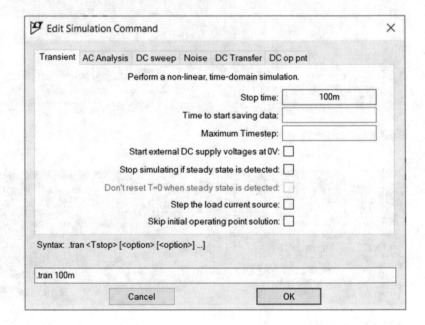

Fig. 1.91 Entering the desired simulation settings to the Edit Simulation Command window

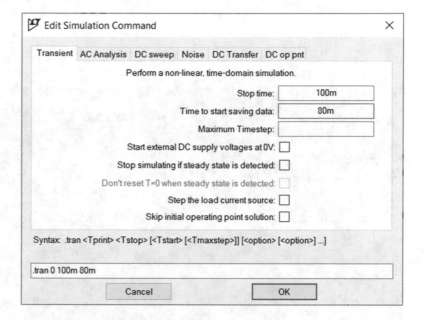

Edit Simulation Command   ×

| Transient | AC Analysis | DC sweep | Noise | DC Transfer | DC op pnt |

Perform a non-linear, time-domain simulation.

Stop time: 100m

Time to start saving data: 80m

Maximum Timestep:

Start external DC supply voltages at 0V: ☐

Stop simulating if steady state is detected: ☐

Don't reset T=0 when steady state is detected: ☐

Step the load current source: ☐

Skip initial operating point solution: ☐

Syntax: .tran <Tprint> <Tstop> [<Tstart> [<Tmaxstep>]] [<option> [<option>] ...]

.tran 0 100m 80m

Cancel   OK

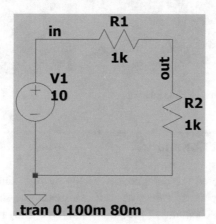

.tran 0 100m 80m

Fig. 1.92 Simulation command is added to the schematic

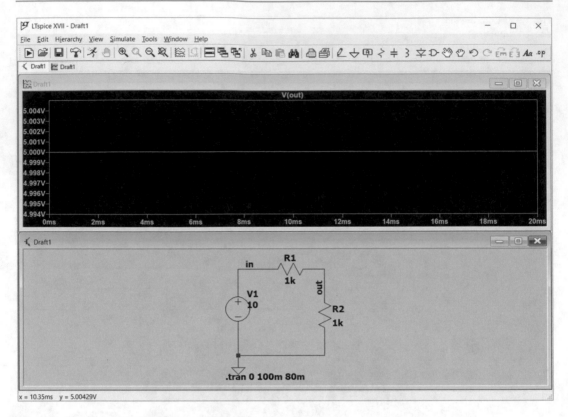

Fig. 1.93 Simulation result for [80 ms, 100 ms] interval

1.4   Example 2: .param Command

In the previous circuit, we entered the components values by right clicking on them. You can use variables to determine the components values as well. This example shows how to use variables to determine components values. Schematic of previous example is shown in Fig. 1.94. Right click on the resistor R1 and R2 and changes the Resistance[Ω] box to {R1} and {R2}, respectively (Figs. 1.95 and 1.96).

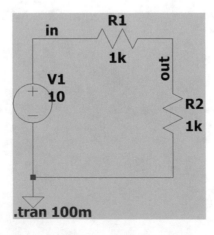

Fig. 1.94 Schematic of Example 1

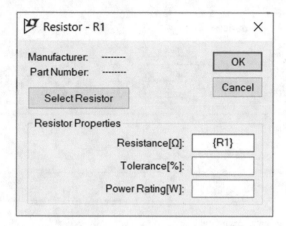

Fig. 1.95 Entering the {R1} to the Resistance[Ω] box of Resistor R1

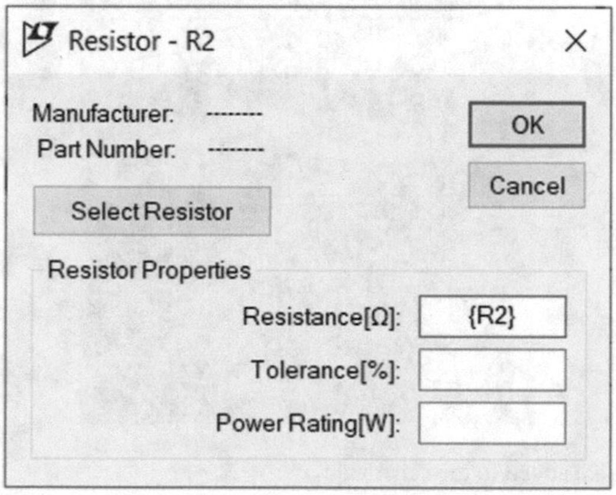

Fig. 1.96 Entering the {R2} to the Resistance[Ω] box of resistor R2

Click the SPICE Directive icon (Fig. 1.97) and enter the commands shown in Fig. 1.98. The commands assign 1 kΩ to R1 and R2. After entering the commands, click the OK button and click on the schematic. After clicking on the schematic, the entered commands are added to it (Fig. 1.99). If you run the simulation, the results shown in previous example are obtained. If you right click on the .param commands shown in Fig. 1.99, the window shown in Fig. 1.98 appears again and permits you to change the values of R1 and R2.

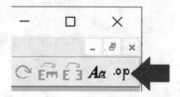

Fig. 1.97 SPICE Directive icon

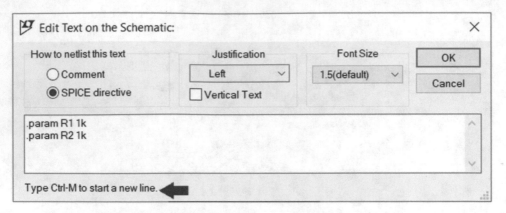

Fig. 1.98 Edit Text on the Schematic window

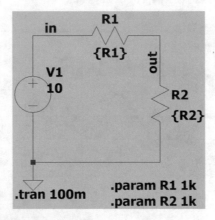

Fig. 1.99 Entered commands are added to the schematic

1.5   Example 3: Potentiometer

You can't find potentiometer in ready to use LTspice blocks. However, you can simulate a potentiometer with two resistors and few lines of code. This example shows how a potentiometer can be simulated in LTspice. Draw the schematic shown in Fig. 1.100. This example simulates a potentiometer with value of 5 kΩ. Variable val determines the position of wiper. If you simulate the circuit, the result shown in Fig. 1.101 appears on the screen. Voltage of node "out1" is 5 V and voltage of node "out2" is 3.5 V.

Fig. 1.100 Schematic of Example 3

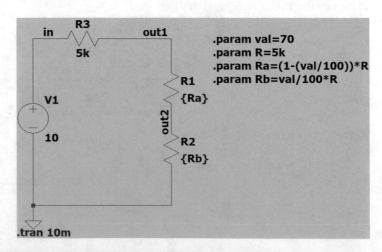

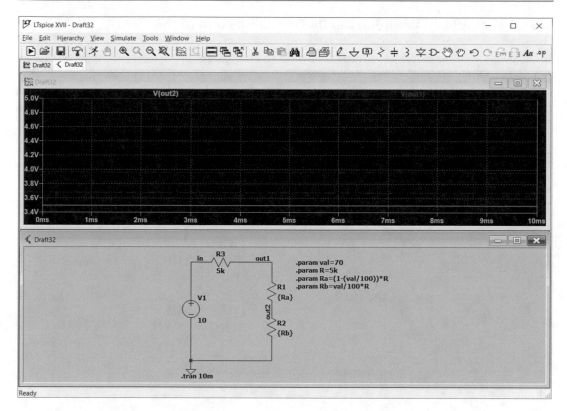

Fig. 1.101 Simulation result

Let's check the obtained results. The calculations shown in Fig. 1.102 show that the LTspice results are correct.

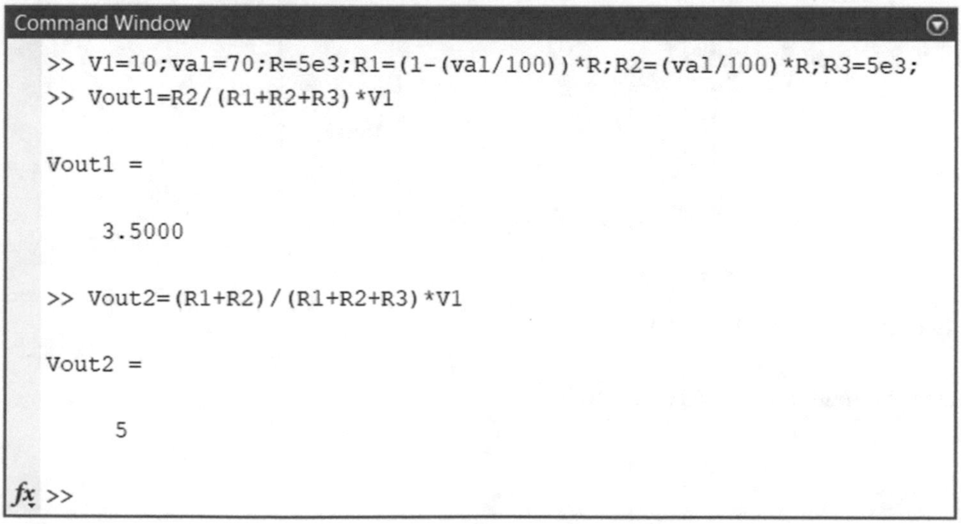

Fig. 1.102 MATLAB calculations

1.6　Example 4: Obtaining the DC Operating Point of the Circuit

In DC steady state, the inductors and capacitors act as short circuit and open circuit, respectively. LTspice can find the steady-state DC voltages/currents easily. This example shows how to obtain the steady-state DC voltages/currents of an electric circuit. DC steady-state analysis is not limited to electric circuits; it can be used to obtain the operating point of electronic circuits as well. For instance, Example 8 in Chap. 2 used the DC steady-state analysis to obtain the operating point of a transistor amplifier.

Consider the circuit shown in Fig. 1.103. The DC steady-state equivalent of this circuit is shown in Fig. 1.104. According to this figure, $V_1 = 12$ V, $V_2 = 6$ V, $V_3 = 0$ V, $V_4 = 0$ V, and $I = 6$ mA.

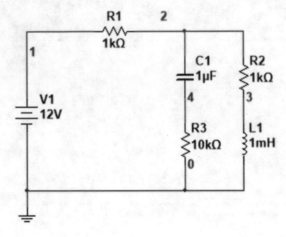

Fig. 1.103　Circuit for Example 4

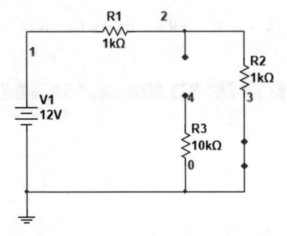

Fig. 1.104　DC equivalent of Fig. 1.103

Draw the circuit in LTspice (Fig. 1.105).

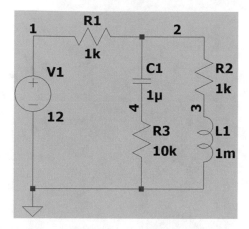

Fig. 1.105 LTspice equivalent of Fig. 1.103

Click the Run icon (Fig. 1.106) and open the DC op pnt tab (Fig. 1.107). After clicking the OK button, the .op command is added to the schematic (Fig. 1.108) and the result shown in Fig. 1.109 appears. The obtained results are quite close to the expected values. Note that e shows power of ten. For instance, node 4 voltage is 6e-14 which means 6×10^{-14} V.

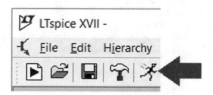

Fig. 1.106 Run icon

Fig. 1.107 DC op pnt tab

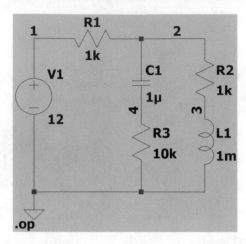

Fig. 1.108 .op command is added to the schematic

```
[LT] * C:\Users\farzinasadi\Documents\LTspiceXVII\DCOperatingPoint.asc                                    ×

       --- Operating Point ---

V(1):           12             voltage
V(2):           6              voltage
V(4):           6e-014         voltage
V(3):           6e-006         voltage
I(C1):          6e-018         device_current
I(L1):          0.006          device_current
I(R3):          6e-018         device_current
I(R2):          0.006          device_current
I(R1):          0.006          device_current
I(V1):          -0.006         device_current
```

Fig. 1.109 Simulation result

You can run the DC operating point analysis by clicking the SPICE Directive icon (Fig. 1.110) as well. After clicking the SPICE Directive icon, enter the ".op" to text box (Fig. 1.111) and click the OK button. Now the schematic looks like Fig. 1.108. If you run the simulation, the result shown in Fig. 1.109 appears.

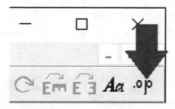

Fig. 1.110   SPICE Directive icon

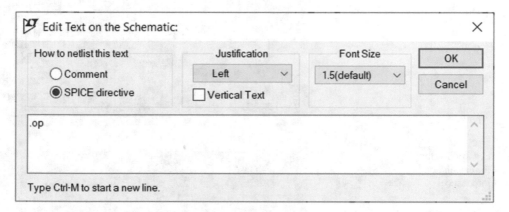

Fig. 1.111   Entering the .op to the text box

1.7   Example 5: DC Transfer Analysis

The DC transfer function analysis calculates the DC gain and input impedance and output impedance of a circuit. Let's study an example. Consider the circuit shown in Fig. 1.112.

Fig. 1.112   Schematic for Example 5

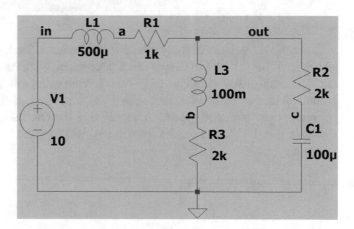

Let's do some hand analysis. The DC steady-state equivalent circuit is shown in Fig. 1.113 (The inductors are replaced with short circuit and capacitors are replaced with open circuit). According to this figure, $V_{in} = 10\,\text{V}, V_{out} = 6.6667\,\text{V}, I = \dfrac{10}{3}\,\text{mA}$ and input impedance seen by V1 is 3 kΩ. The gain of circuit is $\dfrac{V_{out}}{V_1} = 0.66667$. The output impedance seen from node "out" can be calculated with the aid of circuit shown in Fig. 1.114. The output impedance seen from node "out" is $\dfrac{2 \times 1}{2+1} = 0.66667\,k\Omega$.

Remember that independent sources must be killed (i.e., voltage sources must be replaced with short circuit, and current sources must be replaced with open circuit) in the calculation of output impedance.

Fig. 1.113 DC equivalent circuit for Fig. 1.112

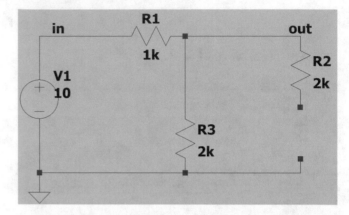

Fig. 1.114 Equivalent circuit for calculation of output impedance

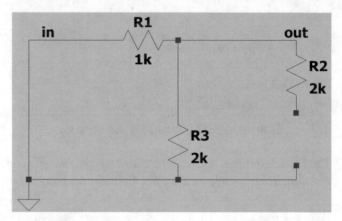

Let's check our hand analysis results with LTspice. Click the Run icon and open the DC Transfer tab. Enter V(out) and V1 to Output and Source boxes, respectively (Fig. 1.115) and click the OK button. The settings shown in Fig. 1.115 calculate the gain (voltage difference between node "out" and ground divided by value of voltage source V1), input impedance seen by V1 and out impedance seen between node "out" and ground. After clicking the OK button, the schematic changes to what is shown in Fig. 1.116, and result shown in Fig. 1.117 appears. Obtained results are the same as hand analysis.

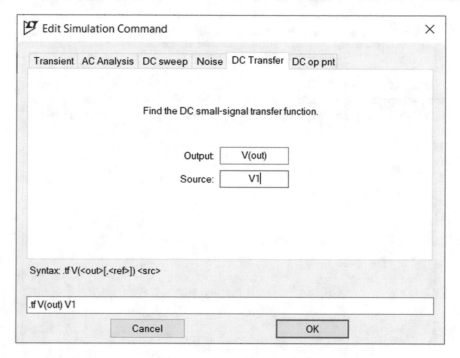

Fig. 1.115 DC Transfer tab

Fig. 1.116 .tf command
is added to the
schematic

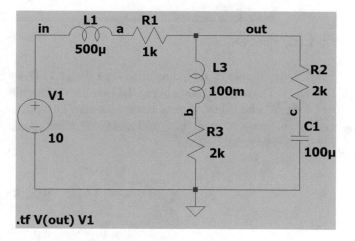

Fig. 1.117 Simulation result

Click the Simulate> Edit Simulation Cmd (Fig. 1.118) and change the settings to what is shown in Fig. 1.119. The settings shown in Fig. 1.119 calculate the gain (voltage difference between node "out" and node "a" divided by value of voltage source V1), input impedance seen by V1 and output impedance seen between node "out" and node "a." After running the simulation, the result shown in Fig. 1.120 appears.

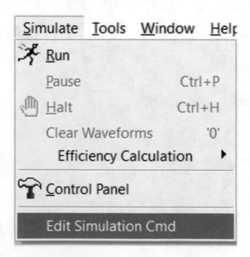

Fig. 1.118 Simulate> Edit Simulation Cmd

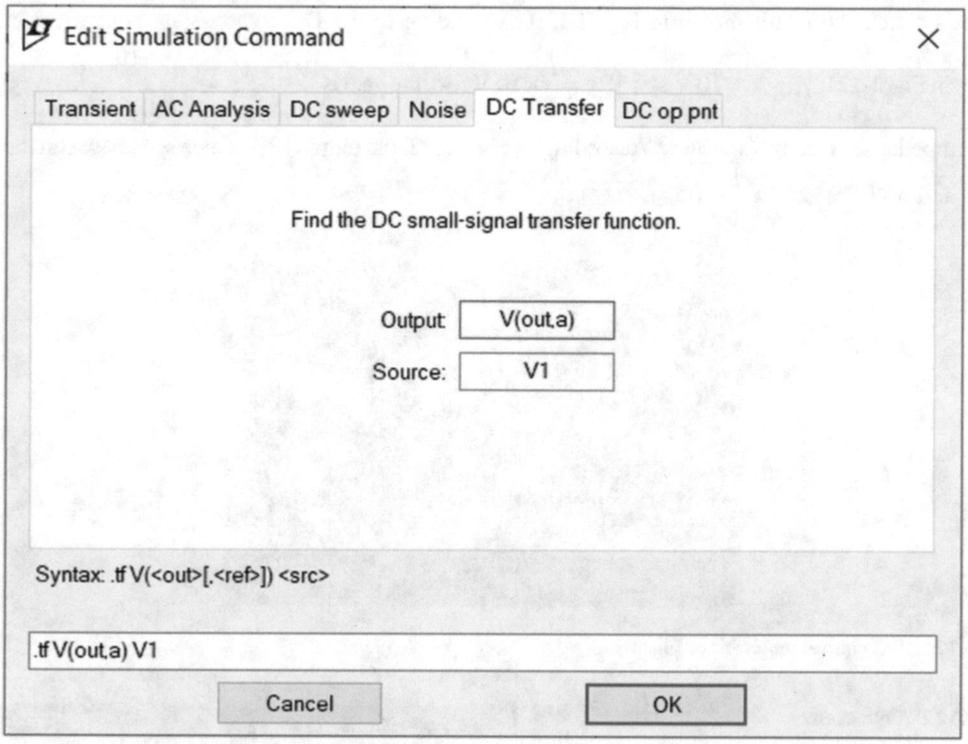

Fig. 1.119 Simulation settings

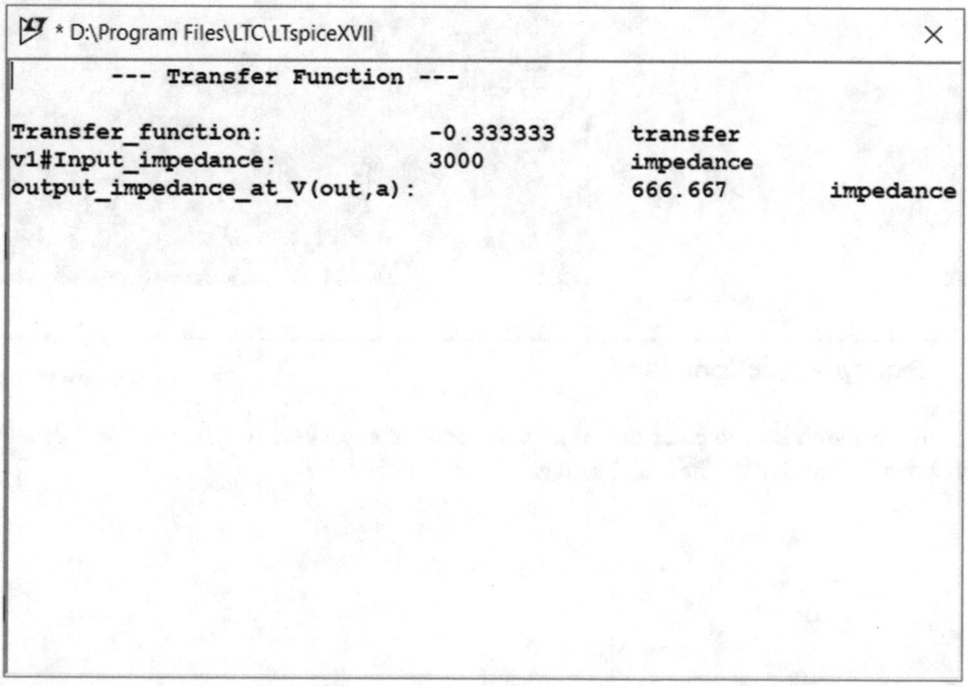

Fig. 1.120 Simulation result

Let's check the result shown in Fig. 1.120. According to the DC steady-state equivalent circuit shown in Fig. 1.121, $V_{in} = V_a = 10$ V and $V_{out} = \frac{2}{3} \times 10$ V. So, the gain is $\dfrac{V_{out} - V_a}{V_1} = \dfrac{\dfrac{20}{3} - 10}{10} = -0.3333$. The input impedance seen by V1 is 3 kΩ. According to Fig. 1.122, the output impedance seen between the node "out" and node "a" is $\dfrac{2 \times 1}{2 + 1} = 0.66667\,k\Omega$.

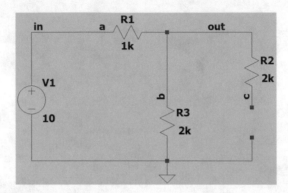

Fig. 1.121 DC equivalent circuit of Fig. 1.112

Fig. 1.122 Equivalent circuit for calculation of output impedance of Fig. 1.121

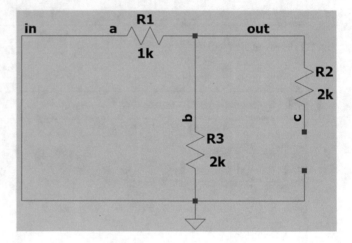

1.8   Example 6: .ic Command

This example shows how to enter the initial conditions of capacitors to LTspice. Consider the RC circuit shown in Fig. 1.123. The initial voltage of capacitor is 2.5 V.

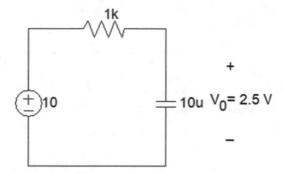

Fig. 1.123 Circuit for Example 6

Draw the schematic shown in Fig. 1.124.

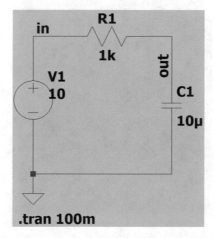

Fig. 1.124 LTspice equivalent of Fig. 1.123

Click the SPICE Directive icon (Fig. 1.125) and enter the single line code shown in Fig. 1.126. Then click the OK button and click on the schematic to add the entered code to it (Fig. 1.127). The .ic V(out)=2.5 V tells the LTspice that the initial voltage node "out" is 2.5 V.

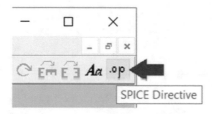

Fig. 1.125 SPICE Directive icon

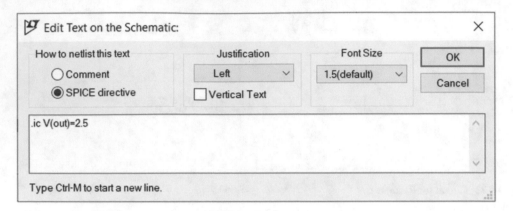

Fig. 1.126 Initial condition directive

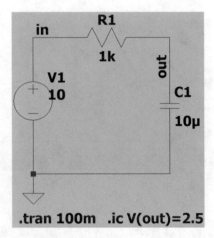

Fig. 1.127 SPICE commands are added to the schematic

Now run the simulation. The result is shown in Fig. 1.128. Note that the voltage of node "out" is 2.5 V at t=0.

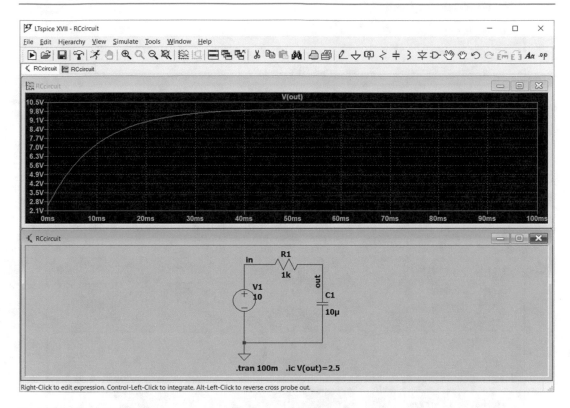

Fig. 1.128 Simulation result

Right click on the .ic command in the schematic. The window shown in Fig. 1.129 appears. Click the Cancel button. The window shown in Fig. 1.130 appears. If you check the Comment button (Fig. 1.131), the command .ic V(out)=2.5 will be ignored (it has no effect on the simulation). The LTspice uses blue color to show the comments on schematic (Fig. 1.132).

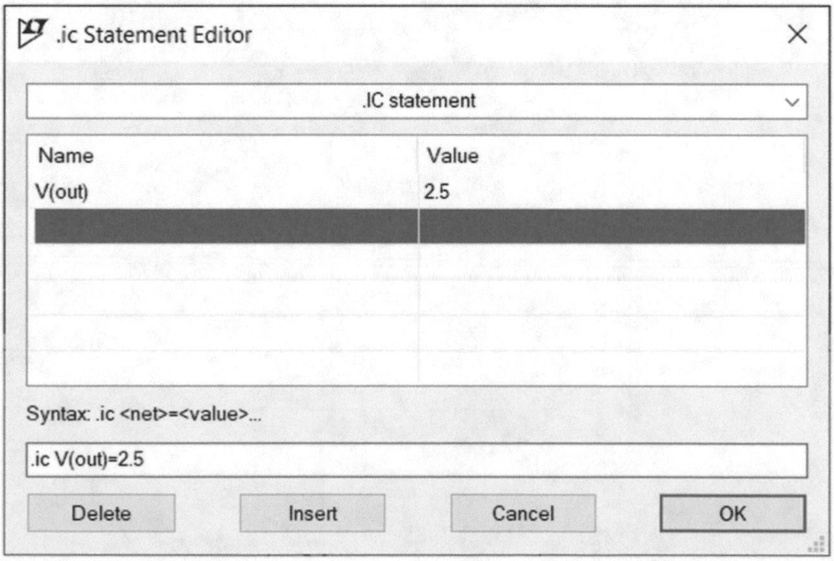

Fig. 1.129 .ic Statement Editor window

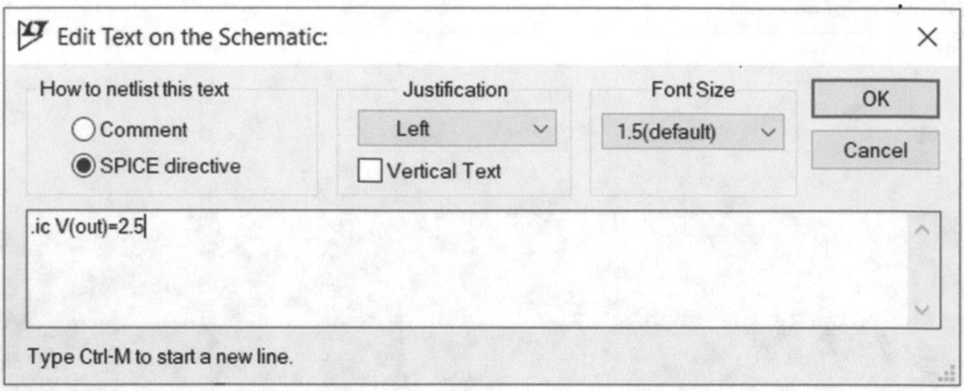

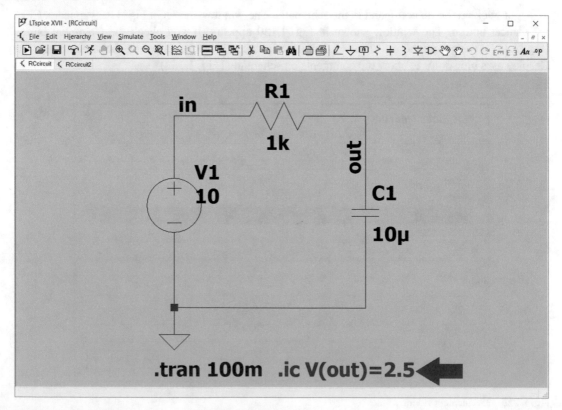

Fig. 1.130 Edit Text on the Schematic window

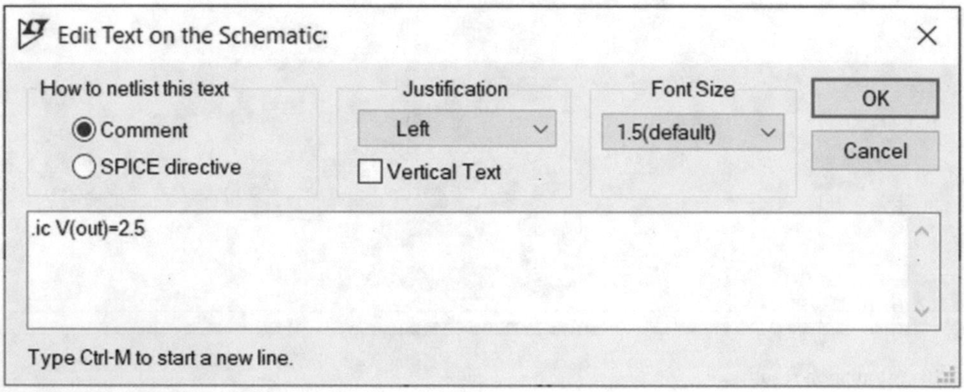

Fig. 1.131 Comment radio button is selected

Fig. 1.132 .ic V(out)=2.5 directive is converted into comment

Left click on the V(out) (Fig. 1.133). After left clicking on V(out), a vertical line (Fig. 1.134) and a window (Fig. 1.135) are added to the graph. You can move the vertical line (cursor) with your mouse and the coordinate of intersection of vertical line and graph is shown in the opened window. If you close the window shown in Fig. 1.135, the vertical line is removed from the graph.

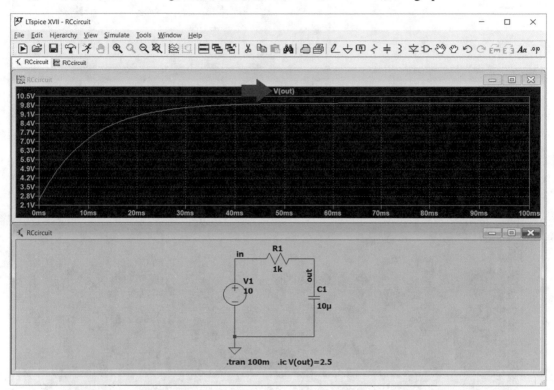

Fig. 1.133 Clicking on V(out) opens the window shown in Fig. 1.134

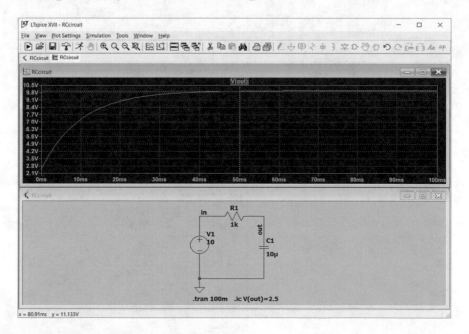

Fig. 1.134 A vertical line is added to the simulation output

Fig. 1.135 Coordinates
of the cursor

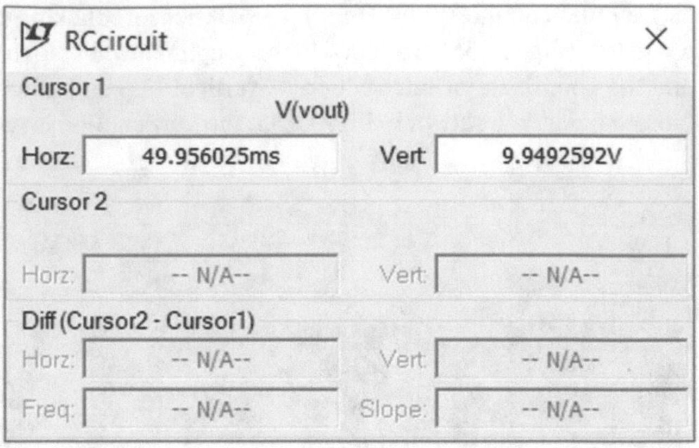

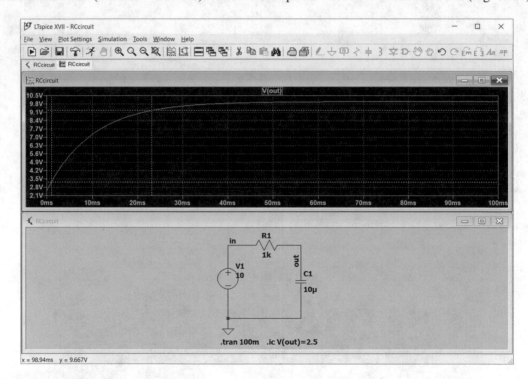

If you click the V(out) for second time, second cursor is added to the graph (Fig. 1.136). The Cursor 2 and Diff (Cursor2 − Cursor1) section of the opened window is activated as well (Fig. 1.137).

Fig. 1.136 Addition of second cursor to the output

Fig. 1.137 Coordinates
of the cursors

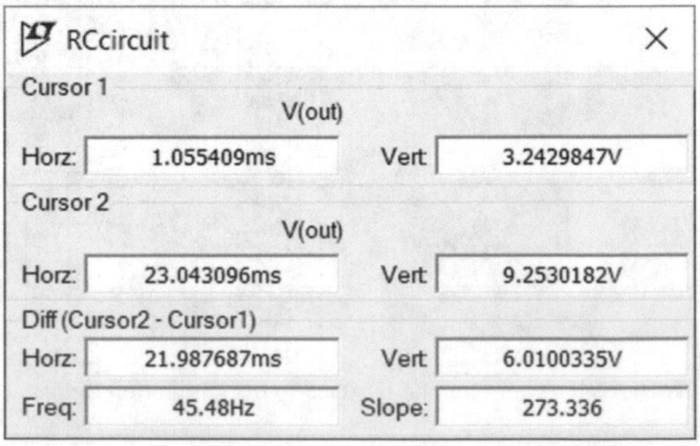

1.9   **Example 7: RL Circuit Analysis**

In this example, we want to analyze the RL circuit shown in Fig. 1.138. The initial current of inductor is 100 mA.

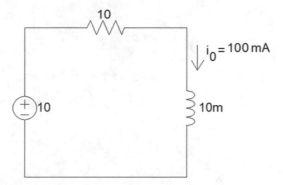

Fig. 1.138 Circuit for Example 7

Draw the schematic shown in Fig. 1.139. The .ic I(L1)=100 m line tells the LTspice that 100 mA current enters to the + terminal of inductor. The positive terminal of un rotated vertical inductors is the upper terminal (Fig. 1.83).

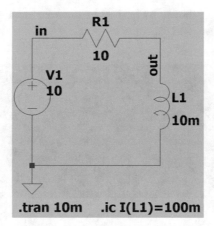

Fig. 1.139 LTspice equivalent of Fig. 1.138

Right click on the inductor. Note that the inductor has a default series resistance of 1 mΩ. When Series Resistance [Ω] box is empty, the LTspice uses the default value of 1 mΩ (Fig. 1.140).

Fig. 1.140 The
inductor has default
series resistance of 1
mΩ

⚑ Inductor - L1	✕

Manufacturer: --------
Part Number: --------

| Select Inductor |

| OK |
| Cancel |

Show Phase Dot ☐

Inductor Properties

Inductance[H]:   `10m`

Peak Current[A]:

Series Resistance[Ω]:

Parallel Resistance[Ω]:

Parallel Capacitance[F]:

(Series resistance defaults to 1mΩ) ◀

Run the simulation. The result shown in Fig. 1.141 shows the inductor current. Note that the inductor current starts from +100 mA= +0.1 A.

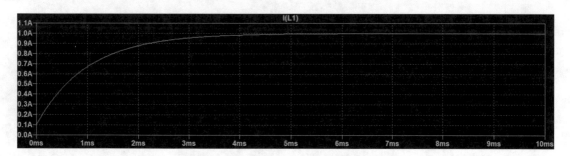

Fig. 1.141 Inductor current

The voltage of node "out" is shown in Fig. 1.142. Note that the voltage at t = 0 is 10 V − 10 Ω × 100 mA = 9 V.

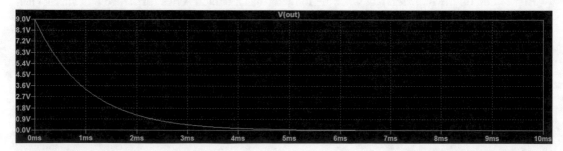

Fig. 1.142 Voltage of node "out"

1.10   Example 8: Switch Block

In this example, we want to simulate the circuit shown in Fig. 1.143. The switch S2 is closed for [0, 50 µs] interval. Switch S1 is open during this interval. At t = 50 µs, switch S2 is opened and switch S1 is closed. Initial current of inductor is zero.

The switch S1 and S2 in Fig. 1.143 can be simulated with the aid of Voltage controlled switch block (Fig. 1.144). The sw block is a voltage controlled switch. LTspice has current controlled switches as well (Fig. 1.145). We use the voltage controlled switch in this example.

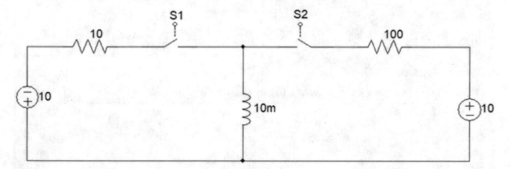

Fig. 1.143   Circuit for Example 8

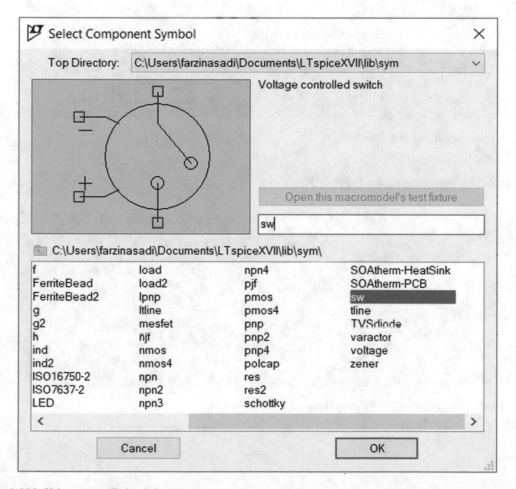

Fig. 1.144   Voltage controlled switch

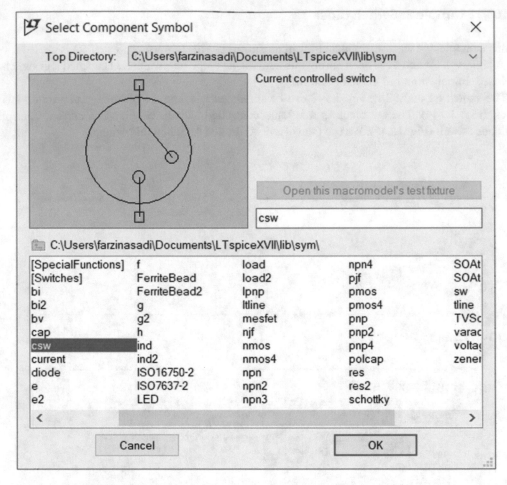

Fig. 1.145 Current controlled switch

Draw the schematic shown in Fig. 1.146.

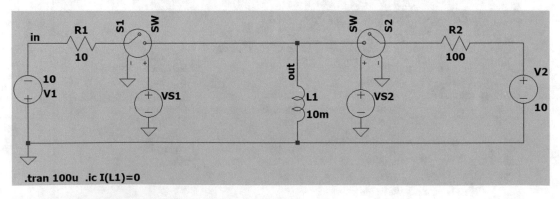

Fig. 1.146 First sketch for circuit shown in Fig. 1.143

Right click on the VS1. The window shown in Fig. 1.147 appears. Click the Advanced button.

Fig. 1.147 Voltage
Source window

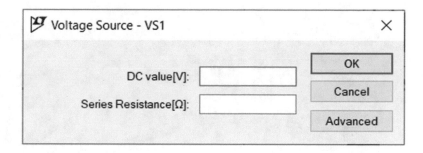

After clicking the advanced button, the window shown in Fig. 1.148 appears. Change the settings to what is shown in Fig. 1.149 and click the OK button. The settings shown in Fig. 1.149 produce the waveform shown in Fig. 1.150.

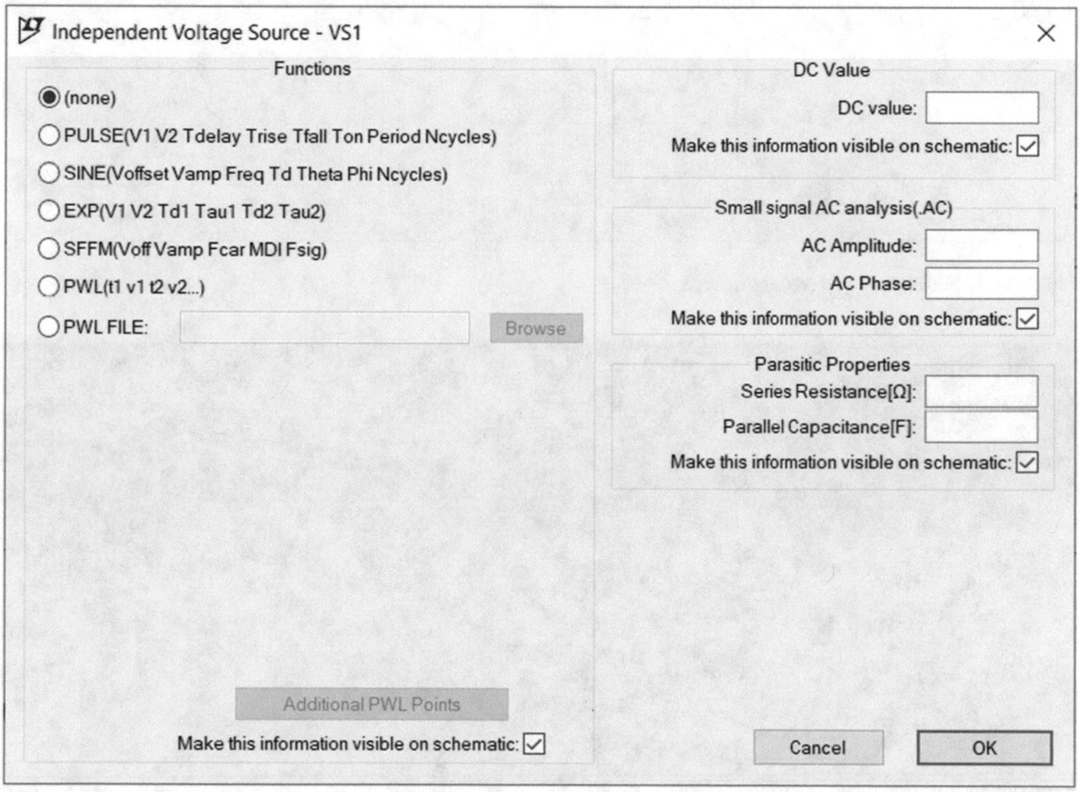

Fig. 1.148 Independent Voltage Source – VS1 window

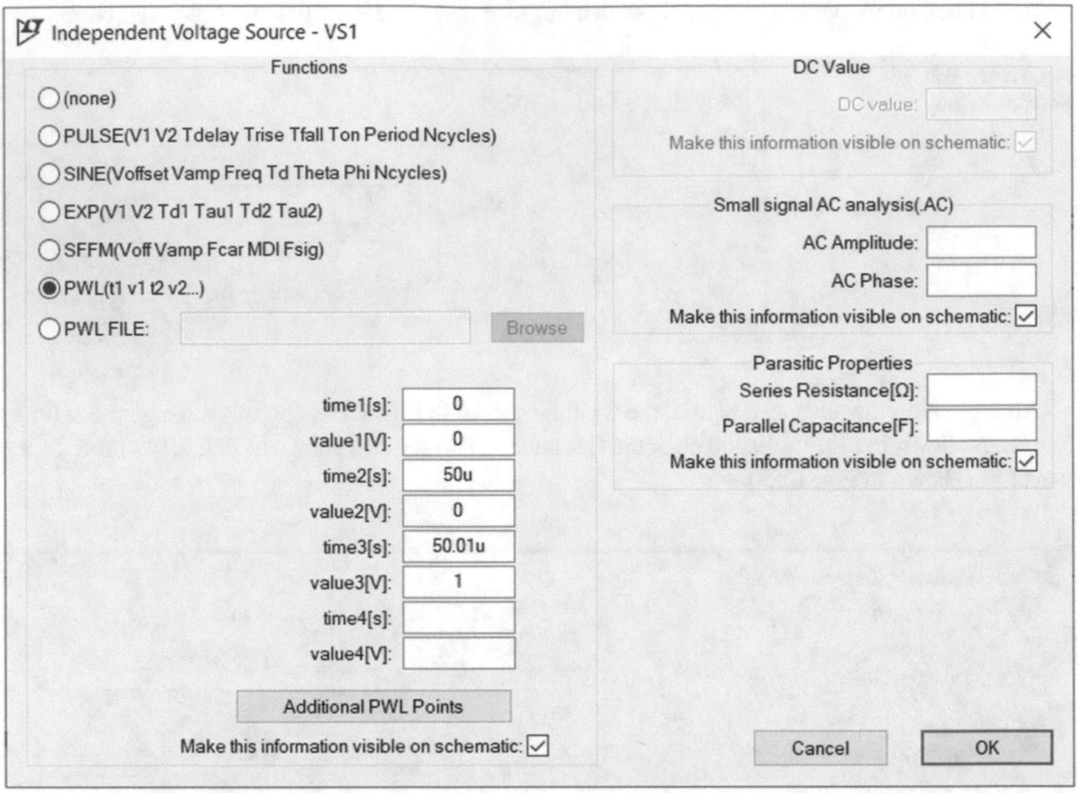

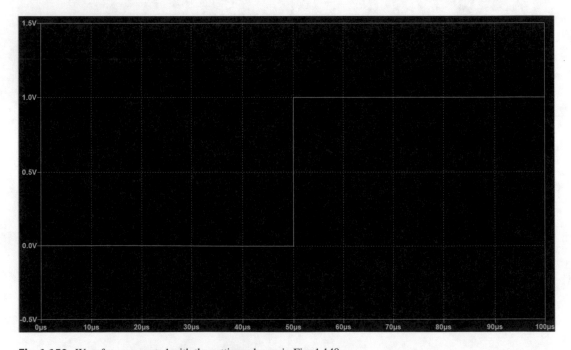

Fig. 1.149 Settings of VS1 voltage source

Fig. 1.150 Waveform generated with the settings shown in Fig. 1.149

Right click on the VS2, then click the advanced button and change the settings to what is shown in Fig. 1.151. The settings shown in Fig. 1.151 produce the waveform shown in Fig. 1.152.

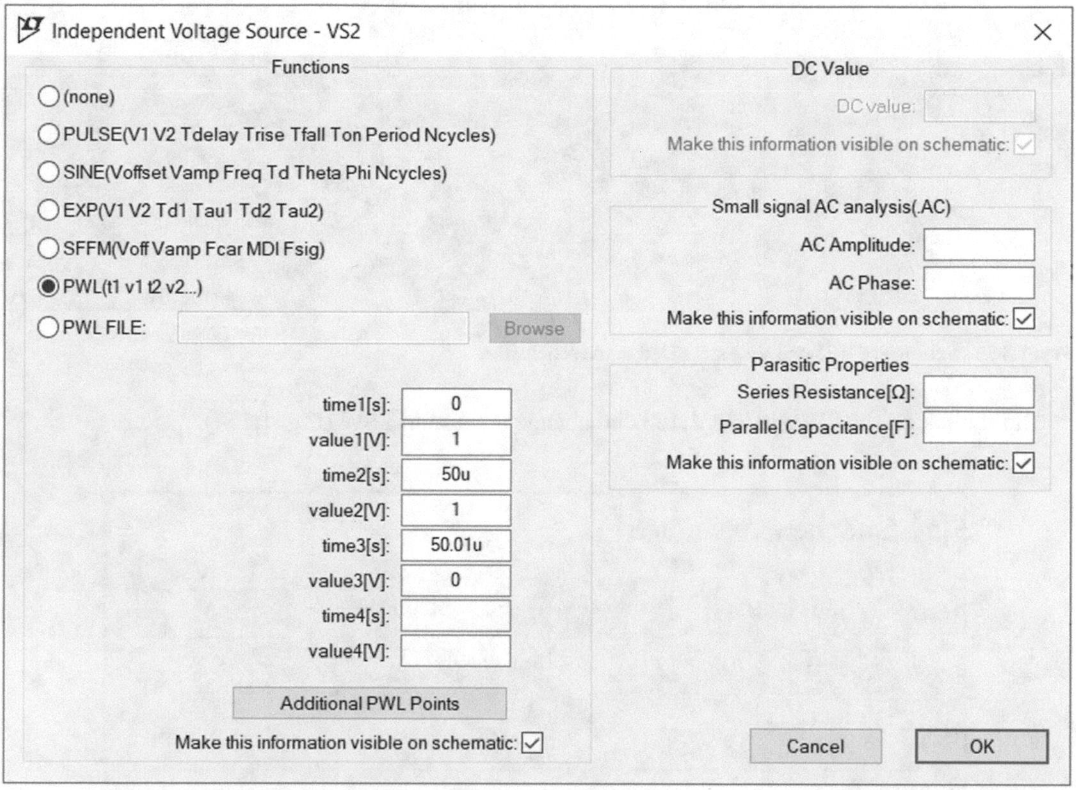

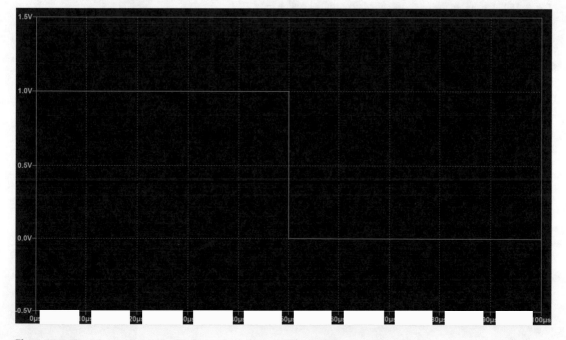

Fig. 1.151 Settings of VS2 voltage source

Fig. 1.152 Waveform generated with the settings shown in Fig. 1.151

Now your schematic looks like Fig. 1.153.

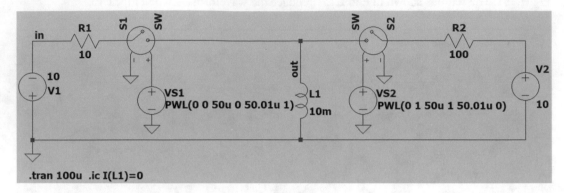

Fig. 1.153 Settings of VS1 and VS2 are added to the schematic

Right click on the SW's in Fig. 1.153 and change them to MYSW (Fig. 1.154).

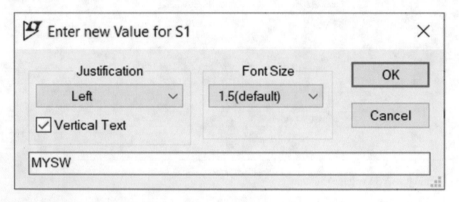

Fig. 1.154 Changing the name of voltage controlled switch to MYSW

Now your schematic looks like Fig. 1.155.

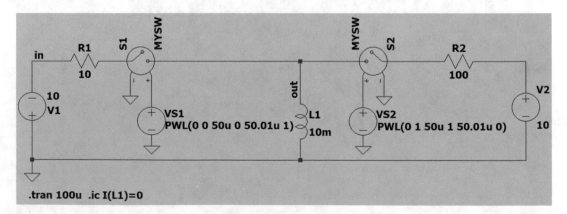

Fig. 1.155 New name of voltage controlled switches are added to the schematic

If you run the simulation, the error message shown in Fig. 1.156 appears. The model for the voltage switches is not given to the LTspice. That is why the error message shown in Fig. 1.156 appears.

Fig. 1.156 Model of voltage controlled switches are not defined

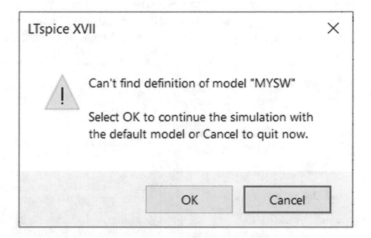

We need to use the .model command in order to determine the simulation model of switch S1 and S2. Add the .model MYSW SW(Ron=1m Roff=1Meg Vt=.5 Vh=0) line to the schematic (Fig. 1.157).

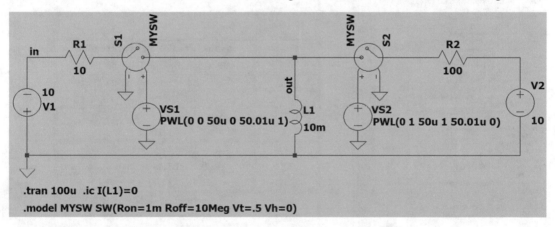

Fig. 1.157 Model of voltage controlled switches are defined with .model command

The schematic shown in Fig. 1.158 can be used to simulate the circuit as well.

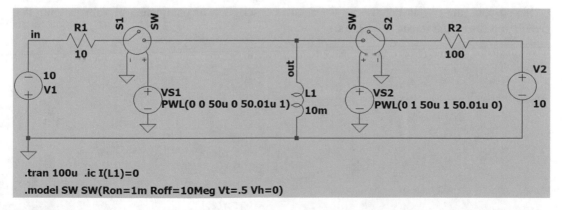

Fig. 1.158 LTspice equivalent of Fig. 1.143

The .model MYSW SW(Ron=1m Roff=1Meg Vt=.5 Vh=0) line tells the LTspice that on state resistance (resistance between switch terminals when the switch is closed) is 1 mΩ, off state resistance (resistance between switch terminals when the switch is opened) is 10 MΩ. When the control voltage is less than 0.5 V, the switch is off and when the control voltage is greater than 0.5 V, the switch is on. .model section of LTspice help gives more information about the .model command (Fig. 1.159).

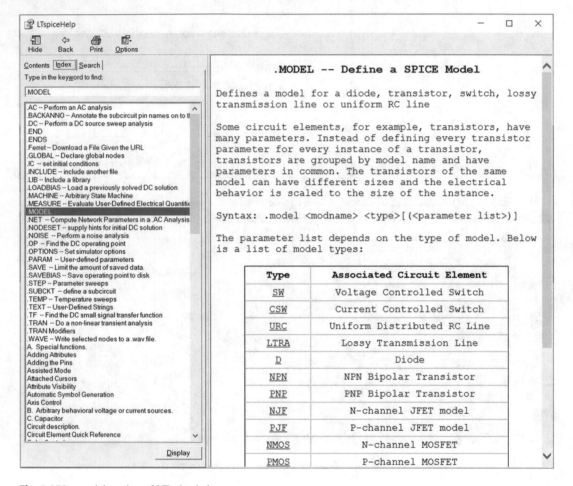

Fig. 1.159   .model section of LTspice help

If you click the <u>SW</u> in Fig. 1.159, the detailed description of parameters of voltage controlled switch appears on the screen (Fig. 1.160).

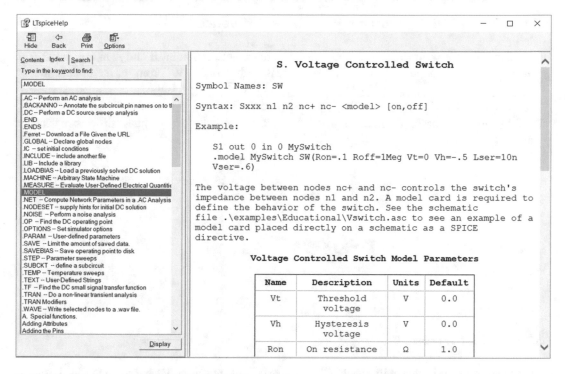

Fig. 1.160 Voltage controlled switch section of LTspice help

Run the simulation. The result shown in Fig. 1.161 appears.

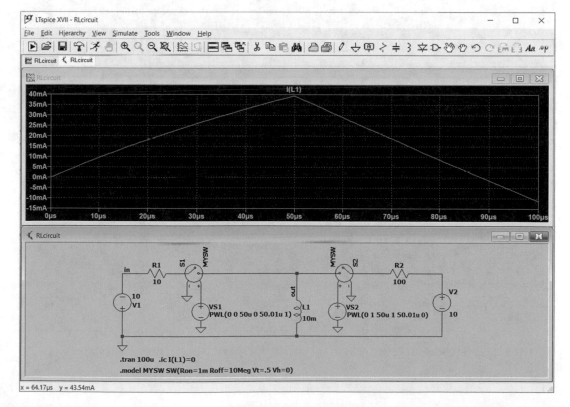

Fig. 1.161 Simulation result

1.11   Example 9: Measurement of Phase Difference

In this example, we want to measure the phase difference between point B and A in Fig. 1.162. The input voltage has frequency of 60 Hz and peak value of 1 V. From basic circuit theory,

$$V_B = \frac{j \times L \times w}{R + j \times L \times w} V_A = \frac{j \times L \times 2pf}{R + j \times L \times 2pf} V_A = \frac{j \times 5\text{m} \times 377}{4 + j \times 5\text{m} \times 377} V_A = \frac{1.885j}{4 + 1.885j} V_A = 0.426 e^{j64.76°} V_A.$$

V_A and V_B show the phasor of voltage of nodes A and B, respectively. So, the phase difference between point B and A is 64.76°.

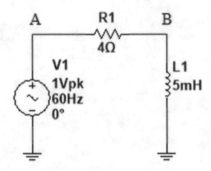

Fig. 1.162 Circuit for Example 9

Draw the schematic shown in Fig. 1.163. Settings of voltage source V1 is shown in Fig. 1.164.

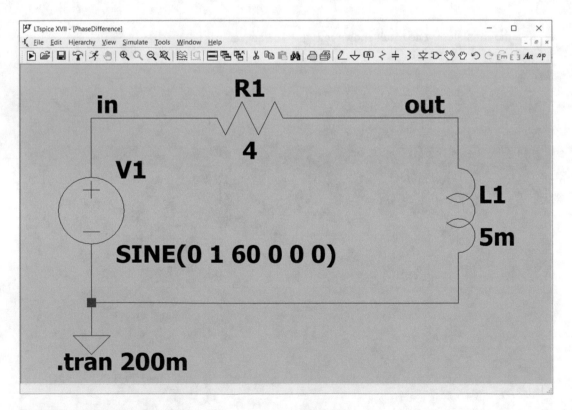

Fig. 1.163 LTspice equivalent of Fig. 1.162

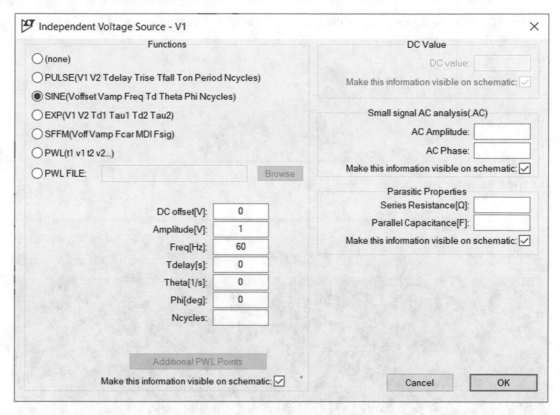

Fig. 1.164 Settings of voltage source V1

Run the simulation and click on the node "in" and node "out" to see their voltages (Fig. 1.165).

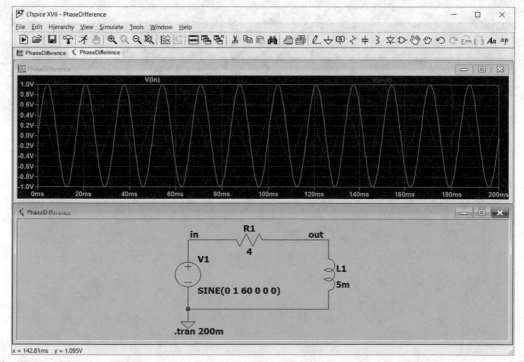

Fig. 1.165 Simulation result

Maximize the graph page and use the magnifier icon (Fig. 1.166) to zoom into the steady-state portion of the graph (Fig. 1.167).

Fig. 1.166 Magnifier icon

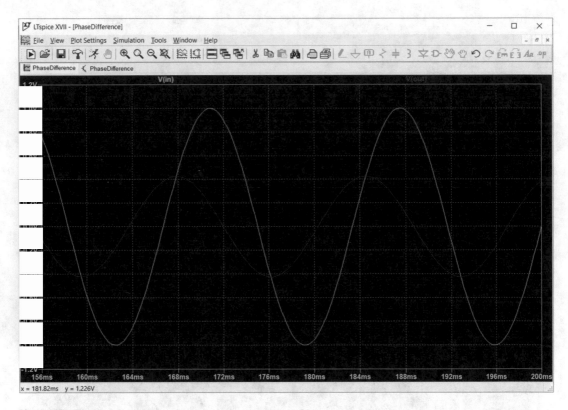

Fig. 1.167 Zoomed output

Add two cursors to the graph and measure the time difference between the starting points of the waveforms (Fig. 1.168). According to Fig. 1.169, the time difference between the starting points of the waveforms is about 3 ms.

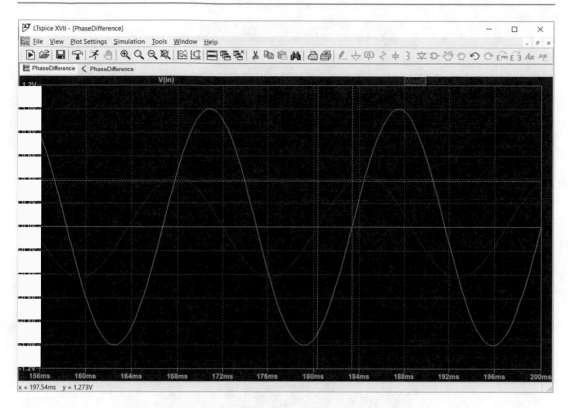

Fig. 1.168 Measurement of time difference between the two waveforms

Fig. 1.169 Time difference between the two waveforms is around 3 ms

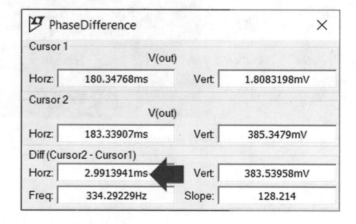

The phase difference is calculated with the aid of commands shown in Fig. 1.170. The obtained result is quite close to the correct value.

```
Command Window                              ⊙
   >> T=1/60;
   >> Delta=2.9913941e-3;
   >> DeltaPhi=Delta/T*360


   DeltaPhi =


       64.6141

fx >>
```

Fig. 1.170 MATLAB calculations

1.12   Example 10: Calculation of RMS and Average Values

This example shows how to calculate the RMS and average values in LTspice. Draw the schematic shown in Fig. 1.171. Settings of voltage source V1 is shown in Fig. 1.172. The settings shown in Fig. 1.172 produce $10 + 20 \sin (2\pi \times 60 \times t + 30°)$ V.

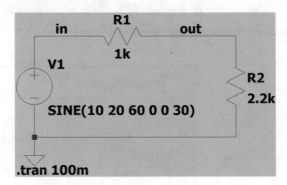

Fig. 1.171 LTspice schematic of Example 10

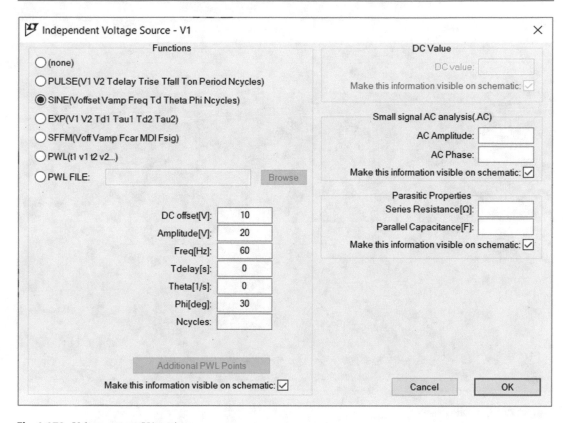

Fig. 1.172 Voltage source V1 settings

Hold down the Ctrl key and click on the I(R2) (Fig. 1.173). After clicking, the window shown in Fig. 1.174 appears.

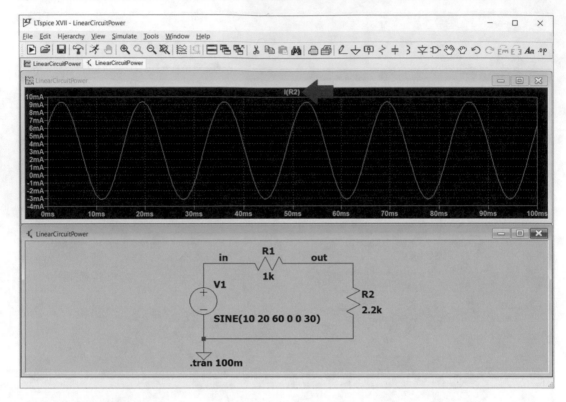

Fig. 1.173   Simulation result

Fig. 1.174   Average and
RMS values for [0,
100 ms] interval

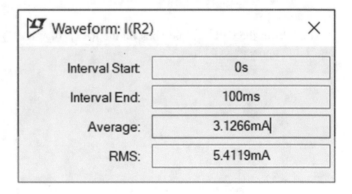

According to Fig. 1.174, the average and RMS values for interval [0, 100 ms] are 3.1266 mA and
5.4119 mA, respectively. The MATLAB commands shown in Fig. 1.175 shows that obtained results
are correct.

Fig. 1.175 MATLAB
calculations

```
Command Window                                                    ⊙
  >> format long
  >> R1=1e3;R2=2.2e3;f=60;w=2*pi*f;phi0=pi/6;
  >> syms t
  >> V1=10+20*sin(w*t+phi0);
  >> I=V1/(R1+R2);
  >> T1=0;T2=100e-3;
  >> eval(1/(T2-T1)*int(I,t,T1,T2))

  ans =

      0.003125000000000

  >> eval(sqrt(1/(T2-T1)*int(I^2,t,T1,T2)))

  ans =

      0.005412658773653

fx >> |
```

Right click on the time axis and enter 69 ms to the Left box (Fig. 1.176). After clicking the OK button, the [69 ms, 100 ms] portion of graph is shown on the screen (Fig. 1.177).

```
Horizontal Axis                                                          ×

   Quantity Plotted:  time                                    |  Eye Diagram  |

                            Axis Limits
   Left:  [  69ms  ]      tick:  [  3ms  ]        Right:  [  100ms  ]

   □ Logarithmic           |  Cancel  |                     |  OK  |
```

Fig. 1.176 Horizontal Axis window

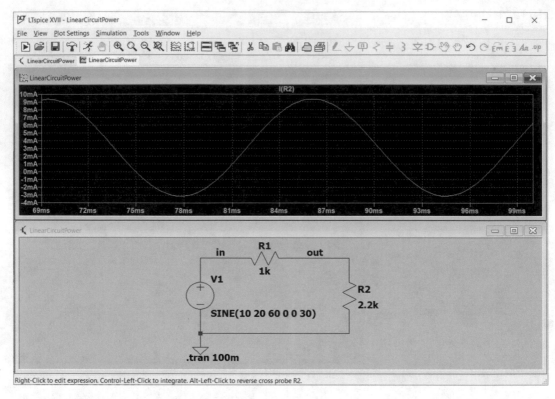

Fig. 1.177 Output graph is shown for [69 ms, 100 ms] interval

Hold down the control and click on the I(R2). After clicking, the window shown in Fig. 1.178 appears.

Fig. 1.178 Average and RMS values for [69 ms, 100 ms] interval

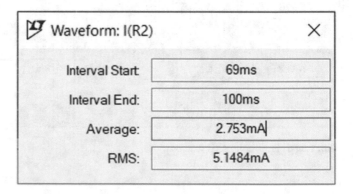

According to Fig. 1.178, the average and RMS values for interval [69 ms, 100 ms] are 2.753 mA and 5.1484 mA, respectively. The MATLAB commands shown in Fig. 1.179 shows that obtained results are correct. Note that LTspice uses the portion of graph that is shown on the screen to calculate the average and RMS values. The portion that is not shown on the screen has no effect on the calculation of average and RMS values.

If you want to calculate the average and RMS values accurately, you need to select an integer number of cycles from the steady-state portion of waveform. For instance, assume that the frequency of waveform is 50 Hz and the circuit is in steady state for t > 10 ms. In this case, you can right click on the time axis and enter 11 ms to the Left box and 31 ms to the Right box. Since the time difference between these two values is 20 ms, one full cycle is shown on the screen. Now you can hold down the Ctrl key and click the title of waveform to see its average and RMS values.

Fig. 1.179 MATLAB
calculations

```
Command Window                                                    ⊙

  >> format long
  >> R1=1e3;R2=2.2e3;f=60;w=2*pi*f;phi0=pi/6;
  >> syms t
  >> V1=10+20*sin(w*t+phi0);
  >> I=V1/(R1+R2);
  >> T1=69e-3;T2=100e-3;
  >> eval(1/(T2-T1)*int(I,t,T1,T2))

ans =

    0.002751041151000

  >> eval(sqrt(1/(T2-T1)*int(I^2,t,T1,T2)))

ans =

    0.005148861353919

fx >> |
```

You can measure the average power as well. After running the simulation, hold down the Alt key and click on the component that you want to measure its average power. After clicking, the power waveform of the component is shown on the screen. If you hold down the Ctrl key and click on the waveform label, the average value of power waveform will be shown.

Let's measure the average power of voltage source V1. The power waveform of voltage source V1 is shown in Fig. 1.180. Hold down the Ctrl key and click on the V(in)*I(V1). According to Fig. 1.181, the average value of power drawn from voltage source V1 is 93.745 mW.

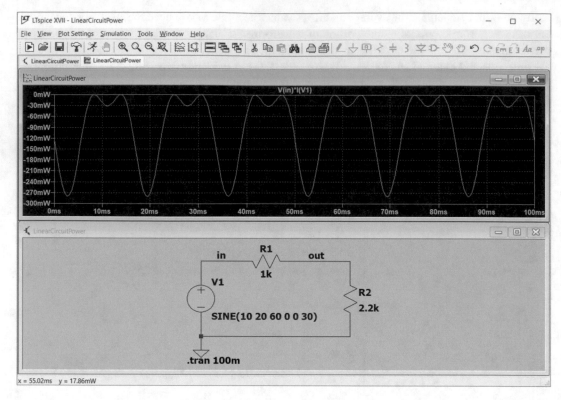

Fig. 1.180   Graph of V(in)*I(V1)

Fig. 1.181   Average and
integral of Fig. 1.180

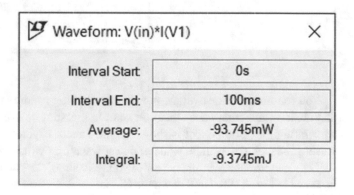

The MATLAB calculations shown in Fig. 1.182 proves that the obtained result is correct.

Fig. 1.182 MATLAB calculations

```
Command Window                                          ⊙
>> format long
>> R1=1e3;R2=2.2e3;f=60;w=2*pi*f;phi0=pi/6;
>> syms t
>> V1=10+20*sin(w*t+phi0);
>> I=V1/(R1+R2);
>> T1=0;T2=100e-3;
>> Pavg=eval(1/(T2-T1)*int(V1*I,t,T1,T2))

Pavg =

    0.093750000000000

>> E=eval(int(V1*I,t,T1,T2))

E =

    0.009375000000000

fx >>
```

1.13 Example 11: .meas Command

In the previous example we introduced a method to measure the average and RMS values of wave-forms shown on the screen. You can measure the average and RMS values with the aid of .meas command as well. This example uses the .meas command to measure the average and RMS values of previous example. Add the .meas IRMS RMS –I(V1) to the schematic of previous example (Fig. 1.183). This command measures the RMS of –I(V1) and put the result in a variable named IRMS. I(V1) is the current that enters the + terminal of voltage source V1. So, –I(V1) is the current leaves the + terminal of voltage source V1.

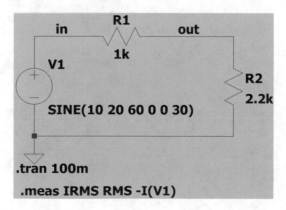

Fig. 1.183 LTspice schematic of Example 11

Run the simulation (Fig. 1.184).

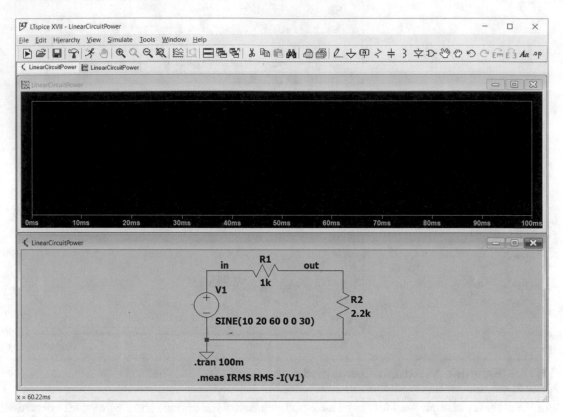

Fig. 1.184 Schematic shown in Fig. 1.183 is run

Now press the Ctrl+L or click the View> SPICE Error Log (Fig. 1.185) to open the SPICE Error Log window.

Fig. 1.185 View> SPICE Error Log

The result is shown in Fig. 1.186. According to Fig. 1.186, the RMS of current for [0, 0.1 s] time interval is 5.41156 mA. The obtained result is the same as the result of previous example.

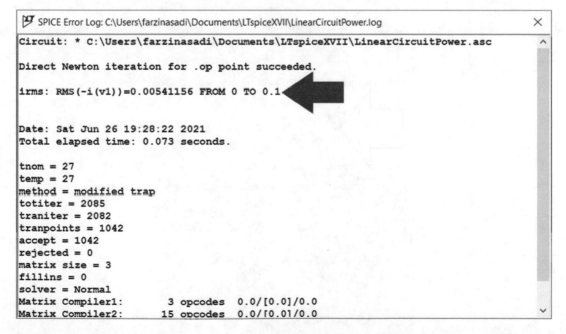

Fig. 1.186 Simulation result

Click the Simulate> Edit Simulation Cmd (Fig. 1.187).

Fig. 1.187 Simulate> Edit Simulation Cmd

Enter 69m to the Time to start saving data box (Fig. 1.188) and click the OK button. The schematic shows the changes that you applied to the simulation command (Fig. 1.189).

Fig. 1.188 Simulation settings

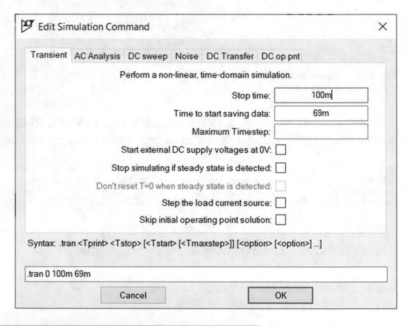

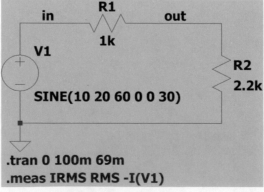

Fig. 1.189 Schematic for settings shown in Fig. 1.188

Run the simulation. Result of simulation is shown in Fig. 1.190. According to Fig. 1.190, the RMS for [69 ms, 100 ms] time interval is 5.14752 mA (Note that 0 and 31 ms in Fig. 1.190 represent the t = 69 ms and t = 100 ms, respectively). The obtained result is the same as the result of previous example.

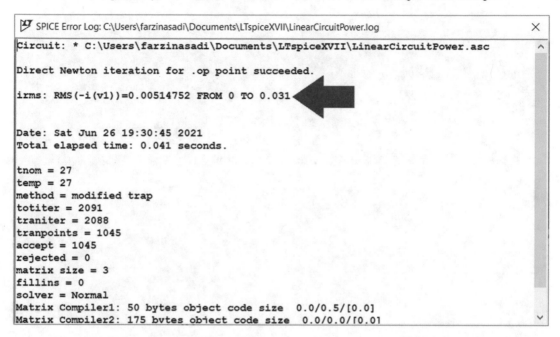

Fig. 1.190 Simulation result

Let's measure the average power of input voltage source. Add the .meas AveragePower AVG −I(V1)*V(V1) command to the schematic (Fig. 1.191). This command measures the average value of −I(V1)*V(V1) on the [0, 100 ms] time interval and put the result in a variable named AveragePower.

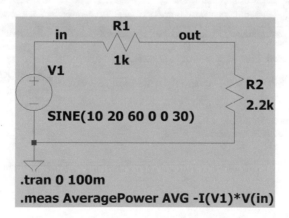

Fig. 1.191 Addition of .meas command to the schematic

Run the simulation (Fig. 1.192).

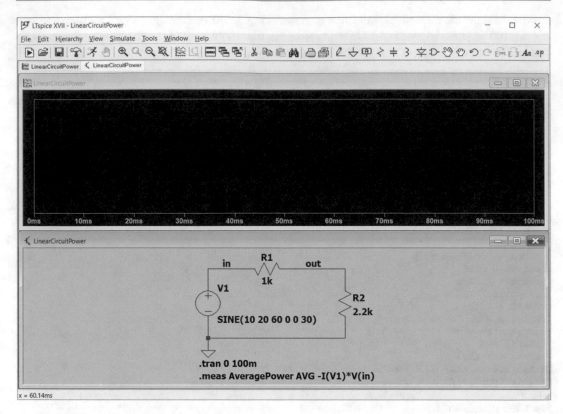

Fig. 1.192 Schematic shown in Fig. 1.191 is run

Press the Ctrl+L (or click the View> SPICE Error Log). The result shown in Fig. 1.193 appears on the screen. According to Fig. 1.193, the average value of power drawn from the source V1 on the [0, 0.1 s] time interval is 93.7329 mA. The obtained result is the same as the result of previous example.

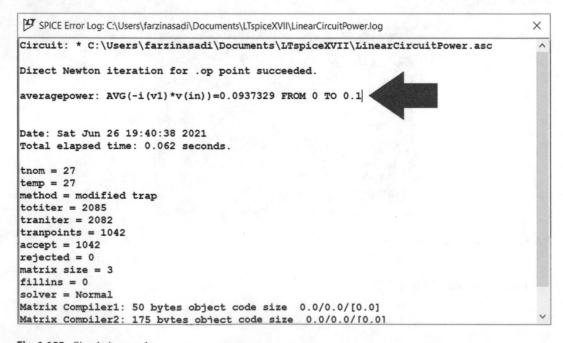

Fig. 1.193 Simulation result

You can use the .meas command to measure the maximum and minimum values as well. For instance, the commands shown in Fig. 1.194 measure the maximum and minimum values of power drawn from the input source. Result is shown in Fig. 1.195.

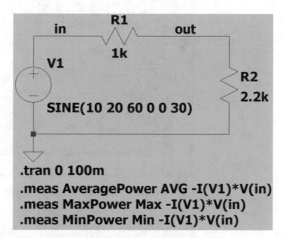

Fig. 1.194 Measurement of maximum and minimum with .meas command

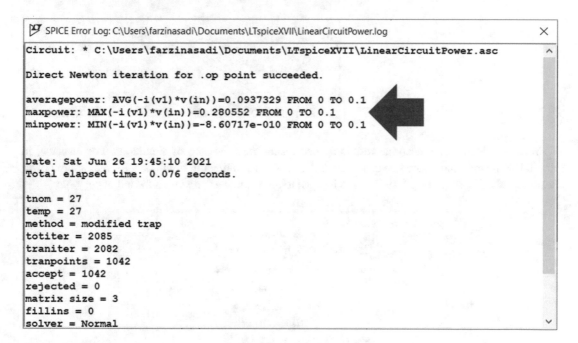

Fig. 1.195   Simulation result

According to the result obtained in Fig. 1.195, the maximum power drawn from the input source is 280.552 mW and minimum is about zero. Let's check the result with the aid of cursors. According to Fig. 1.196 the maximum of power waveform is about 281.25 mW which is a little bit bigger than the value suggested by .meas command. According to Fig. 1.197 the minimum is 122.9 nW which can be considered as zero.

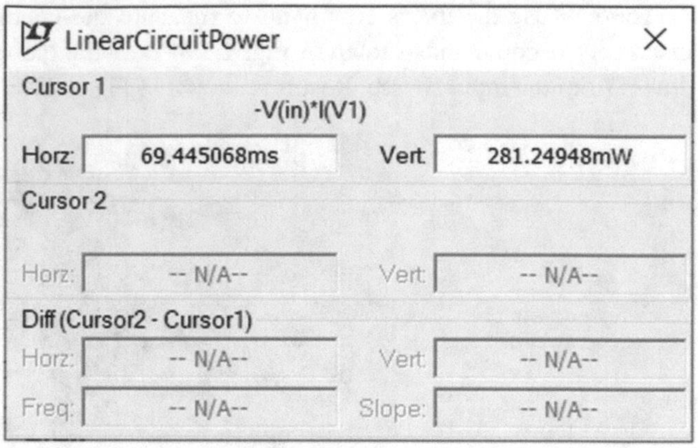

Fig. 1.196 Maximum
of graph is around
281.25 mW

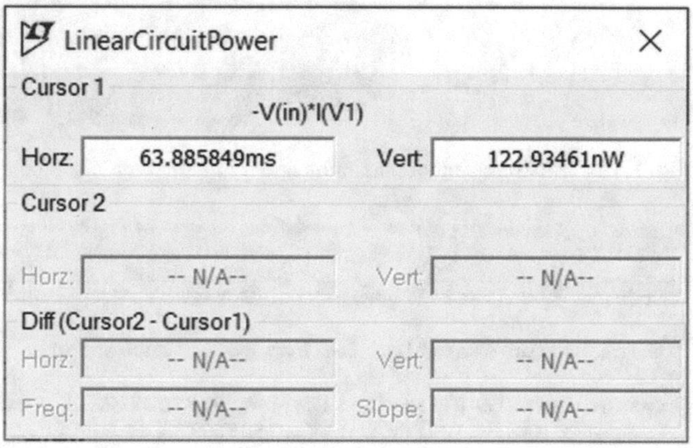

Fig. 1.197 Minimum
of graph is around
122.93 mW

You can obtain more accurate results by decreasing the time step of simulation. For instance, in Fig. 1.198 the maximum step size is limited to 100 ns. After running the simulation, the maximum power becomes 281.249 mW (Fig. 1.199) which is equal to the value we found with the aid of cursors.

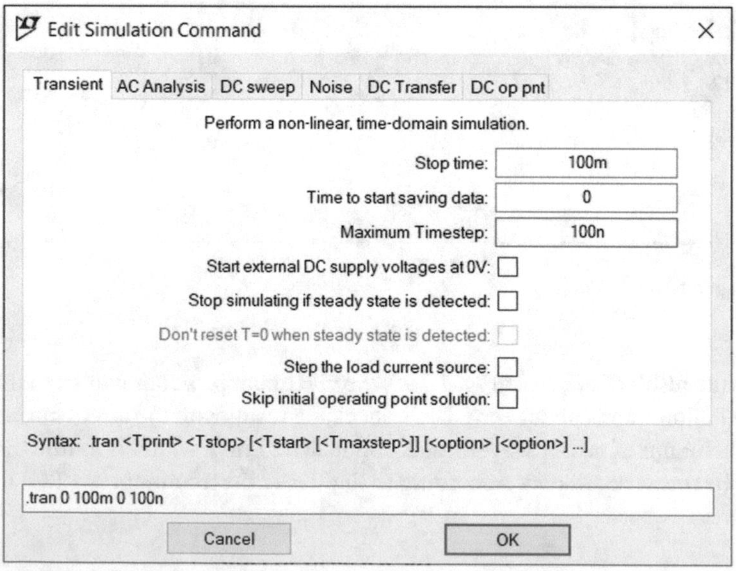

Fig. 1.198 Step size of simulation is set to 100 ns

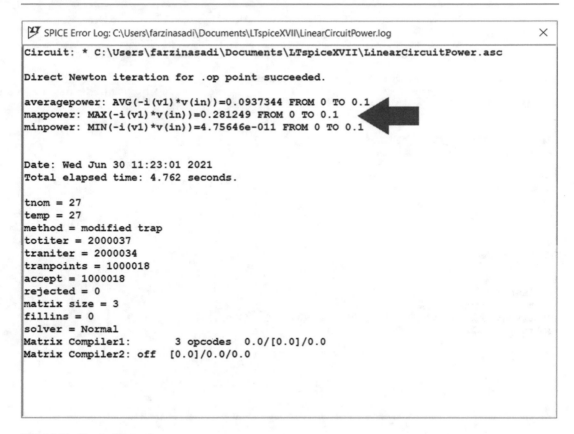

Fig. 1.199 Simulation result

You can calculate the integrals of a waveform with the aid of .meas commands as well. For instance, the command in Fig. 1.200, calculate the integral of power waveform in the [69 ms, 89 ms] interval (Fig. 1.201).

Fig. 1.200 Measurement of energy with .meas command

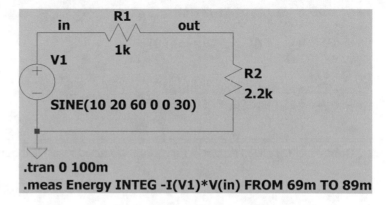

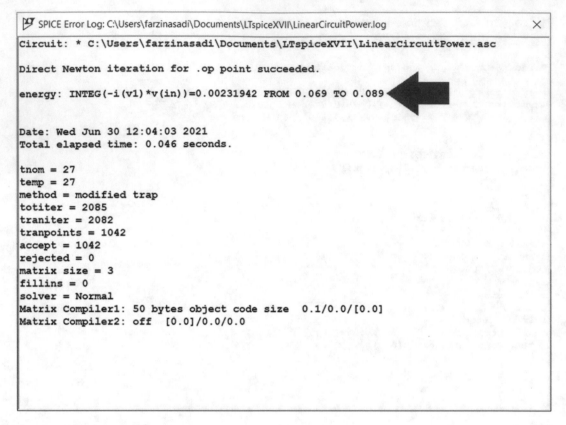

Fig. 1.201 Simulation result

Let's check the obtained result. The following MATLAB code calculates the integral of power waveform on the [69 ms, 89 ms] time interval.

```
format long
f=60;phi0=pi/6;R1=1e3;R2=2.2e3;
syms t
Vin=10+20*sin(2*pi*f*t+phi0);
I=Vin/(R1+R2);
p=Vin*I;
time=linspace(0,0.1,2500);
Energy=eval(int(p,t,69e-3,89e-3))
```

Output of the code is shown in Fig. 1.202. Obtained result is the same as Fig. 1.201.

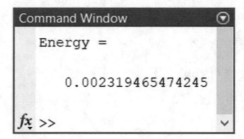

Fig. 1.202 Output of MATLAB code

Peak-Peak values can be calculated with .meas command as well. For instance, the command shown in Fig. 1.203 calculates the peak-peak of circuit current. According to the result shown in Fig. 1.204, the peak-peak current is 12.4851 mA.

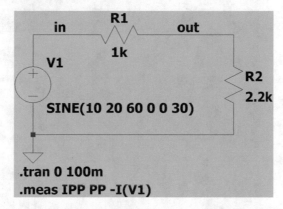

Fig. 1.203 Measurement of peak-peak with .meas command

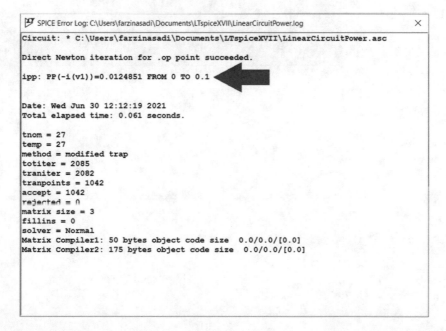

Fig. 1.204 Simulation result

Let's check the obtained result. The following MATLAB code draws the graph of circuit current.

```
f=60;phi0=pi/6;R1=1e3;R2=2.2e3;
syms t
Vin=10+20*sin(2*pi*f*t+phi0);
I=Vin/(R1+R2);
ezplot(I, [0, 0.1])
```

Output of this code is shown in Fig. 1.205. You can click on the graph to read the peak values (Fig. 1.206). According to Fig. 1.207, the peak-peak value is 12.5 mA which is quite close to LTspice result.

Fig. 1.205 Output of MATLAB code

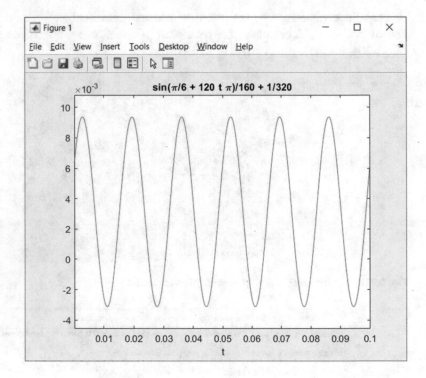

Fig. 1.206 Maximum and minimum of graph shown in Fig. 1.205

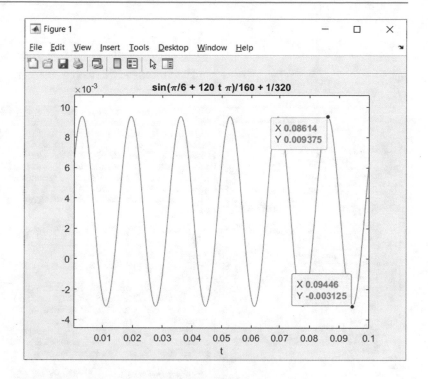

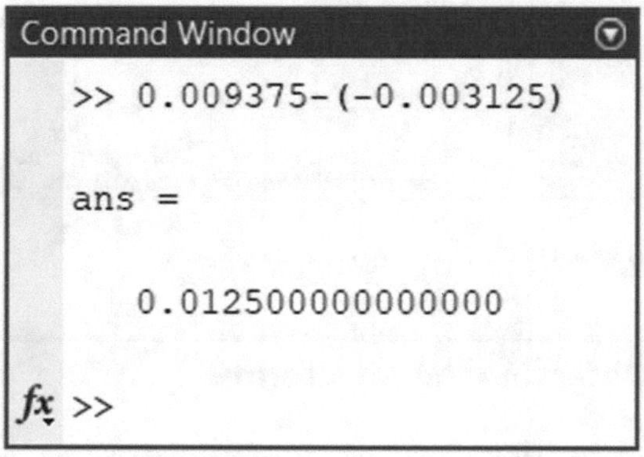

Fig. 1.207 Peak-peak of Fig. 1.206 is around 12.5 mA

The help page of .meas command (Fig. 1.208) is the best reference to learn more details about this command.

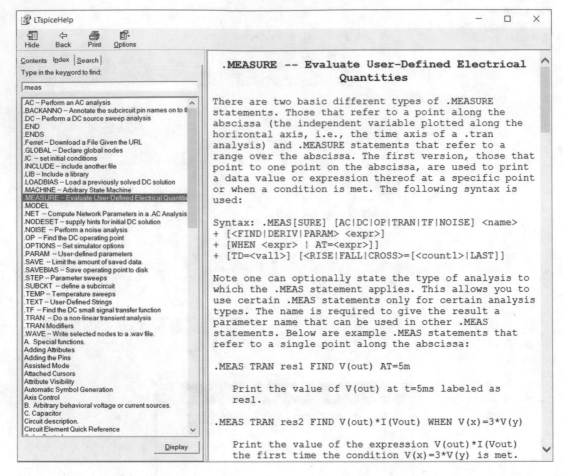

Fig. 1.208 .meas section of LTspice Help

1.14   Example 12: Observing the Waveform

In the previous examples we learned that after running the simulation, holding down the Alt key and clicking on the component shows the power waveform of the component. In this example, we learn another way to draw the power waveform of a component. In this example, we will draw the power waveforms of the circuit shown in Fig. 1.209. The input voltage of this circuit is $120\sqrt{2}\sin\left(2\times\pi\times60\times t+45°\right)\cong169.7\sin\left(2\times\pi\times60\times t+45°\right)$. The initial current of the inductor is zero.

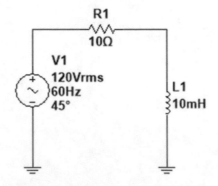

Fig. 1.209 Circuit of Example 12

Draw the schematic shown in Fig. 1.210. The voltage source is used to measure the circuit current. It measures the current that enter its + terminal. Settings of voltage source V2 is shown in Fig. 1.211.

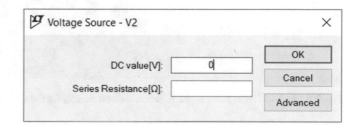

Fig. 1.210 LTspice equivalent of Fig. 1.209

Fig. 1.211 Settings of voltage source V2

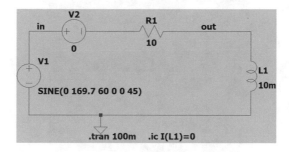

Run the simulation (Fig. 1.212).

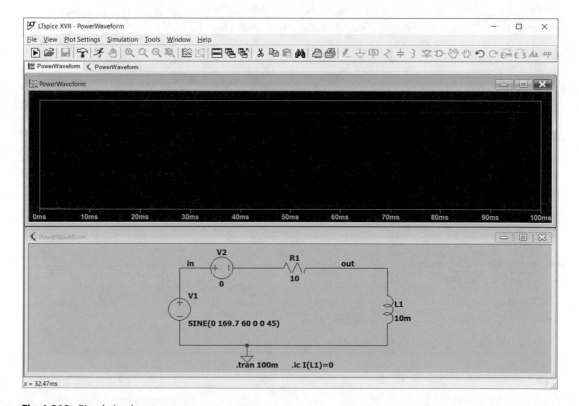

Fig. 1.212 Simulation is run

Right click on the black window and click the Add Traces (Fig. 1.213).

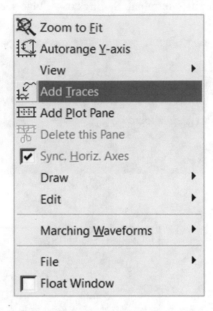

Fig. 1.213 Add Traces permits you to define desired outputs for the simulation

Enter −I(V2)*V(in) to the Expression(s) to add box (Fig. 1.214) and click the OK button. After clicking the OK button, the graph shown in Fig. 1.215 appears. −I(V2)*V(in) is the instantaneous power of voltage source V1. You can obtain the instantaneous power graph of voltage source V1 by entering I(V1)*V(in) to the Expression(s) to add box as well (Fig. 1.216).

Fig. 1.214 Add Traces to Plot window

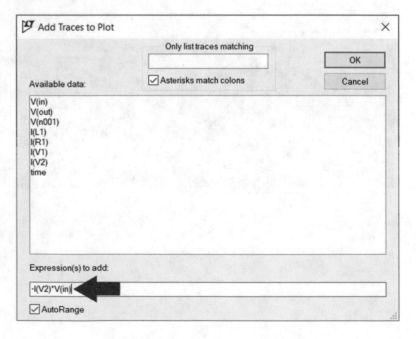

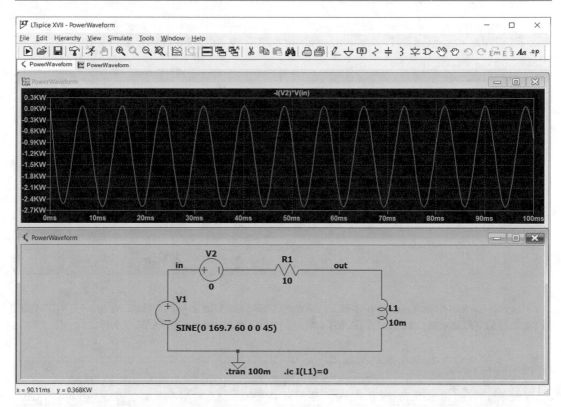

Fig. 1.215   Graph of −I(V2)*V(in)

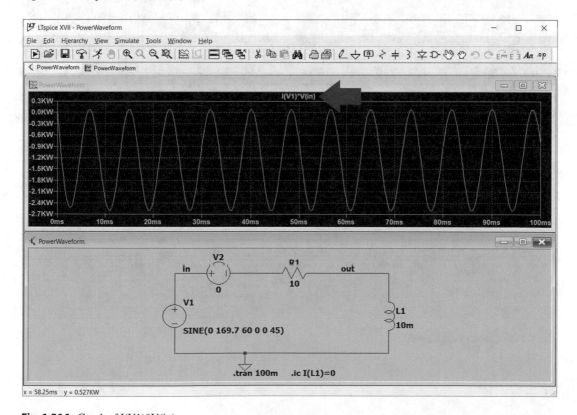

Fig. 1.216   Graph of I(V1)*V(in)

Let's measure the maximum and minimum of the waveform shown in Fig. 1.215. According to Fig. 1.217, the maximum and minimum are about 86.333 W and −2.603 kW, respectively.

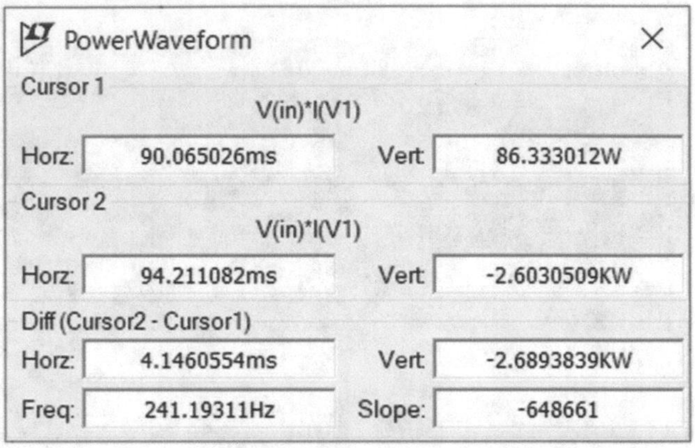

Fig. 1.217 Maximum and minimum of waveform shown in Fig. 1.215

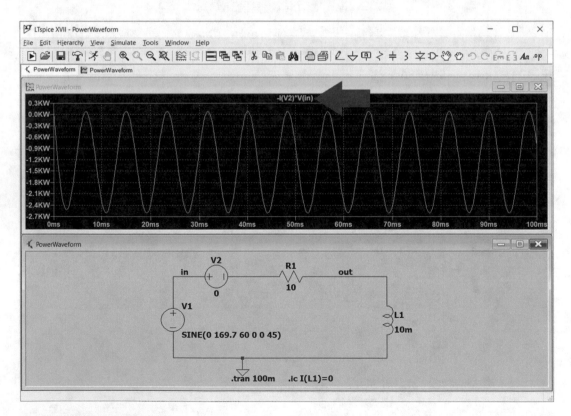

You can measure the average power by holding down the Ctrl key and click on the −I(V2)*V(in) (Fig. 1.218). According to Fig. 1.219, the average power is about −1.266 kW.

Fig. 1.218 Graph of −I(V2)*V(in)

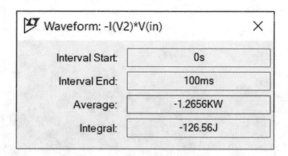

Fig. 1.219 Average and integral of −I(V2)*V(in) for interval of [0, 100 ms]

Right click on the −I(V2)*V(in) (Fig. 1.218). Expression editor window appears (Fig. 1.220). Clear the −I(V2)*V(in) and enter I(V2)*(V(in)-V(out)) and click the OK button (Fig. 1.221). After clicking the OK button, graph of I(V2)*(V(in)-V(out)) appears on the screen (Fig. 1.222). I(V2)*(V(in)-V(out)) is the instantaneous power of resistor.

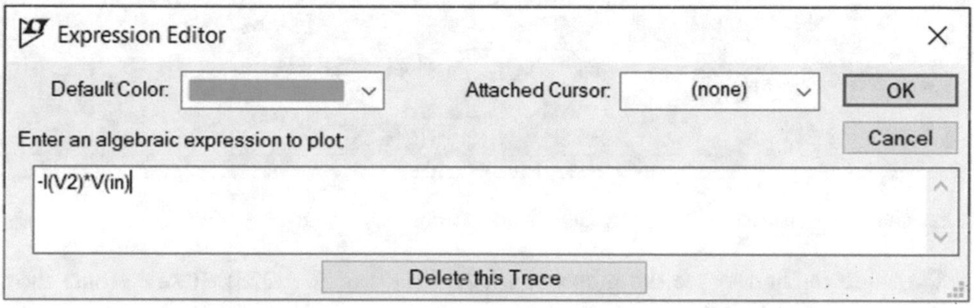

Fig. 1.220 Expression Editor window

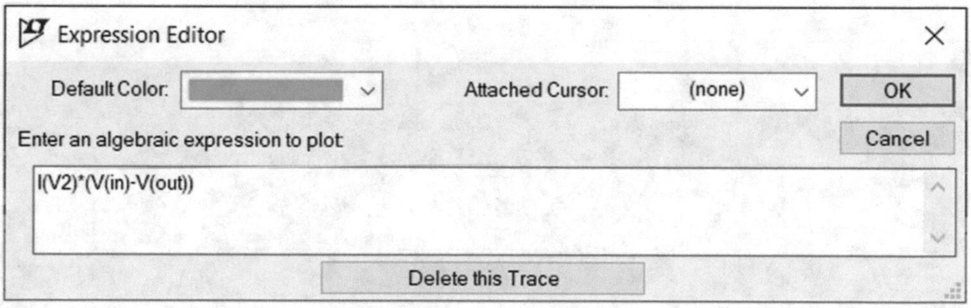

Fig. 1.221 I(V2)*(V(in)-V(out)) is entered to the box

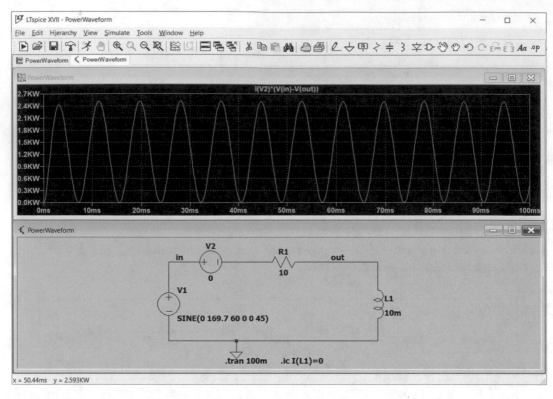

Fig. 1.222 Graph of instantaneous power dissipated in the resistor

Use the same method to see the graph of I(V2)*V(out) (Fig. 1.223). I(V2)*V(out) shows the instantaneous power of inductor.

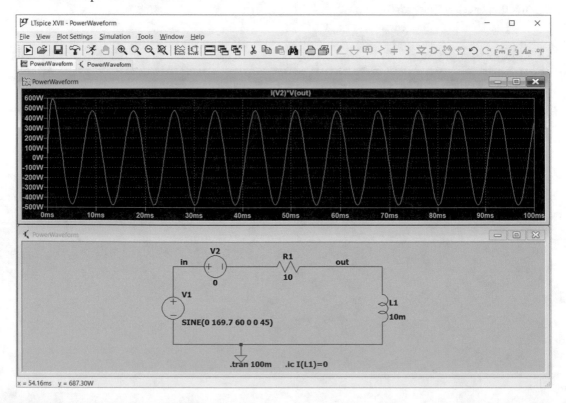

Fig. 1.223 Instantaneous power of inductor

Let's check the obtained results. The following MATLAB code draws the instantaneous power of circuit components.

```
clc
clear all

R1=10;L1=10e-3;f=60;T=1/f;w=2*pi*f;phi0=pi/4;Vm=169.7;
syms i(t) V1(t)
V1=Vm*sin(w*t+phi0);
ode=L1*diff(i,t)+R1*i==V1;
cond=i(0)==0;
iSol(t)=dsolve(ode,cond);

pV1=simplify(-Vm*sin(w*t+phi0)*iSol);
figure(1)
ezplot(pV1,[0 0.1])
title('Instantaneous power of AC source')
grid minor
pR1=simplify(R1*iSol^2);
figure(2)
ezplot(pR1,[0 0.1])
title('Instantaneous power of resistor')
grid minor

VL1=V1-R1*iSol;
pL1=simplify(VL1*iSol);
figure(3)
ezplot(pL1,[0 0.1])
title('Instantaneous power of inductor')
grid minor

% Average power is calculated in the steady state region of
% currennt waveform. t0 must big enough to enter the steady
% state region.
t0=65e-3;
Paverage=eval(1/T*int(V1*iSol,t0,t0+T))
```

After running the code, the results shown in Figs. 1.224, 1.225, 1.226, and 1.227 are obtained.

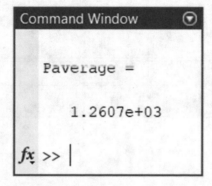

Fig. 1.224 Average value of power drawn from the source

Fig. 1.225 Graph of instantaneous power drawn from the AC source (Min: −2608, Max: 86.59)

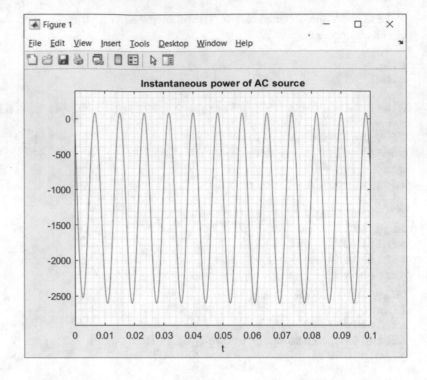

Fig. 1.226 Graph of instantaneous power dissipated in the resistor (Min: 0.0256, Max: 2521)

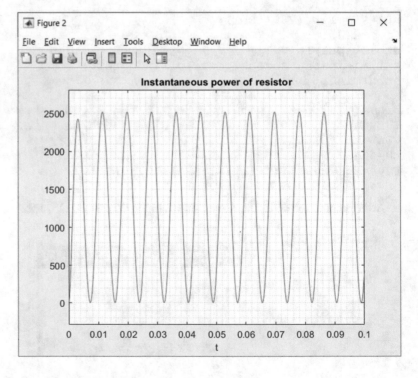

Fig. 1.227 Graph of instantaneous power of inductor (Min: −475.3, Max: 475.3)

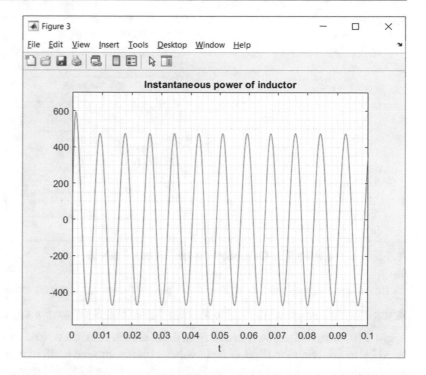

You can use cursors to compare the MATLAB results with LTspice. For instance, according to MATLAB output (Fig. 1.225), the minimum and maximum of instantaneous power of V1 is −2.608 kW and 86.59 W, respectively. According to LTspice result shown in Fig. 1.217, the minimum and maximum of instantaneous power of V1 is −2.603 kW and 86.333 W, respectively. The LTspice results are quite close to the MATLAB results.

If you decrease the step size of LTspice simulation, the minimum and maximum of LTspice graph become closer to the MATLAB graph. For instance, if we decrease the maximum step size to 100 ns (Fig. 1.228), the minimum and maximum of the LTspice graph become −2.608 kW and 86.53 W, respectively (Fig. 1.229).

Fig. 1.228 Simulation settings

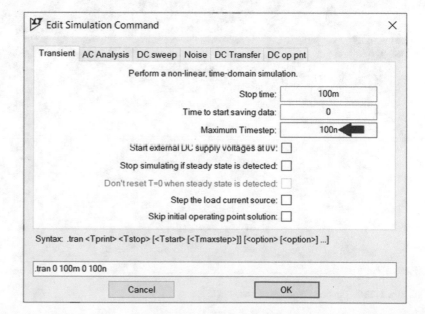

Fig. 1.229 Minimum
and maximum of the
$-I(V2)*V(in)$ graph

PowerWaveform			✕
Cursor 1			
	-I(V2)*V(in)		
Horz:	94.20639ms	Vert:	-2.6076683KW
Cursor 2			
	-I(V2)*V(in)		
Horz:	40.070609ms	Vert:	86.52943W
Diff (Cursor2 - Cursor1)			
Horz:	-54.135781ms	Vert:	2.6941977KW
Freq:	18.472071Hz	Slope:	-49767.4

1.15   Example 13: Calculation of Power Factor

In this example, we want to measure the power factor of previous example. The power factor is defined as $\dfrac{P}{S}$, where P shows the average power and S shows the apparent power. According to Fig. 1.219, the average power drawn from the source is 1.2656 kW. The apparent power can be calculated by multiplying the RMS of input voltage source into the RMS of current drawn from it. So, we need to measure the RMS of input voltage source and the RMS of current drawn from it. Let's measure the RMS of input voltage source. In order to do this, run the schematic of Example 12 and draw the voltage of node "in" (Fig. 1.230).

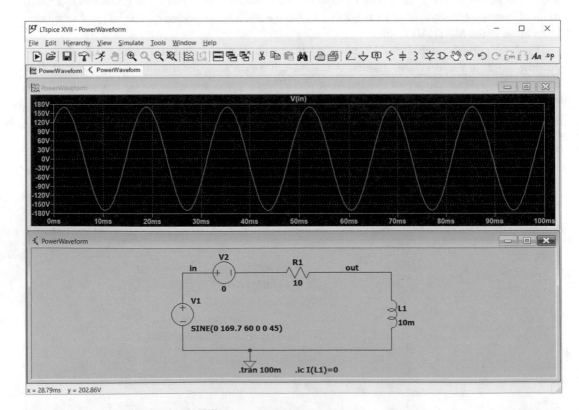

Fig. 1.230 Graph of voltage of node "in"

According to Fig. 1.231, the RMS of voltage of node "in" is 119.88 V.

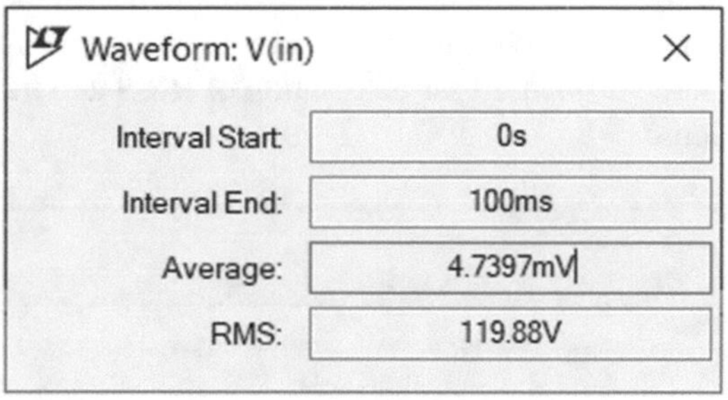

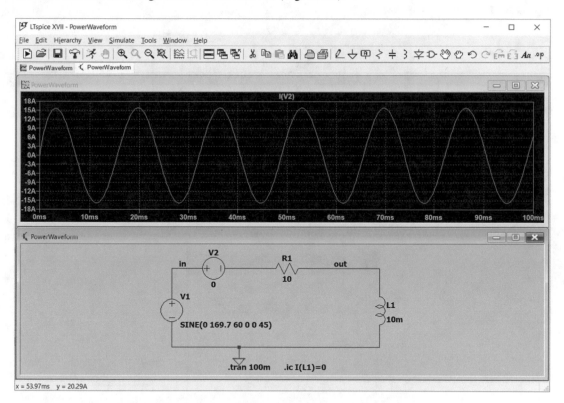

Fig. 1.231 Average and RMS values of node "in" voltage for interval of [0, 100 ms]

Draw the graph of current drawn from voltage source V1. According to Fig. 1.232, the RMS of current drawn from voltage source V1 is 11.27 A (Fig. 1.233).

Fig. 1.232 Graph of current drawn from source V1 (=current enter to the + terminal of V2)

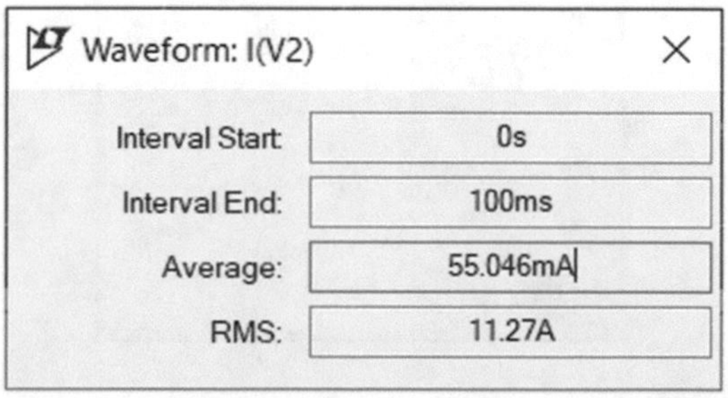

Fig. 1.233 Average and RMS values of current drawn from voltage source V1 for [0, 100 ms] interval

If you want to measure the RMS of current accurately, right click on the time axis and enter 83.333 ms to the Left box (Fig. 1.234) and click OK. Then hold down the Ctrl key and click on the I(V2). Result is shown in Fig. 1.235. With these settings, only the last cycle of the waveform (i.e., [83.333 ms, 100 ms] time interval) is used for calculation of RMS and the transient region of graph does not enter the calculations.

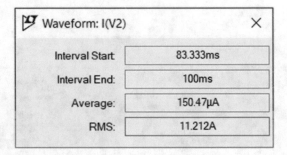

Fig. 1.234   Horizontal Axis window

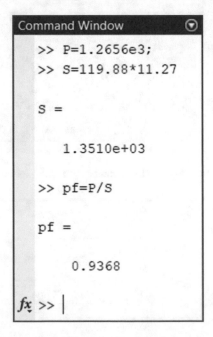

Fig. 1.235   Average and RMS values of current drawn from voltage source V1 for [83.333 ms, 100 ms] interval

The power factor is calculated with the aid of commands shown in Fig. 1.236.

```
>> P=1.2656e3;
>> S=119.88*11.27

S =

    1.3510e+03

>> pf=P/S

pf =

    0.9368

fx >> |
```

Fig. 1.236   MATLAB calculation

Let's check the obtained result. The following MATLAB code calculates the power factor of the circuit.

```
clc
clear all

R1=10;L1=10e-3;f=60;T=1/f;w=2*pi*f;phi0=pi/4;Vm=169.7;

syms i(t) V1(t)
V1=Vm*sin(w*t+phi0);
ode=L1*diff(i,t)+R1*i==V1;
cond=i(0)==0;
iSol(t)=dsolve(ode,cond);

% value of t0 is arbitrary
% however it must be selected
% from the steady state region
% of the graph
t0=65e-3;
Paverage=1/T*int(V1*iSol,t0,t0+T);
IRMS=sqrt(1/T*int(iSol^2,t,t0,t0+T));
VRMS=sqrt(1/T*int(V1^2,t,t0,t0+T));
S=IRMS*VRMS;
pf=eval(Paverage/S)
```

Output of the code is shown in Fig. 1.237. The obtained result is quite close to the result obtained in Fig. 1.236.

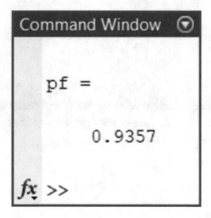

Fig. 1.237 Output of MATLAB code

Since the circuit is linear, the power factor can be calculated with the aid $\cos(\Delta\varphi)$ formula. $\Delta\varphi$ shows the phase difference between the voltage of the source and the current drawn from it. Figure 1.238 uses this method to calculate the power factor of the circuit. Result of this method is the same as Fig. 1.237.

```
Command Window                                    ⊙
>> R1=10;L1=10e-3;f=60;w=2*pi*f;
>> pf=cos(atan(L1*w/R1))

pf =

    0.9357

fx >> |
```

Fig. 1.238 MATLAB code

1.16 Example 14: Thevenin Equivalent Circuit

In this example, we want to find the Thevenin equivalent circuit with respect to the terminals "a" and "b" for the circuit shown in Fig. 1.239.

Fig. 1.239 Circuit of
Example 14

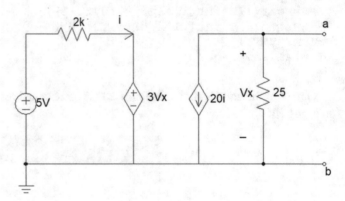

Figure 1.239 contains voltage-dependent voltage source and current-dependent current source. The voltage-dependent voltage source and current-dependent current source can be simulated with the aid of blocks shown in Figs. 1.240 and 1.241, respectively.

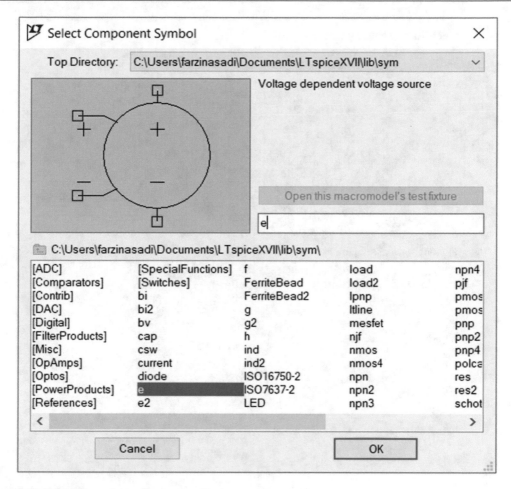

Fig. 1.240 Voltage-dependent voltage source block

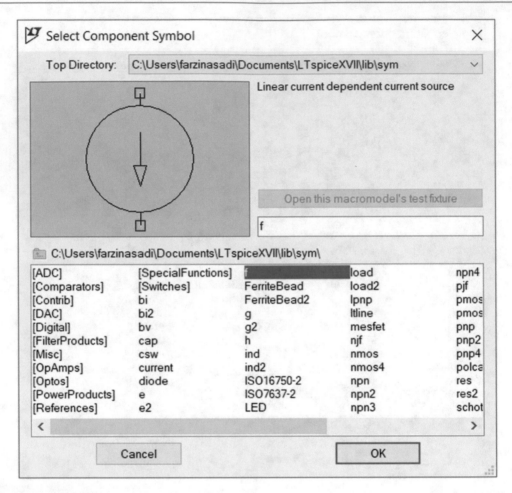

Fig. 1.241 Linear current-dependent current source block

Draw the schematic shown in Fig. 1.242.

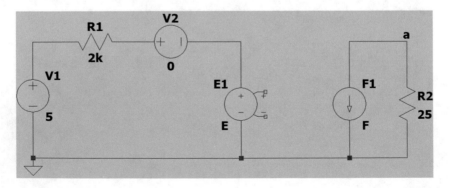

Fig. 1.242 First sketch for circuit shown in Fig. 1.239

Connect the negative control terminal of E1 to ground. Connect the positive control terminal of E1 to a label with name "a" (Fig. 1.243).

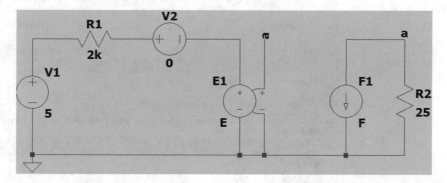

Fig. 1.243   Negative control terminal of E1 is connected to ground and positive control terminal is connected to node "a"

Figure 1.244 is equivalent to the circuit shown in Fig. 1.243. You can use Fig. 1.244 if you prefer.

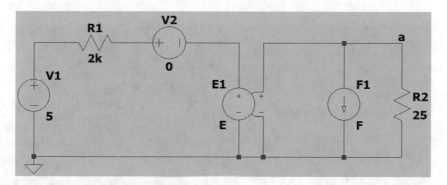

Fig. 1.244   This schematic is equivalent to Fig. 1.243

Right click on the voltage-dependent voltage source E1 and enter 3 to the Value box (Fig. 1.245). Then click the OK button. After clicking the OK button, the schematic changes to what is shown in Fig. 1.246. Gain of the voltage-dependent voltage source E1 is shown behind it.

Fig. 1.245 Value of
voltage gain for
voltage-dependent
voltage source E1 is
shown on the schematic

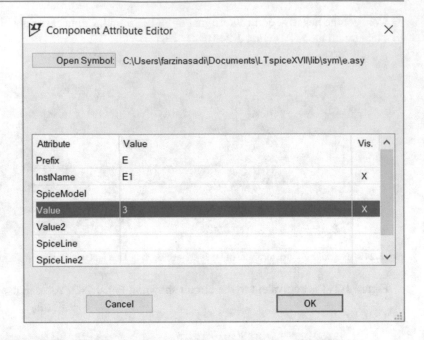

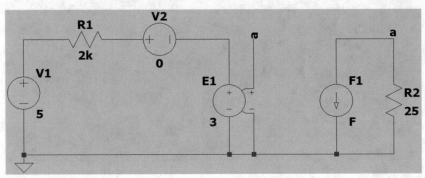

Fig. 1.246 Settings of E1 is shown on the schematic

Let's simulates the current-dependent current source of Fig. 1.239. In order to do this, right click on the current-dependent current source F1 and enter V2 20 to the Value box (Fig. 1.247). Then click the OK button. After clicking the OK button, the schematic changes to what is shown in Fig. 1.248. Note that voltage source V2 is used to determine the current that controls the current-dependent current source F1.

Fig. 1.247 Settings of
current-dependent
current source F1

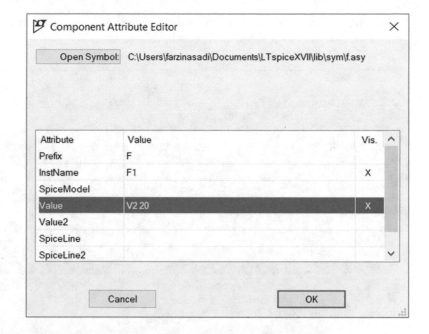

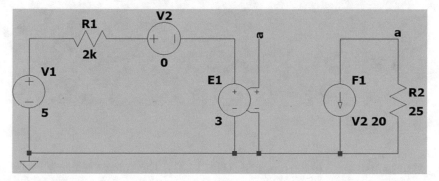

Fig. 1.248 Settings of current-dependent current source F1 are shown on the schematic

Add the command shown in Fig. 1.249 to the schematic.

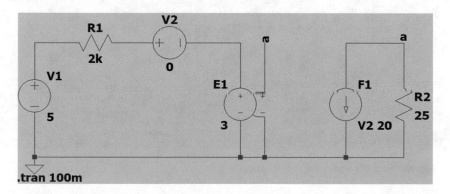

Fig. 1.249 Transient command is added to the schematic

Run the simulation and measure the voltage of node "a." According to Fig. 1.250, the open circuit voltage of node "a" is −5 V. So, the Thevenin voltage is −5 V.

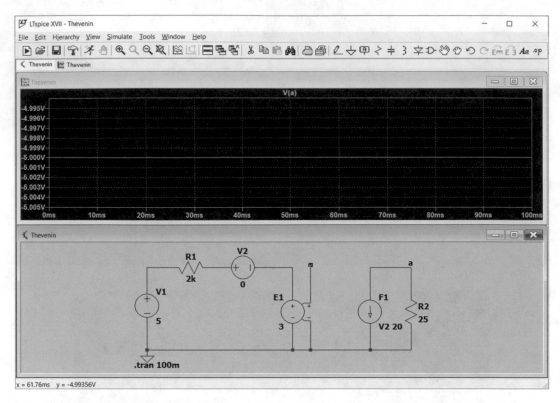

Fig. 1.250 Simulation result

Add the voltage source V3 to the schematic (Fig. 1.251). The voltage source V3 short circuit the output of circuit. The ratio of Thevenin voltage to the current pass through V3 gives the Thevenin resistor.

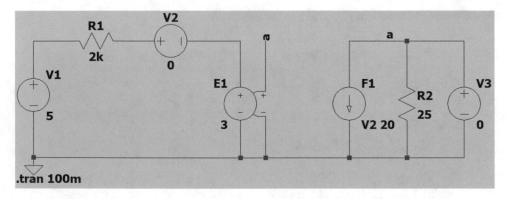

Fig. 1.251 Measurement of output short circuit current

Run the simulation and measure the current of voltage source V3. According to Fig. 1.252, the current of voltage source V3 is −50 mA. So, the Thevenin resistor is $\dfrac{-5\,\text{V}}{-50\,\text{mA}} = 100\,\Omega$.

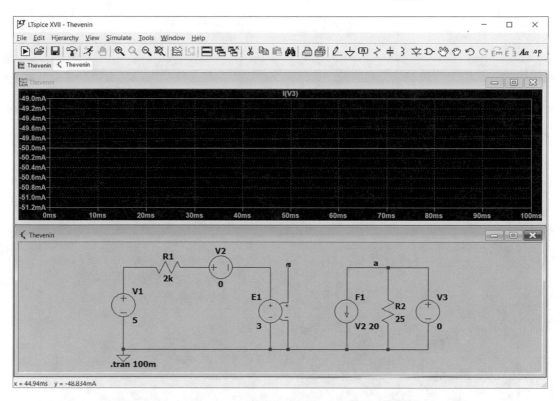

Fig. 1.252 Simulation result

The Thevenin equivalent circuit of Fig. 1.239 is shown in Fig. 1.253.

Fig. 1.253 Thevenin equivalent circuit of studied example (circuit shown in Fig. 1.239)

1.17   Example 15: Current-Dependent Voltage Source

We used current-dependent current source and voltage-dependent voltage source in the previous example. This example shows how to simulate circuits contain current-dependent voltage sources. Consider the circuit shown in Fig. 1.254. We want to use LTspice to measure the voltage of node "out." From basic circuit theory, we expect the voltage of node "out" to be 10 V.

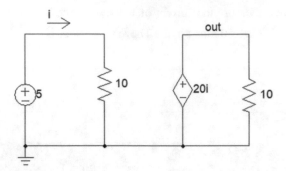

Fig. 1.254 Circuit for Example 15

In LTspice, current-dependent voltage sources can be simulated with the aid of block shown in Fig. 1.255.

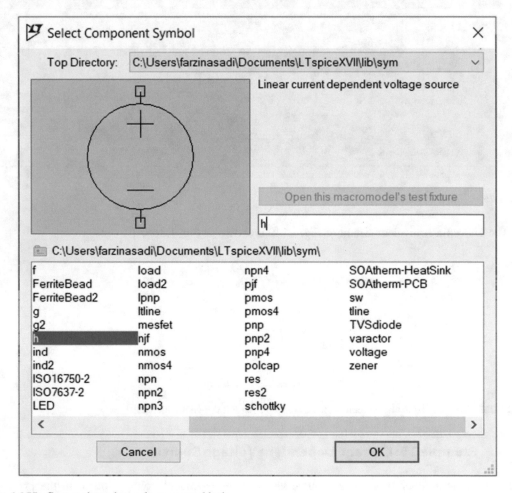

Fig. 1.255 Current-dependent voltage source block

Draw the schematic shown in Fig. 1.256. Settings of current-dependent voltage source H1 is shown in Fig. 1.257.

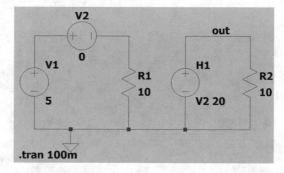

Fig. 1.256 LTspice equivalent of Fig. 1.254

Fig. 1.257 Current-
dependent voltage
source settings

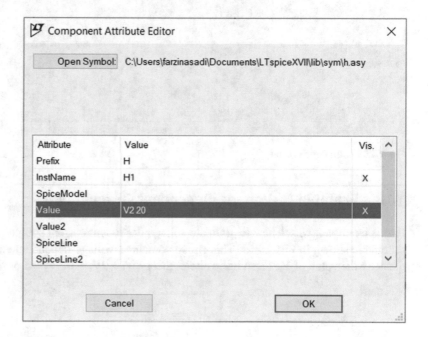

Run the simulation and draw the voltage of node "out." According to Fig. 1.258, the voltage of node "out" is 10 V.

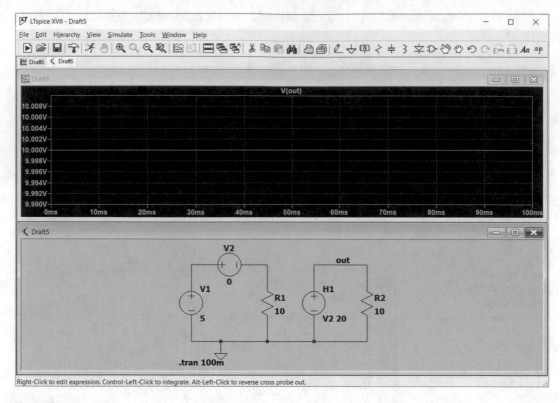

Fig. 1.258 Simulation result

1.18   Example 16: Voltage-Dependent Current Source

In this example, we see how to simulate voltage-dependent current sources in LTspice. Consider the schematic shown in Fig. 1.259. This circuit contains a voltage-dependent current source. We want to measure the value of V_x. From basic circuit theory, $V_x + 2V_x - 0.3V_x = 5$ or $V_x = 1.8518$ V.

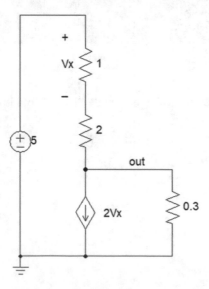

Fig. 1.259 Circuit for Example 16

The voltage-dependent current source can be simulated with the aid of block shown in Fig. 1.260.

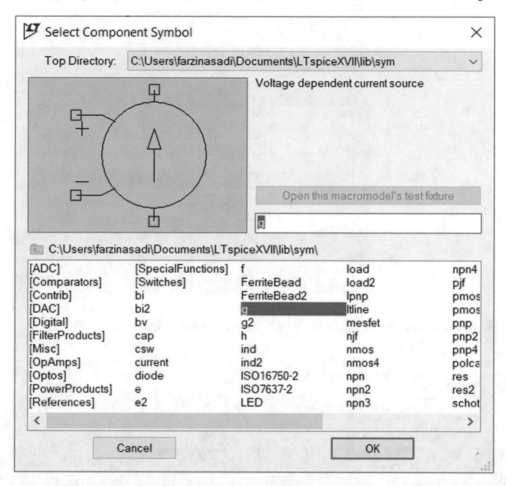

Fig. 1.260 Voltage-dependent current source block

Draw the schematic shown in Fig. 1.261. Settings of voltage-dependent current source G1 is shown in Fig. 1.262.

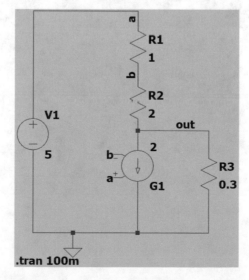

Fig. 1.261 LTspice equivalent for Fig. 1.259

Fig. 1.262 Settings of voltage-dependent current source

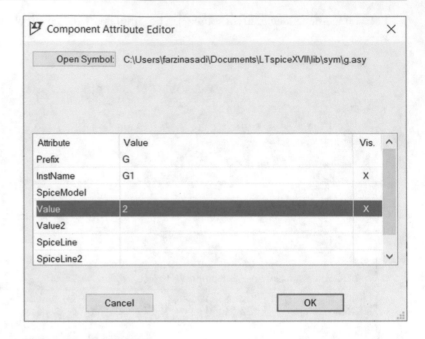

Run the simulation. According to Fig. 1.263, the voltage difference between nodes "a" and "b" is 1.8518 V.

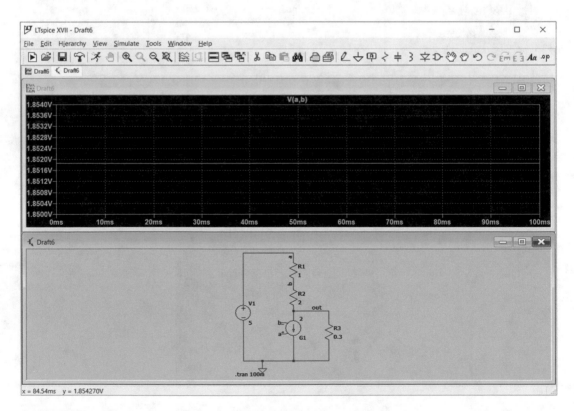

Fig. 1.263 Simulation result

1.19 Example 17: Three-Phase Circuits

Three-phase circuits can be simulated in LTspice easily. A delta connected three-phase source with frequency of 60 Hz and line-line voltage of 120 V is simulated in Fig. 1.264. You need to put small resistors in series with the sources for convergence issues. If you try to simulate the schematic shown in Fig. 1.265, the error message shown in Fig. 1.266 appears.

The schematic shown in Fig. 1.267 generated no error as well. However, it is not balanced. So, it is recommended to use the schematic shown in Fig. 1.264.

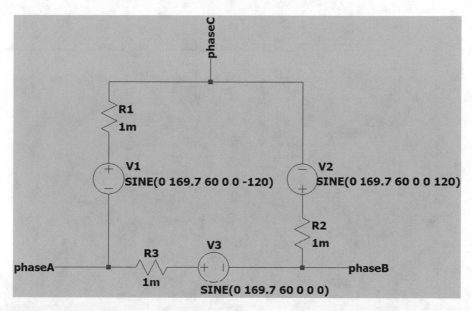

Fig. 1.264 Delta connected three-phase source

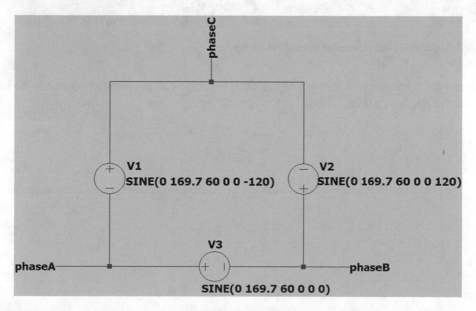

Fig. 1.265 Connection of three sources without any internal resistances leads to an error

Fig. 1.266 Error
message for connection
shown in Fig. 1.265

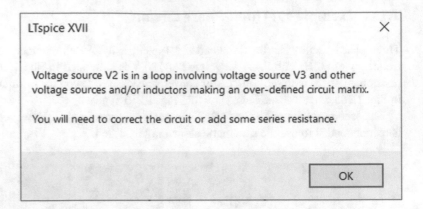

Fig. 1.267 Unbalanced delta connected three-phase source

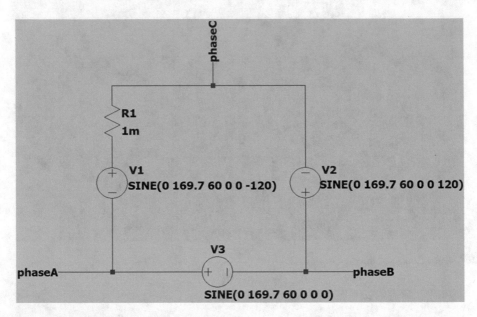

You can put some resistors (or resistor-inductor) in series with the phases to simulate the internal impedance of the source (Fig. 1.268).

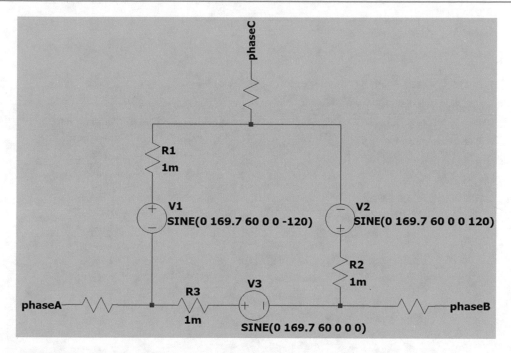

Fig. 1.268 Delta connected three-phase source with internal resistances

Settings of V1, V2 and V3 are shown in Figs. 1.269, 1.270, and 1.271.

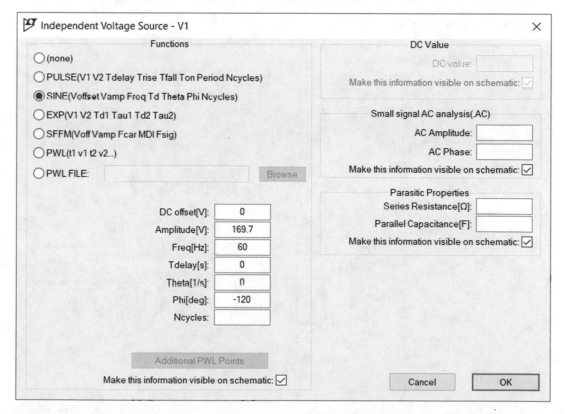

Fig. 1.269 Settings of source V1

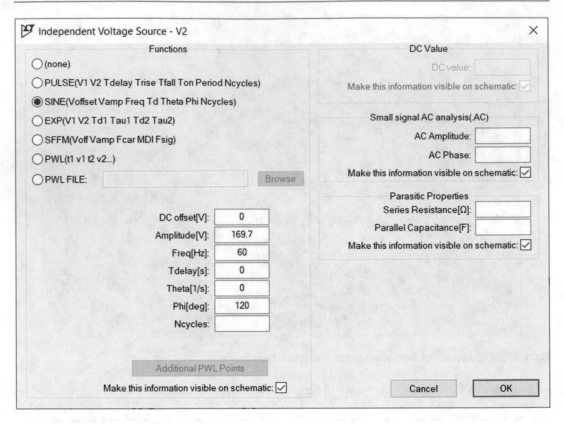

Fig. 1.270 Settings of source V2

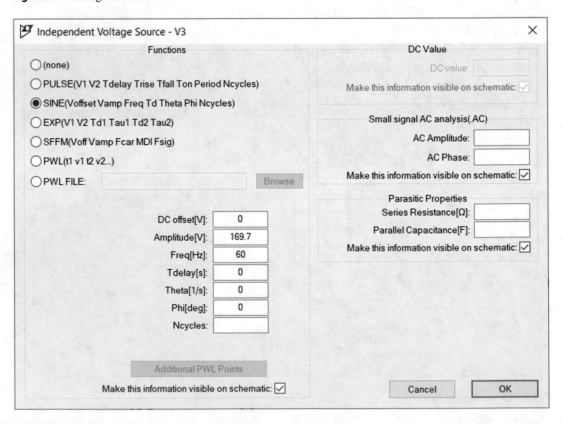

Fig. 1.271 Settings of source V3

A Y connected three-phase source with line-line voltage of 207.8 V and frequency of 60 Hz is simulated in Fig. 1.272. Settings of the V1, V2, and V3 are shown in Figs. 1.269, 1.270, and 1.271, respectively.

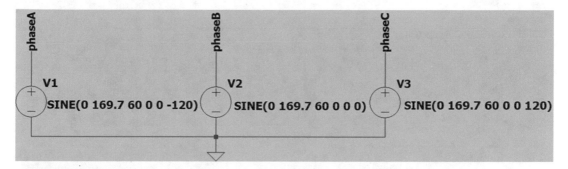

Fig. 1.272 Y connected three-phase source

Let's study an example. Consider the schematic shown in Fig. 1.273.

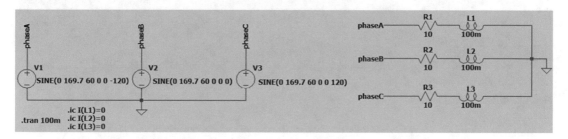

Fig. 1.273 Sample simulation

Run the simulation. After simulation is done, hold down the Alt key and click on the resistor R1. This draws the instantaneous power of resistor R1 for you (Fig. 1.274). Note that that the frequency of instantaneous power is 120 Hz which is two times the frequency of input AC source. In other words, the period of instantaneous power is 8.333 ms. Hold down the Ctrl key and click on the V(phaseA,N001)*I(R1) to measure the average power of resistor R1. According to Fig. 1.275, the average power of resistor R1 is 97.271 W.

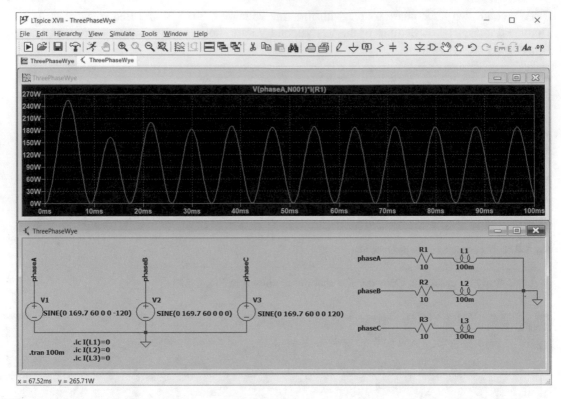

Fig. 1.274 Simulation result

Fig. 1.275 Average value and integral of V(phaseA,N001)*I(R1) for [0, 100 ms] interval

If you want to measure the average power accurately, you need not to allow the transient region affect the calculations. Let's use the last cycle of the waveform for calculation of average power. Right click on the time axis and change the Left box to 91.667 ms and click the OK button (Fig. 1.276). The [91.667 ms, 100 ms] portion of the waveform appears on the screen (Fig. 1.277). Hold down the Ctrl key and click on the V(phaseA,N001)*I(R1). This calculates the average power on the [91.667 ms, 100 ms] time interval. According to the result shown in Fig. 1.278, the average power is 94.411 W.

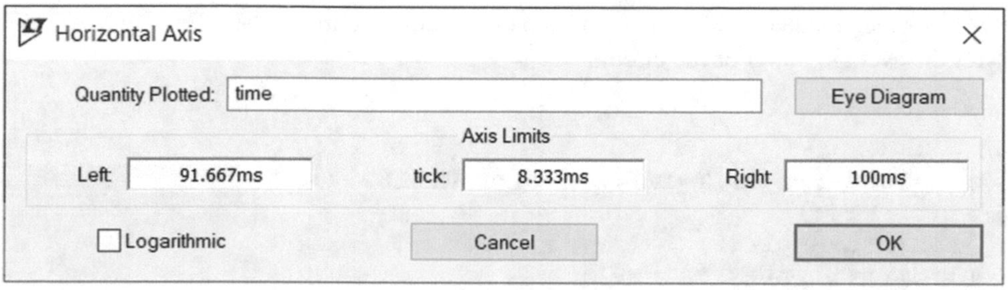

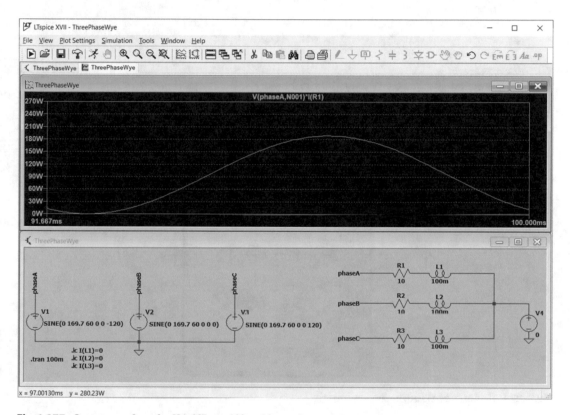

Fig. 1.276 Horizontal Axis window

Fig. 1.277 Output waveform for [91.667 ms, 100 ms] interval

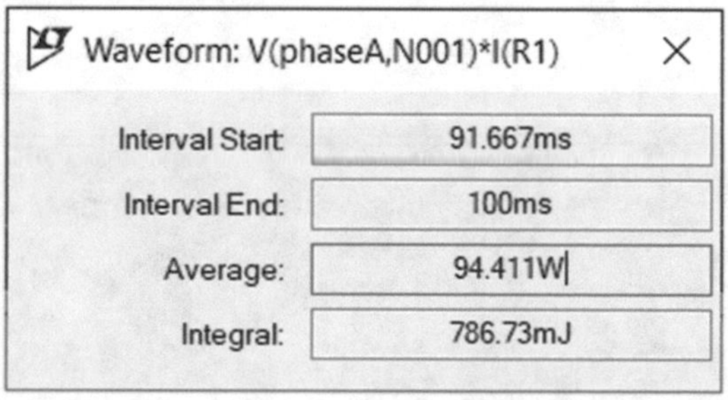

Fig. 1.278 Average value and integral of V(phaseA,N001)*I(R1) for [91.667 ms, 100 ms] interval

Let's check the obtained results and see which one is more accurate. The following MATLAB code calculates the average power of the circuit.

```
clc
clear all
R1=10;L1=100e-3;f=60;T=1/f;w=2*pi*f;Vm=169.7;
syms i1(t) V1(t)
V1=Vm*sin(w*t-2*pi/3);
ode1=L1*diff(i1,t)+R1*i1==V1;
cond=i1(0)==0;
iPhaseA(t)=dsolve(ode1,cond);
p=V1*iPhaseA;
% t0 must be big enough to use the steady state
% region of current waveform for average power
% calculation
t0=300e-3;
Paverage=eval(1/T*int(p,t0,t0+T))
```

After running the code, the result shown in Fig. 1.279 is obtained. So, the measurement which ignores the transient region is more accurate.

Fig. 1.279 Output of MATLAB code

Let's measure the current in the neutral wire. It changes the schematic to what is shown in Fig. 1.280.

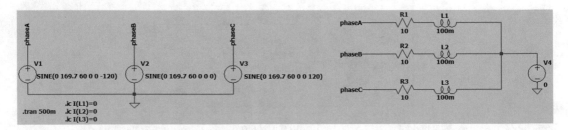

Fig. 1.280 Measurement of current in neutral line

Run the simulation and click on the voltage source V4 to see its current. The current of voltage source V4 is shown in Fig. 1.281. Its maximum is about 120×10^{-15} A. So, we can deduce that the current in the neutral wire is zero. This is expected since the load is balanced.

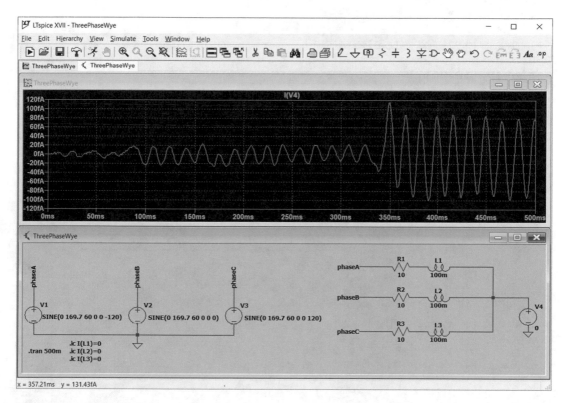

Fig. 1.281   Simulation result

1.20   Example 18: .step Command

In this example, we want to find the value of resistor Rload (Fig. 1.282) which consumes the maximum power.

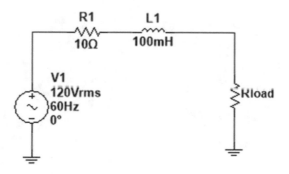

Fig. 1.282   Circuit for Example 18

Let's solve this problem with MATLAB. The following MATLAB code draws the graph of average resistor power as a function of resistor value.

```
syms Rload
R=10;L=100e-3;f=60;w=2*pi*f;
XL=j*L*w;
I=120*exp(j*0)/(R+Rload+XL);
P=Rload*abs(I)^2;
ezplot(P,[10 80])
```

Output of the above code is shown in Fig. 1.283. Zoom in the curve to measure the maximum. According to Fig. 1.284 the maximum occurred at 39 Ω.

Fig. 1.283 Output of MATLAB code

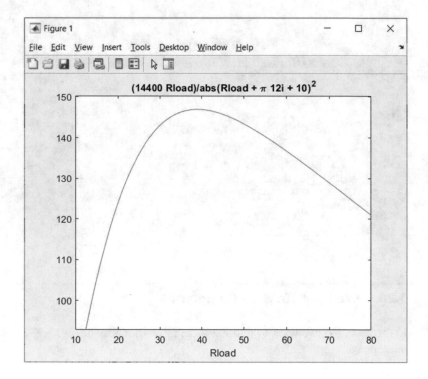

Fig. 1.284 Maximum occurs at around 39 Ω

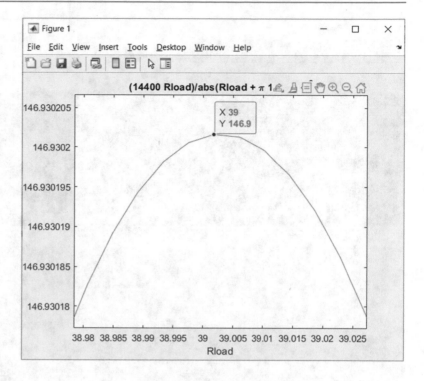

Let's solve this problem with LTspice. Draw the schematic shown in Fig. 1.285. The .step command changes the value of variable Rload from 10 Ω to 80 Ω with 5 Ω steps.

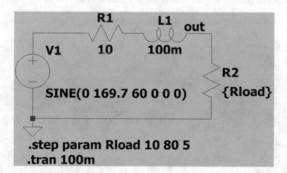

Fig. 1.285 Value of resistor R2 changes from 10 Ω to 80 Ω with 5 Ω steps

Run the simulation (Fig. 1.286).

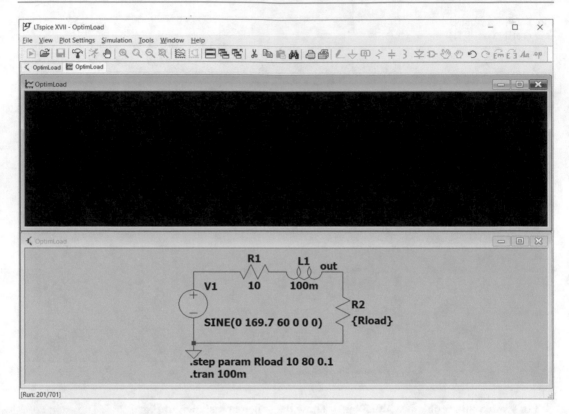

Fig. 1.286 Simulation is run

Hold down the Alt key and click on the resistor R2. The result shown in Fig. 1.287 appears.

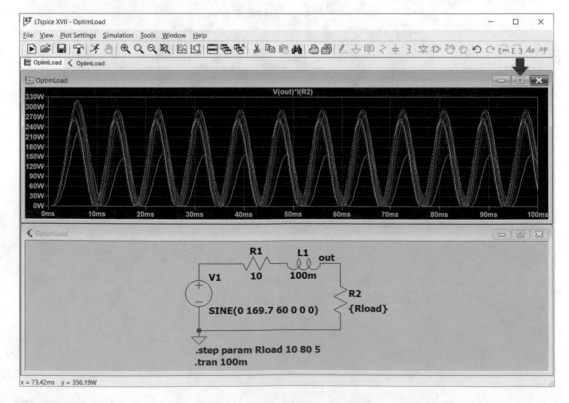

Fig. 1.287 Simulation result

Maximize the graph window (Fig. 1.288).

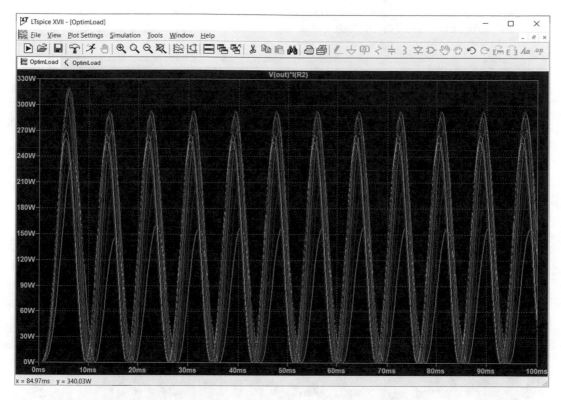

Fig. 1.288 Output window is maximized

Zoom into the steady-state region (Fig. 1.289).

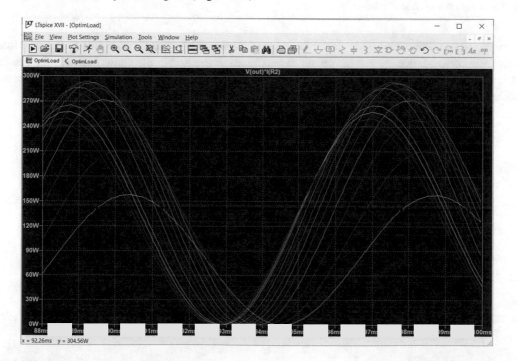

Fig. 1.289 Steady-state region of result

Click on the V(out)*I(R2). This activates the cursor. Use the mouse pointer to put the cursor in the time instant that maximum occurs. Now press the keyboard up and down arrow keys to move the horizontal line to the maximum (Fig. 1.290).

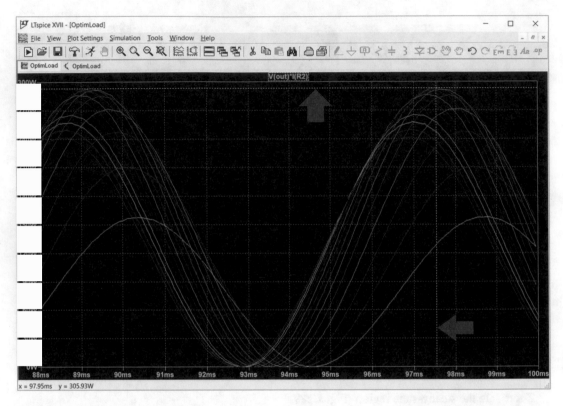

Fig. 1.290 A cursor is added to the output screen

Right click on the cursor. This shows the value of Rload which is associated with the current location of the cursor. Since we put the cursor at maximum of the graph, the shown value is the resistor value which consumes the maximum power. According to Fig. 1.291, the resistor which consumes maximum power is 40 Ω.

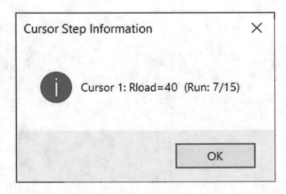

Fig. 1.291 Rload = 40 Ω consumes the maximum power

You can calculate the value of load resistor which consumes the maximum power with higher accuracy by decreasing the increase in the load resistor. The schematic shown in Fig. 1.292 increase the Rload from 10 Ω to 80 Ω with 0.5 Ω steps.

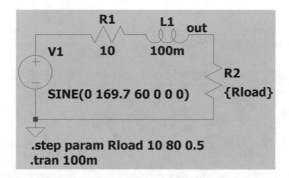

Fig. 1.292 Value of resistor R2 changes from 10 Ω to 80 Ω with 0.5 Ω steps

Run the simulation. Hold down the Alt key and click on the Rload. The graph shown in Fig. 1.293 appears.

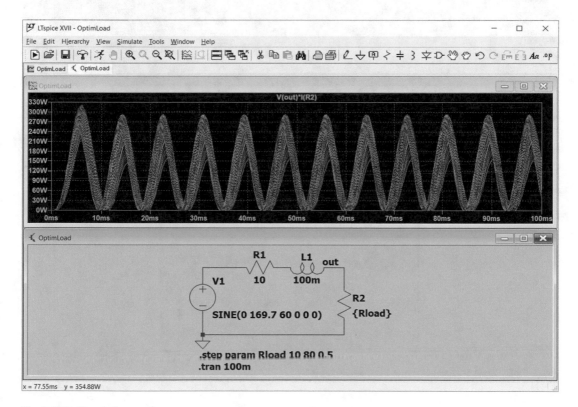

Fig. 1.293 Simulation result

Zoom into the steady-state region of the graph and use the aforementioned method to find the maximum of the graph (Fig. 1.294). According to Fig. 1.295, the load resistor which consumes the maximum power is 39 Ω.

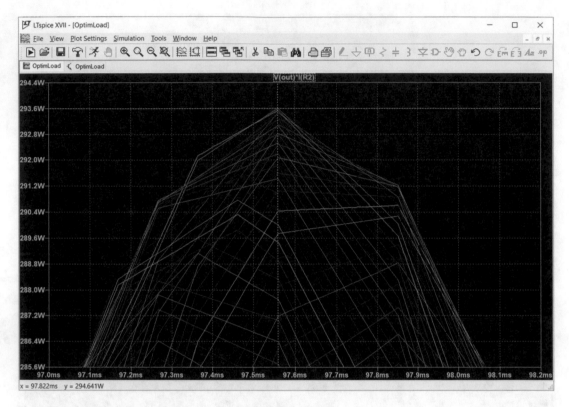

Fig. 1.294   A cursor is used to find the maximum of graph

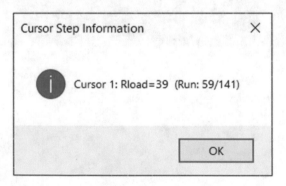

Fig. 1.295   Rload = 39 Ω consumes the maximum power

1.21   **Example 19: Coupled Inductors**

The circuits which contain coupled inductors can be simulated easily in LTspice. As an example, let's simulate the circuit shown in Fig. 1.296. Vin is a step voltage and M is the mutual inductance between L1 and L2. The coupling coefficient between the two coils is $k = \dfrac{M}{\sqrt{L_1 L_2}} = \dfrac{0.9\,\text{m}}{\sqrt{1\text{m} \times 1.1\text{m}}} = 0.8581$.

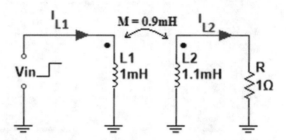

Fig. 1.296   Circuit for Example 19

From basic circuit theory,

$$\begin{cases} L_1 \dfrac{di_{L1}}{dt} - M \dfrac{di_{L2}}{dt} = V_{in}(t) \\[2mm] Ri_{L2} + L_2 \dfrac{di_{L2}}{dt} - M \dfrac{di_{L1}}{dt} = 0 \end{cases}$$

Take the Laplace transform of both side:

$$\begin{bmatrix} L_1 s & -Ms \\ -Ms & R + L_2 s \end{bmatrix} \times \begin{bmatrix} I_{L1}(s) \\ I_{L2}(s) \end{bmatrix} = \begin{bmatrix} V_{in}(s) \\ 0 \end{bmatrix}$$

So,

$$\begin{bmatrix} I_{L1}(s) \\ I_{L2}(s) \end{bmatrix} = \begin{bmatrix} L_1 s & -Ms \\ -Ms & R + L_2 s \end{bmatrix}^{-1} \times \begin{bmatrix} V_{in}(s) \\ 0 \end{bmatrix}$$

$V_{in}(s) = \dfrac{1}{s}$, so

$$\begin{bmatrix} I_{L1}(s) \\ I_{L2}(s) \end{bmatrix} = \begin{bmatrix} \dfrac{(11s + 10000) \times 10000}{s^2 \times (29s + 100000)} \\[4mm] \dfrac{90000}{s(29s + 100000)} \end{bmatrix}$$

You can use the following MATLAB commands to see the time domain graph of I_{L1} and I_{L2}. Output of the code is shown in Figs. 1.297 and 1.298.

```
s=tf('s');
I1=(11*s+10000)*10000/s/(29*s+100000);
I2=90000/(29*s+100000);
figure(1)
step(I1,[0:0.06/100:0.06]),grid on
figure(2)
step(I2), grid on
```

Fig. 1.297 Output of
MATLAB code

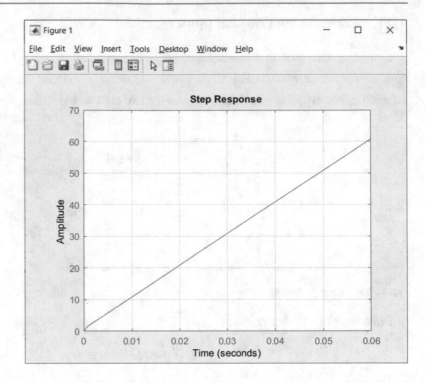

Fig. 1.298 Output of
MATLAB code

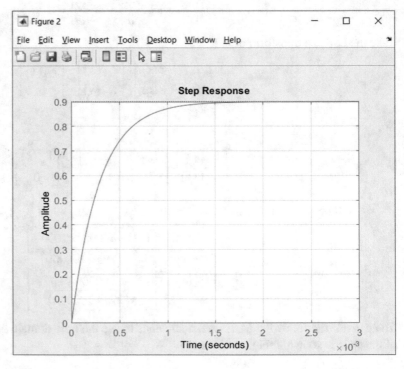

Let's solve this problem with LTspice. Draw the schematic shown in Fig. 1.299. V2 and V3 are
used to measure the current I_{L1} and I_{L2}, respectively.

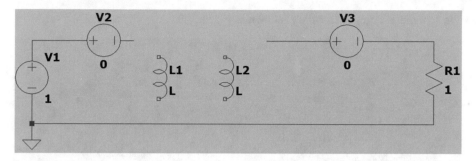

Fig. 1.299 Uncompleted LTspice schematic

Right click on the L1 and L2 and do the settings similar to Figs. 1.300 and 1.301 and click the OK button.

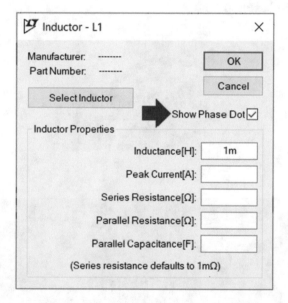

Fig. 1.300 Inductor L1 properties

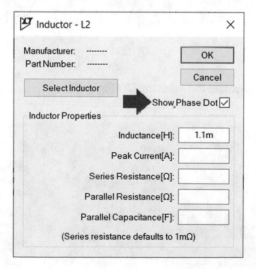

Fig. 1.301 Inductor L2 properties

Now the schematic changes to what is shown in Fig. 1.302.

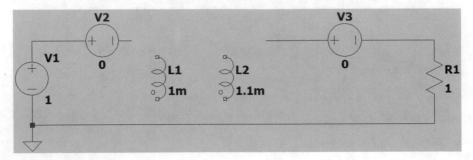

Fig. 1.302 Values of inductors are shown on the schematic

Click on the Drag icon (Fig. 1.303). After clicking the Drag icon, click on the L1 and use Ctrl+R and Ctrl+E to rotate the L1 (Fig. 1.304).

Fig. 1.303 Drag icon

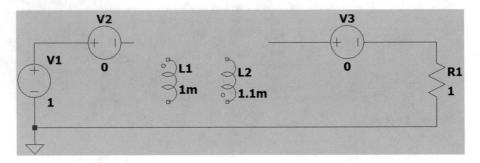

Fig. 1.304 L1 is rotated

Use the same technique to rotate the inductor L2 (Fig. 1.305).

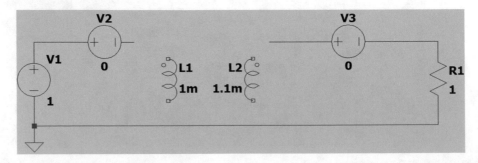

Fig. 1.305 L2 is rotated

Use the drag icon to move the labels to the sides of the inductors (Fig. 1.306).

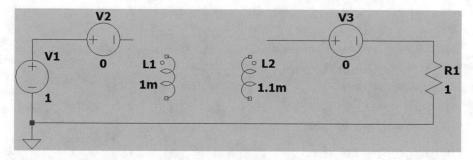

Fig. 1.306 Inductor labels are placed in desired places

Connect the inductor L1 and L2 to the circuit (Fig. 1.307).

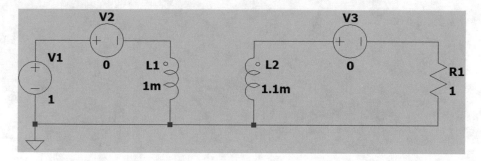

Fig. 1.307 Inductors are connected to the rest of circuit

Right click on V1 and enter its DC value (Fig. 1.308). Click the Advanced button and change the settings to what is shown in Fig. 1.309. Then click the OK button. After clicking the OK button, the schematic changes to what is shown in Fig. 1.310.

Fig. 1.308 Settings of voltage source V1

Voltage Source - V1	✕
DC value[V]: 1	OK
Series Resistance[Ω]:	Cancel
	Advanced

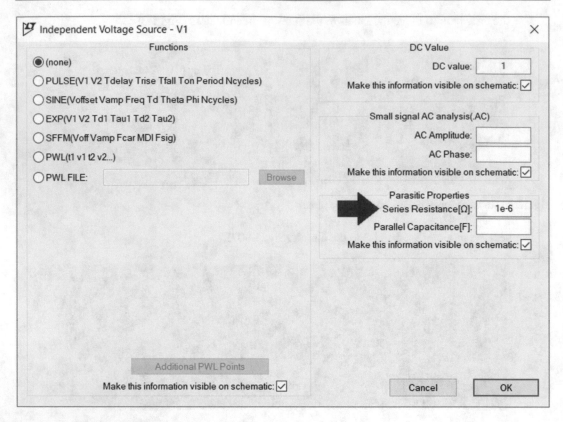

Fig. 1.309 Advanced settings of voltage source V1

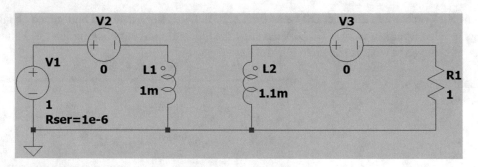

Fig. 1.310 DC source V1 has voltage of 1 V and series resistance of 1 μΩ

Note that you need to put a small resistor between the voltage source and inductor. Otherwise the error message shown in Fig. 1.311 appears and you cannot simulate the circuit. To avoid this error, the Series Resistance[Ω] box is filled with 1e-6 which means 1 μΩ (Fig. 1.309).

Fig. 1.311 Generated
error message

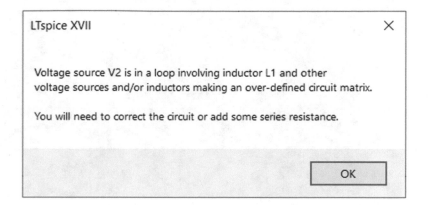

Click the SPICE Directive icon (Fig. 1.312) and enter the commands shown in Fig. 1.313 and click the OK button. The schematic changes to what is shown in Fig. 1.314. The k1 L1 L2 0.8581 line sets the coupling between L1 and L2 to 0.8581. the .ic I(L1)=0 and .ic I(L2)=0 lines set the initial current of inductors to 0 A.

Fig. 1.312   SPICE Directive

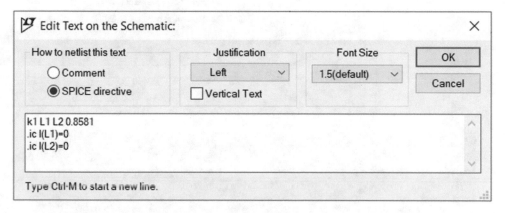

Fig. 1.313   Defining the magnetic coupling between the inductors

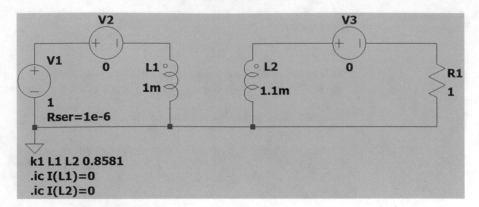

Fig. 1.314 LTspice equivalent of Fig. 1.296

Click the Run icon and set up a transient simulation with Stop time of 100 ms (Fig. 1.315). The schematic changes to what is shown in Fig. 1.316.

Fig. 1.315 Simulation settings

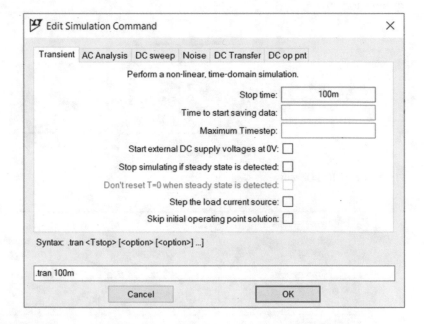

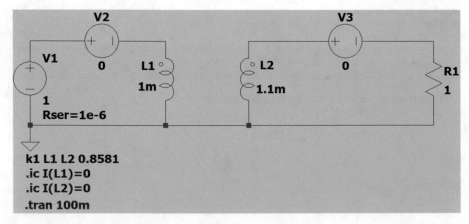

Fig. 1.316 Simulation command is added to the schematic

The schematic is ready for simulation. Click the Run button and draw the current of voltage source V2 and V3. Result is shown in Figs. 1.317 and 1.318. You can use cursors to ensure that the obtained results are the same as the MATLAB result.

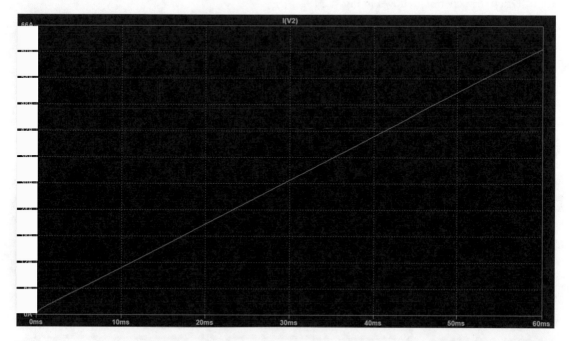

Fig. 1.317 Graph of I(V2)

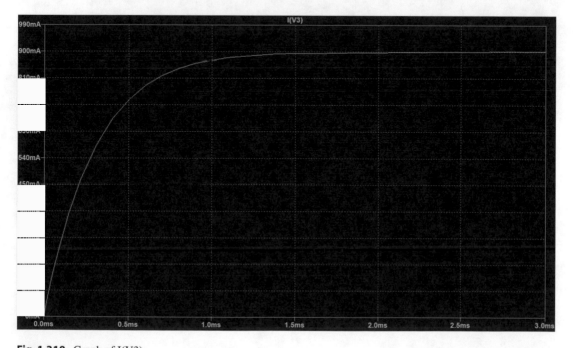

Fig. 1.318 Graph of I(V3)

1.22   Example 20: Step Response of Circuits

In this example, we want to simulate the step response of a simple RLC circuit with zero initial conditions. Draw the schematic shown in Fig. 1.319.

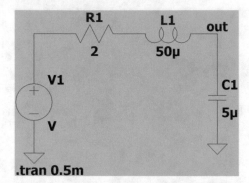

Fig. 1.319 Schematic for Example 20

Right click on the voltage source V1 and click the Advanced button (Fig. 1.320). Then change the settings to what is shown in Fig. 1.321. Settings shown in Fig. 1.321 generate the waveform shown in Fig. 1.322. After clicking the OK button in Fig. 1.321, the schematic changes to what is shown in Fig. 1.323.

Fig. 1.320 Settings for voltage source V1

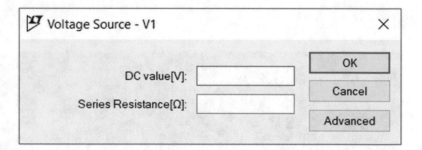

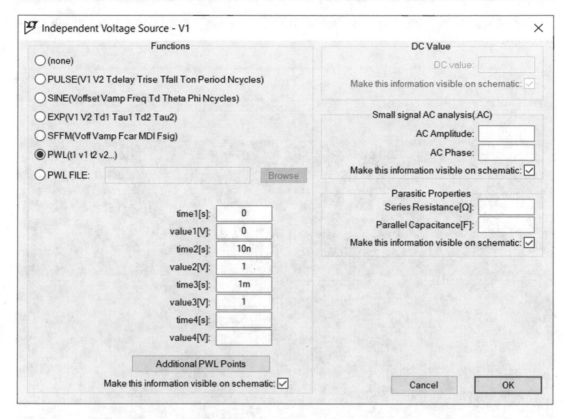

Fig. 1.321 Advanced settings for voltage source V1

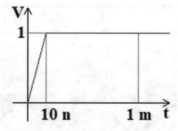

Fig. 1.322 Waveform generated with settings shown in Fig. 1.321

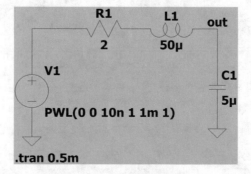

Fig. 1.323 Settings of voltage source V1 are shown on the schematic

Run the simulation. Voltage of node "out" and inductor current are shown in Figs. 1.324 and 1.325, respectively.

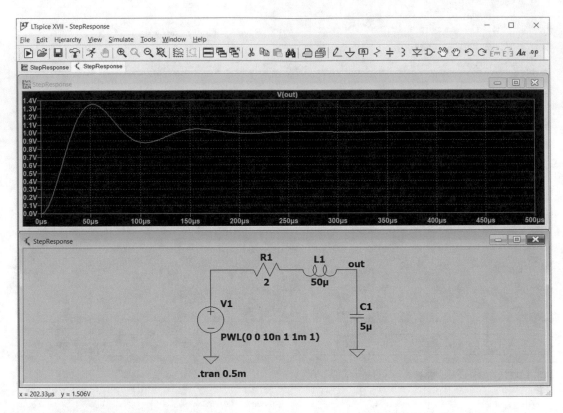

Fig. 1.324   Graph of V(out)

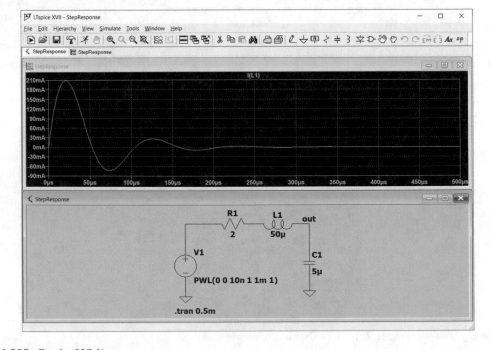

Fig. 1.325   Graph of I(L1)

Let's check the results. The transfer function of the circuit can be written

as $\dfrac{V_{C_1}(s)}{V_1(s)} = \dfrac{\dfrac{1}{C_1 s}}{R_1 + L_1 s + \dfrac{1}{C_1 s}} = \dfrac{1}{L_1 C_1 s^2 + R_1 C_1 s + 1}$. The commands shown in Fig. 1.326 draws the step

response of this transfer function. Output of these commands are shown in Fig. 1.327. The result is the same as the LTspice result.

Fig. 1.326 MATLAB commands

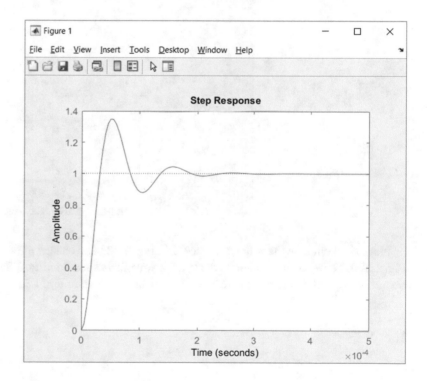

```
Command Window
   >> R1=2;L1=50e-6;C1=5e-6;
   >> H=tf([1],[L1*C1 R1*C1 1]);
   >> step(H,0.5e-3)
fx >>
```

Fig. 1.327 Output of MATLAB commands

The transfer function of the circuit can be written as $\dfrac{I_{L_1}(s)}{V_1(s)} = \dfrac{1}{R_1 + L_1 s + \dfrac{1}{C_1 s}} = \dfrac{C_1 s}{L_1 C_1 s^2 + R_1 C_1 s + 1}$ .The

commands shown in Fig. 1.328 draws the step response of this transfer function. Output of these commands are shown in Fig. 1.329. The result is the same as the LTspice result.

```
Command Window                                          ⊙
  >> R1=2;L1=50e-6;C1=5e-6;
  >> H=tf([C1 0],[L1*C1 R1*C1 1]);
  >> step(H,0.5e-3)
fx >> |
```

Fig. 1.328 MATLAB commands

Fig. 1.329 Output of
MATLAB commands

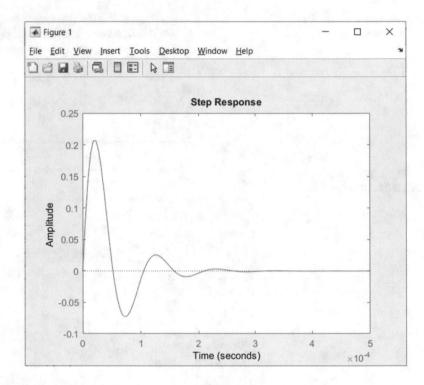

Now, right click on the voltage source V1 in Fig. 1.323 and change the settings to what is shown in
Fig. 1.330. These settings generate the voltage waveform shown in Fig. 1.331. After clicking the OK
button in Fig. 1.330, the schematic changes to what is shown in Fig. 1.332.

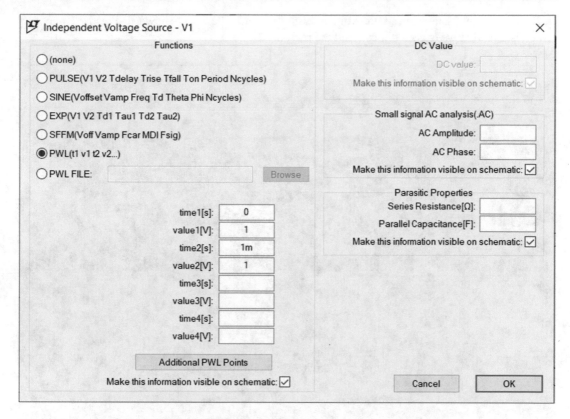

Fig. 1.330 New settings of voltage source V1

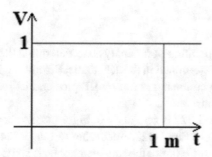

Fig. 1.331 Waveform generated with settings shown in Fig. 1.330

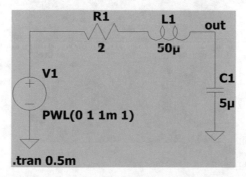

Fig. 1.332 Settings of voltage source V1 are shown on the schematic

Run the simulation and draw the voltage of node "out." The result shown in Fig. 1.333 appears which is not what we expected. Let's find the reason.

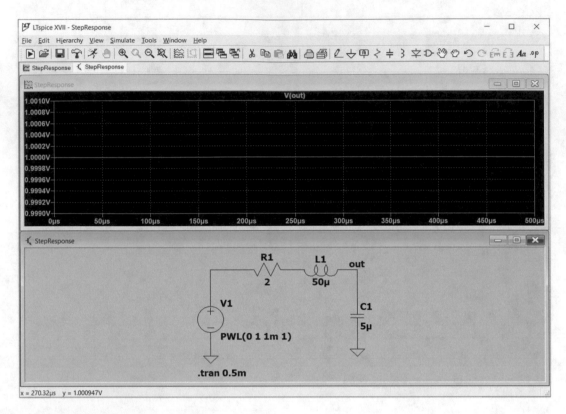

Fig. 1.333 Graph of V(out)

LTspice uses the result of operating point analysis to determine the initial condition of the circuit when no initial conditions are determined in the schematic. So, although we didn't wrote any .op command in the schematic, the .op command is run by LTspice and its results are used to determine the starting point of .trans command.

The waveform shown in Fig. 1.322 is zero at t = 0. Figure 1.334 shows the steady-state DC equivalent circuit of schematic shown in Fig. 1.323. According to Fig. 1.334, voltage of node "out" is zero and current of inductor is zero at t = 0. So, the transient analysis starts from zero initial conditions.

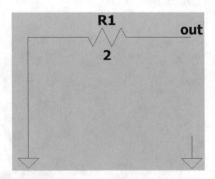

Fig. 1.334 Steady-state DC equivalent circuit of Fig. 1.323

The waveform shown in Fig. 1.331 is one at t = 0. Figure 1.335 shows the steady-state DC equivalent circuit of schematic shown in Fig. 1.332. According to Fig. 1.335, voltage of node "out" is one and current of inductor is zero at t = 0. So, the transient analysis starts from none zero initial conditions. You can force the simulation to start from zero initial conditions with the aid of .ic command (Fig. 1.336).

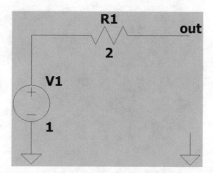

Fig. 1.335 Steady-state DC equivalent circuit of Fig. 1.332

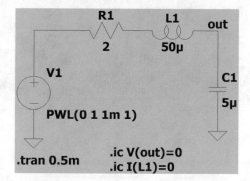

Fig. 1.336 Simulation starts from zero initial conditions

After running the schematic shown in Fig. 1.336, the results shown in Figs. 1.337 and 1.338 are obtained. These are the expected results.

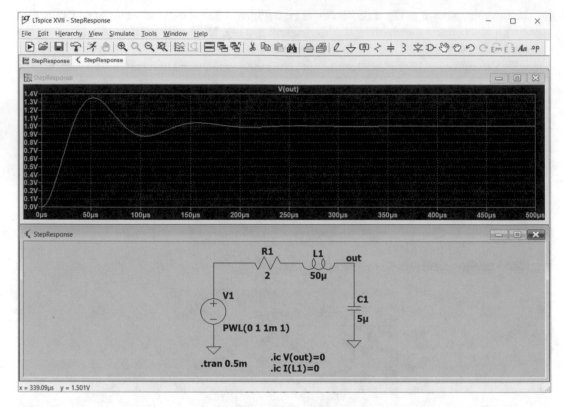

Fig. 1.337 Graph of V(out)

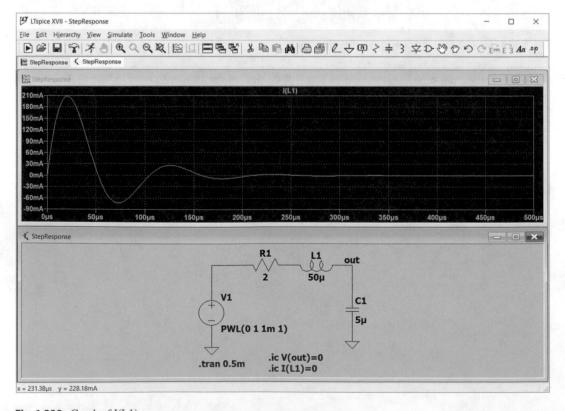

Fig. 1.338 Graph of I(L1)

1.23   Example 21: Impulse Response of the Circuit

In this example, we want to see the impulse response of the RLC circuit of previous example. Change the schematic of previous example to what is shown in Fig. 1.339. Settings of voltage source V1 are shown in Fig. 1.340. These settings generate the voltage waveform shown in Fig. 1.341. Integral of the waveform shown in Fig. 1.341 is 1, and it is short enough to simulate the impulse input.

You can apply a pulse with smaller amplitude to the circuit as well. For instance, you can apply the pulse shown in Fig. 1.342 to the circuit and multiply the output response by 10^7 to obtain the unit impulse response of the circuit.

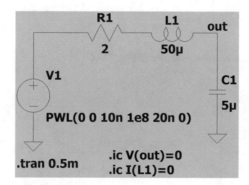

Fig. 1.339   Schematic of Example 21

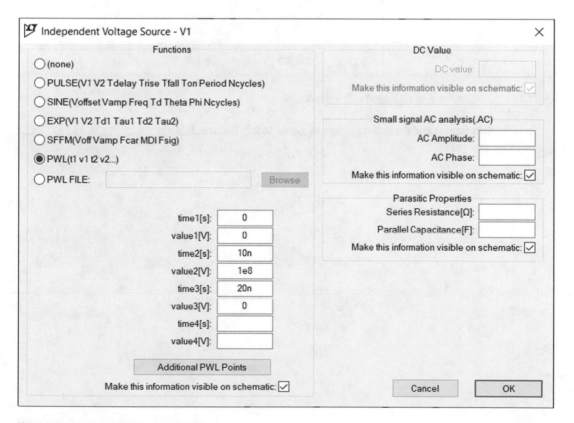

Fig. 1.340   Settings of voltage source V1

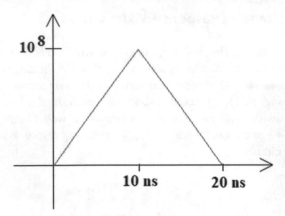

Fig. 1.341   Waveform generated with settings shown in Fig. 1.340

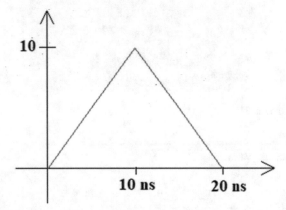

Fig. 1.342   The circuit can be stimulated with a smaller pulse

Run the simulation and draw the voltage of node "out." The result is shown in Fig. 1.343.

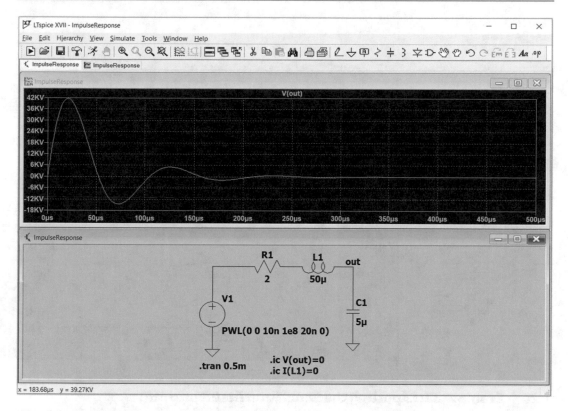

Fig. 1.343   Graph of V(out)

Let's check the result. The MATLAB code shown in Fig. 1.344 draws the impulse response of the

circuit ($\dfrac{V_{C_1}(s)}{V_1(s)} = \dfrac{\dfrac{1}{C_1 s}}{R_1 + L_1 s + \dfrac{1}{C_1 s}} = \dfrac{1}{L_1 C_1 s^2 + R_1 C_1 s + 1}$). Output of this code is shown in Fig. 1.345.

MATLAB result and LTspice result are the same.

```
Command Window                        ⊙
    >> R1=2;L1=50e-6;C1=5e-6;
    >> H=tf([1],[L1*C1 R1*C1 1]);
    >> impulse(H,0.5e-3)
fx >> |
```

Fig. 1.344   MATLAB code

Fig. 1.345 Output of
MATLAB code

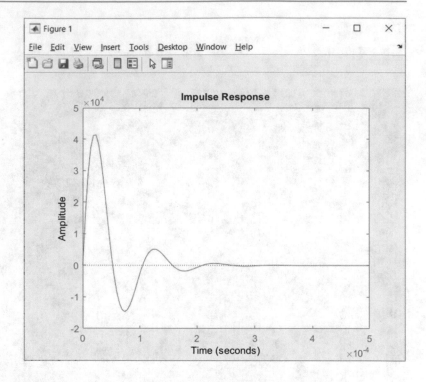

1.24 Example 22: Exporting the Simulation Result into MATLAB

LTspice waveforms can be exported to MATLAB. Importing the waveforms into MATLAB environment permits further analysis with the aid of tools that MATLAB provides.

Let's import the result of previous analysis into the MATLAB environment. Right click on the black area in Fig. 1.343. Then click the File and Export data as text (Fig. 1.346).

Fig. 1.346 Exporting
the data as a text file

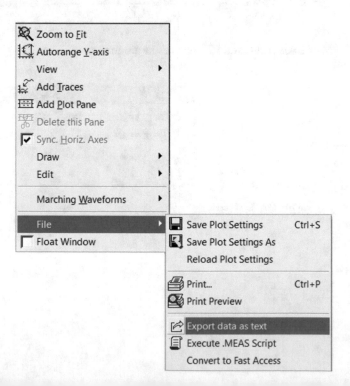

After clicking the Export data as text, Select Traces to Export window appears on the screen (Fig. 1.347). Use the Browse button to determine the path that output file is saved. Select the waveform(s) that you want to be saved in the output file and click the OK button. Note that the output file format is .txt. So, you can read it with Excel as well.

Fig. 1.347 Select Traces to Export window

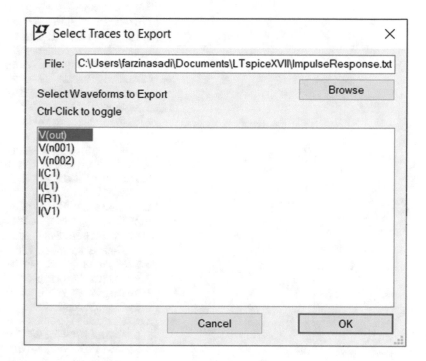

The generated text file is shown in Fig. 1.348. First column is always the simulation time.

Fig. 1.348 Generated
data

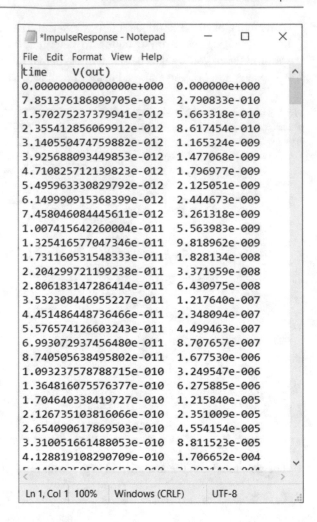

Now open the MATLAB and enter the commands shown in Fig. 1.349. Output of theses commands
are shown in Fig. 1.350. The graph is the same as Fig. 1.343. So, we imported the data into MATLAB
successfully.

```
Command Window                                                          ⊙
  >> F=importdata('C:\Users\farzinasadi\Documents\LTspiceXVII\ImpulseResponse.txt');
  >> time=F.data(:,1);
  >> Vout=F.data(:,2);
  >> plot(time,Vout)
fx >>
```

Fig. 1.349 MATLAB commands

Fig. 1.350 Output of
MATLAB commands

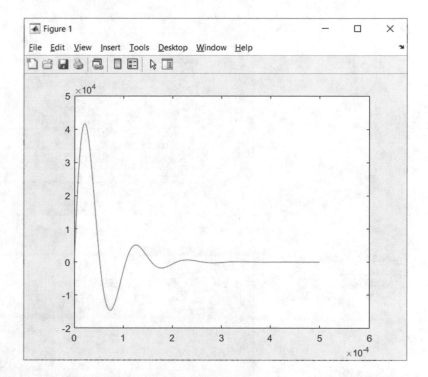

Let's obtain the mathematical equation of the graph shown in Fig. 1.350. From basic circuit theory we know that such a response belongs to function $f(t) = ae^{-bt} \sin(\omega t)$. Value of ω can be found easily. Just zoom in the graph of Fig. 1.350 with the aid of Zoom in icon (Fig. 1.351) to find the location of first intersection with time axis. According to Fig. 1.352, the intersection happened at 1.04744×10^{-4}. So, $\omega = \dfrac{2\pi}{1.04744 \times 10^{-4}} = 5.9986 \times 10^4$. So, $f(t) = ae^{-bt} \sin(5.9986 \times 10^4 t)$. Now we need to find the value of a and b.

Fig. 1.351 Zoom In icon

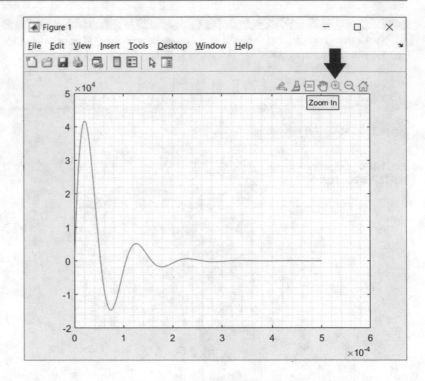

Fig. 1.352 Value of graph is zero at 0.104744 ms

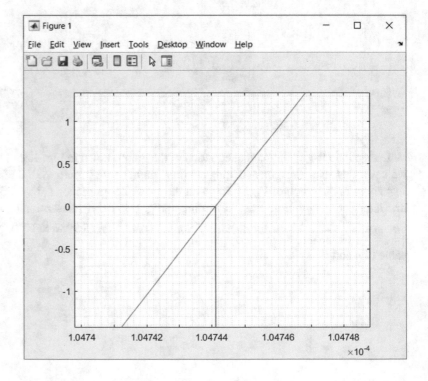

Let's use Curve Fitting Toolbox® to find the best values for a and b. Run the Curve Fitting Toolbox with the aid of cftool command (Fig. 1.353) and do the settings similar to Fig. 1.354. According to Fig. 1.354, $a = 6.663 \times 10^4$ and $b = 2 \times 10^4$. So, equation of impulse response of the circuit is $6.663 \times 10^4 e^{-2 \times 10^4 t} \sin\left(5.9986 \times 10^4 t\right)$.

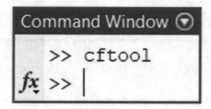

Fig. 1.353   cftool command

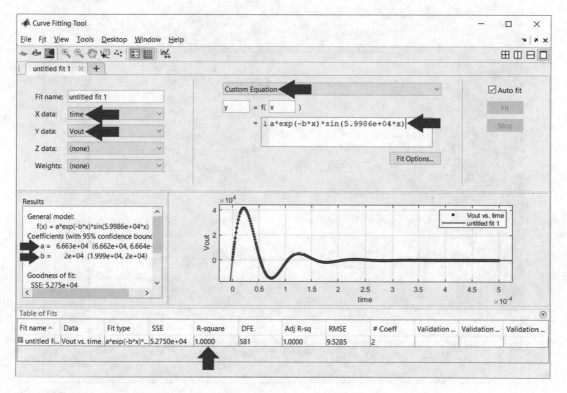

Fig. 1.354   Curve Fitting Tool

The commands shown in Fig. 1.355 show the LTspice result and graph of $f(t) = 6.663 \times 10^4 e^{-2 \times 10^4 t} \sin\left(5.9986 \times 10^4 t\right)$ simultanously. Output of this code is shown in Fig. 1.356. The two graphs overlap. This shows that $f(t) = 6.663 \times 10^4 e^{-2 \times 10^4 t} \sin\left(5.9986 \times 10^4 t\right)$ is a good function to represent our data.

```
Command Window                                                                    ⊙
  >> F=importdata('C:\Users\farzinasadi\Documents\LTspiceXVII\ImpulseResponse.txt');
  >> time=F.data(:,1);
  >> Vout=F.data(:,2);
  >> plot(time,Vout)
  >> hold on
  >> plot(time,6.663e4*exp(-2e4*time).*sin(5.9986e4*time),'r--')
fx >>
```

Fig. 1.355 MATLAB commands

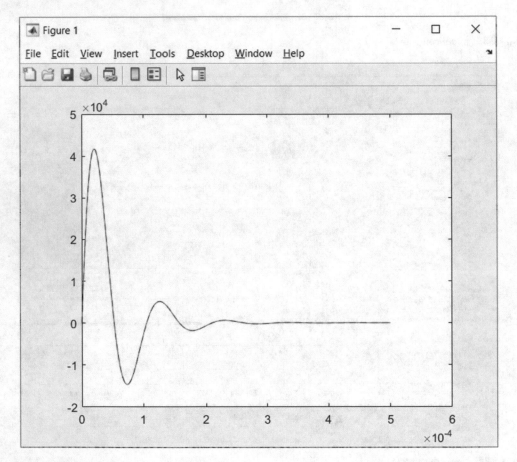

Fig. 1.356 Output of MATLAB commands

The MATLAB code shown in Fig. 1.357 draws the difference between LTspice result and $f(t) = 6.663 \times 10^4 e^{-2 \times 10^4 t} \sin(5.9986 \times 10^4 t)$ function. The output of this code is shown in Fig. 1.358. According to Fig. 1.358, the maximum error is less than 60.

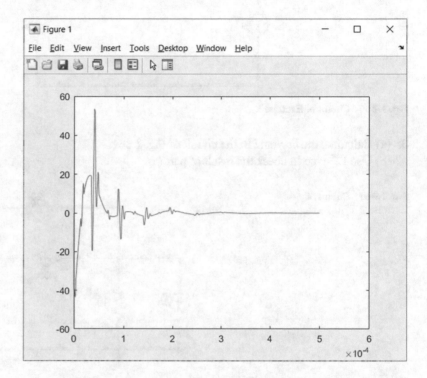

```
Command Window                                                          ⊙
>> F=importdata('C:\Users\farzinasadi\Documents\LTspiceXVII\ImpulseResponse.txt');
>> time=F.data(:,1);
>> Vout=F.data(:,2);
>> plot(time,Vout-6.663e4*exp(-2e4*time).*sin(5.9986e4*time))
fx >>
```

Fig. 1.357 MATLAB commands

Fig. 1.358 Output of MATLAB commands

1.25 Exercises

1. Simulate the circuit shown in Fig. 1.359 with LTspice. Initial conditions are shown on the Figure.

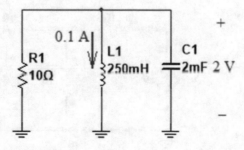

Fig. 1.359 Circuit of Exercise 1

2. In the circuit shown in Fig. 1.360, $V_1 = 10 + 25 \sin (2\pi \times 60t)$. Initial conditions are $V_C = 10\ V$ and $i_L = 0\ A$. Use LTspice to observe the circuit current.

   Hint: Use two .ic commands, i.e., .ic V(a)=0 and .ic V(out)=0.

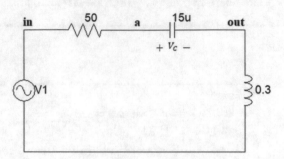

Fig. 1.360 Circuit of Exercise 2

3. (a) Calculate the current i in the circuit of Fig. 1.361.
   (b) Use LTspice to check the result of part (a).

Fig. 1.361 Circuit of
Exercise 3

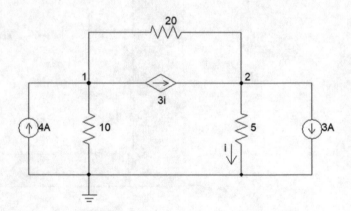

4. (a) Find the Thevenin equivalent circuit with respect to the terminals "a" and "b" for the circuit shown in Fig. 1.362.
   (b) Use LTspice to check the result of part (a).

Fig. 1.362 Circuit of
Exercise 4

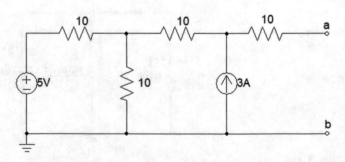

5. Use LTspice to find the Thevenin equivalent circuit with respect to the terminals "a" and "b" for the circuit shown in Fig. 1.363.

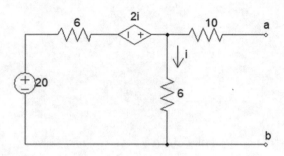

Fig. 1.363 Circuit of Exercise 5

6. Set up an LTspice simulation to measure the RMS of a triangular wave. Use MATLAB or hand calculation to verify the result.
7. Use LTspice to find the value of the load resistor Rload (Fig. 1.364) which consumes the maximum power.

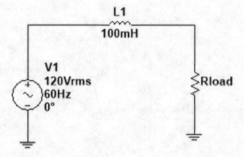

Fig. 1.364 Circuit of Exercise 7

8. Calculate the impulse response of the circuit shown in Fig. 1.339 (output is capacitor voltage). Compare your result with the function $f(t)$ found in Example 22.

References

1. Hayt, W., Kemmerly, J., Durbin, S.: Engineering circuit analysis, 9th edition, McGraw-Hill (2021)
2. Nilsson, J., Riedel, S.: Electric circuits, 11th edition, Pearson (2018)
3. Thomas, R. E., Rosa, A. J., Toussaint G. J.: The Analysis and Design of Linear Circuits, 9th edition, John Wiley and Sons (2020)
4. Alexander, C., Sadiku, M.N.O: Fundamentals of Electric Circuits, 6th edition, McGraw-Hill (2016)

2.1 Introduction

In this chapter, you will learn how to analyze electronic circuits in LTspice. The theory behind the studied circuits can be found in any standard electronic/microelectronic text book [1–3]. Similar to previous chapter, doing some hand calculations for the given circuits and comparing the hand analysis results with LTspice results are recommended.

2.2 Example 1: Transformer

This example shows how to simulate a transformer in LTspice. Consider the schematic shown in Fig. 2.1. We want to add a transformer to this circuit to reduce the input voltage to about 35 V. A transformer with turn ratio of 9:1 can do this for us.

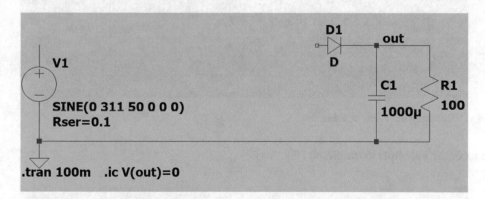

Fig. 2.1 Uncompleted schematic of Example 1

Add two inductors to the schematic (Fig. 2.2).

© The Author(s), under exclusive license to Springer Nature Switzerland AG 2023
F. Asadi, *Essential Circuit Analysis using LTspice®*, https://doi.org/10.1007/978-3-031-09853-6_2

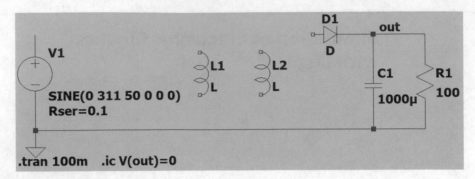

Fig. 2.2 Addition of inductor L1 and L1 to schematic in Fig. 2.1

The relationship between inductances and turn ratio is given by $\frac{L_p}{L_s} = \left(\frac{N_p}{N_s}\right)^2$ formula. L_p, L_s, N_p, and N_s show the primary winding inductance, secondary winding inductance, number of primary winding turns, and number of secondary winding turns, respectively. So, for a 9:1 transformer, $\frac{L_p}{L_s} = \left(\frac{N_p}{N_s}\right)^2 = \left(\frac{9}{1}\right)^2 = 81$ or $L_p = 81L_s$. If we assume $L_s = 100$ μH, then $L_p = 8100$ μH. Enter these values to the software (Fig. 2.3). Note that real transformers have primary and secondary inductances that are much bigger than numbers used in this example (generally in the mH range).

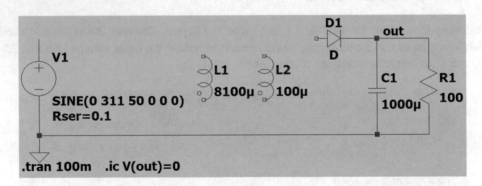

Fig. 2.3 Determining the values of inductors

Connect the inductors to the circuit (Fig. 2.4).

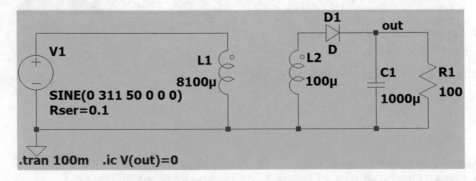

Fig. 2.4 Inductors are connected to the rest of circuit

You can use the Edit > Draw Line (Fig. 2.5) to draw two parallel lines between the windings (Fig. 2.6). These two lines have no effect on the simulation; however, they show that the circuit contains a transfer clearly.

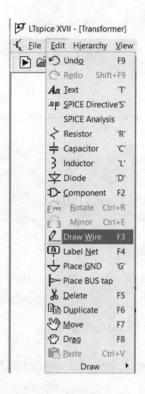

Fig. 2.5   Edit> Draw Wire

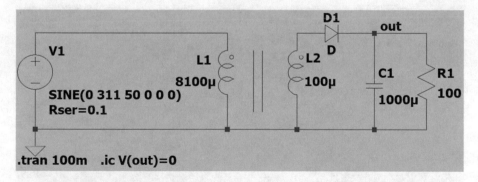

Fig. 2.6   Two parallel lines are drawn between the inductors

Click the SPICE directive icon (Fig. 2.7) and enter the commands shown in Fig. 2.8. The k1 L1 L2 1 line defines a magnetic coupling with coupling factor of 1 between the two windings.

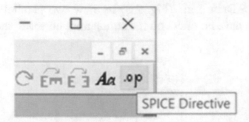

Fig. 2.7 SPICE Directive icon

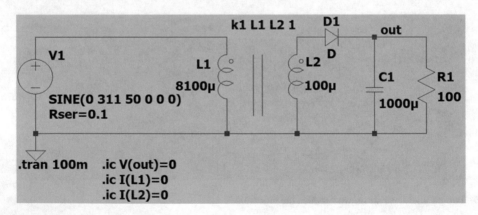

Fig. 2.8 Magnetic coupling factor between L1 and L2 is set to 1

Set the initial conditions of the circuit. Now the schematic of half wave rectifier with output filter capacitor is ready (Fig. 2.9).

Fig. 2.9 Completed schematic of Example 1

Run the simulation. Voltage of node "out" is shown in Fig. 2.10.

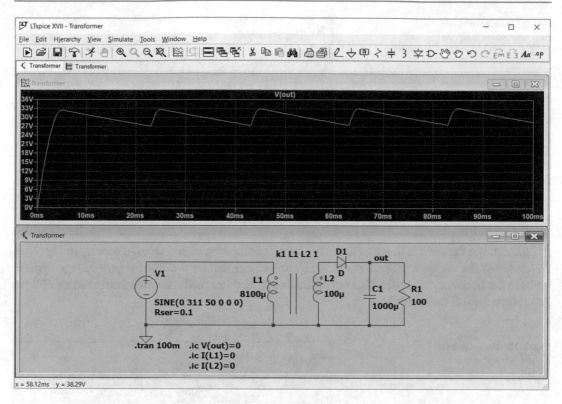

Fig. 2.10 Simulation result

Let's measure the output voltage ripple. Add two cursors to the graph, put the cursor 1 in the minimum of the graph, and put the cursor 2 in the maximum of the graph (Fig. 2.11). According to Fig. 2.12, the ripple of output voltage is about 5.508 V.

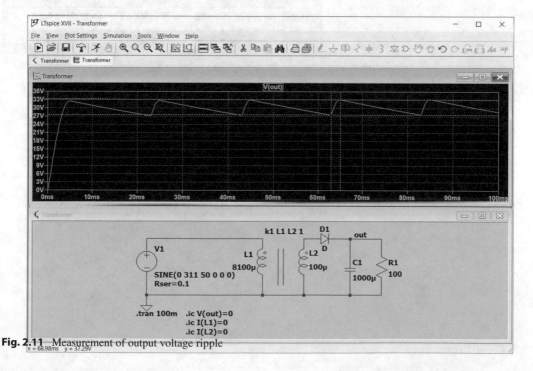

Fig. 2.11 Measurement of output voltage ripple

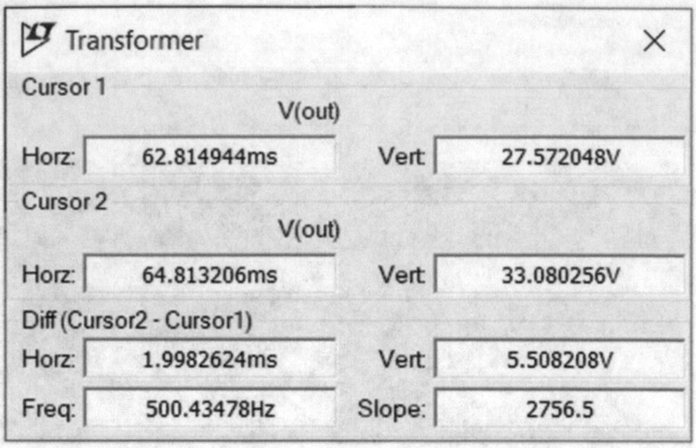

Fig. 2.12 Measurement
of output voltage ripple

🏴 Transformer			✕
Cursor 1	V(out)		
Horz:	62.814944ms	Vert:	27.572048V
Cursor 2	V(out)		
Horz:	64.813206ms	Vert:	33.080256V
Diff (Cursor2 - Cursor1)			
Horz:	1.9982624ms	Vert:	5.508208V
Freq:	500.43478Hz	Slope:	2756.5

You can use the cursors to measure the frequency of output voltage ripple. According to Fig. 2.13, frequency of the output voltage ripple is 50 Hz. Remember that frequency of output voltage tipple is equal to the frequency of input AC source for half wave rectifiers, and it is twice the frequency of input AC source for full wave rectifiers.

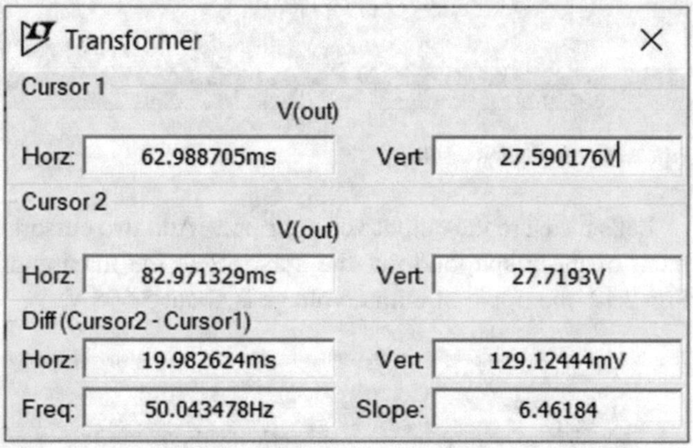

Fig. 2.13 Frequency of
output voltage ripple is
50 Hz

🏴 Transformer			✕
Cursor 1	V(out)		
Horz:	62.988705ms	Vert:	27.590176V
Cursor 2	V(out)		
Horz:	82.971329ms	Vert:	27.7193V
Diff (Cursor2 - Cursor1)			
Horz:	19.982624ms	Vert:	129.12444mV
Freq:	50.043478Hz	Slope:	6.46184

The capacitor and diode currents are shown in Figs. 2.14 and 2.15, respectively.

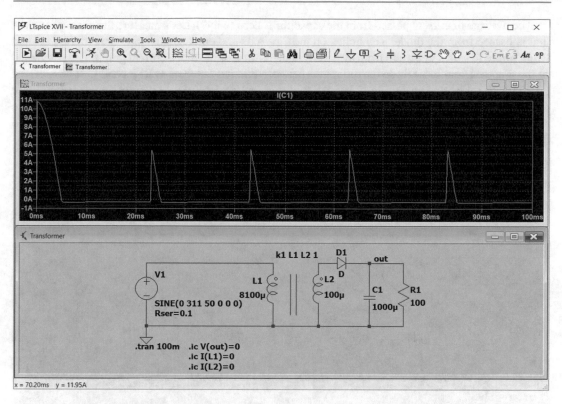

Fig. 2.14 Capacitor current

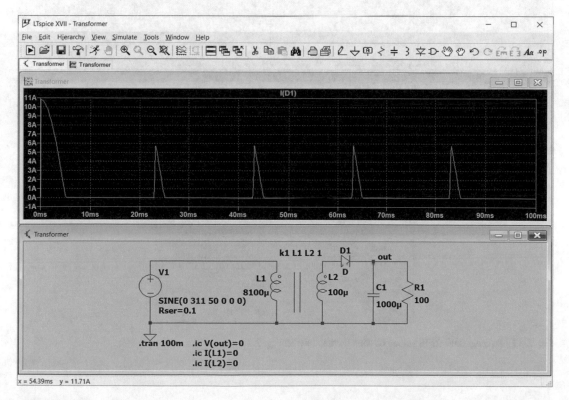

Fig. 2.15 Diode current

The average of current passed from the capacitor is zero (why?). Let's measure the average current that is passed from the diode. Right click on the time axis and enter suitable numbers to the Left and Right boxes to select one cycle from the steady state portion of the diode current graph. The frequency of the diode current is 50 Hz so the difference between entered numbers to Right and Left boxes must be 20 ms to select one cycle (Fig. 2.16). After seeing one cycle of the waveform on the screen, hold down the Alt key and click on the I(D1). This shows the average of current passed from the diode. According to Fig. 2.17, the average current passed from the diode is 311.17 mA. You need to select a diode which is capable to pass such an average current.

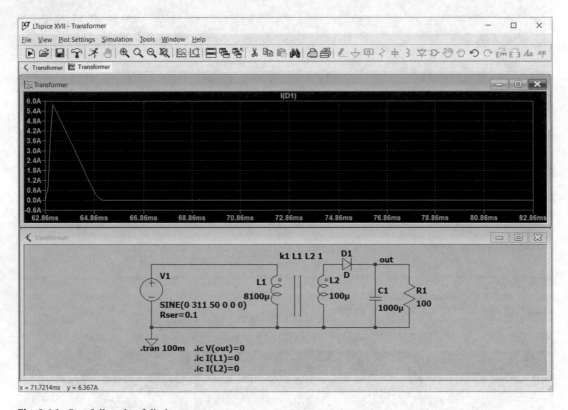

Fig. 2.16 One full cycle of diode current

Fig. 2.17 Average and RMS values of waveform shown in Fig. 2.16

Let's measure the maximum of reverse voltage applied to the diode. The diode voltage (voltage difference between the anode and cathode, i.e., $V_{Anode} - V_{cathode}$) is shown in Fig. 2.18. According to Fig. 2.19, the maximum reverse voltage which is applied to the diode is −65.04 V. So, we need to select a diode which is capable to withstand this reverse voltage.

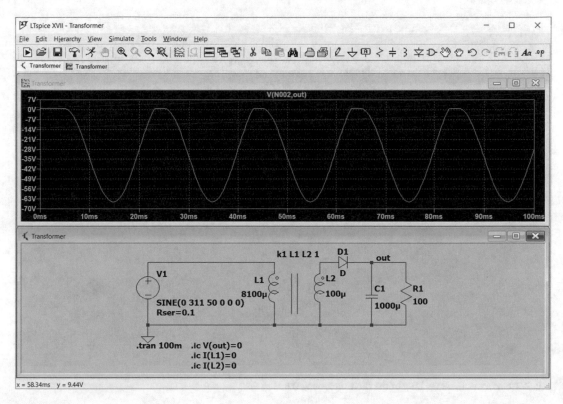

Fig. 2.18 Diode voltage waveform

Fig. 2.19 Maximum reverse voltage applied to the diode is around 65 V

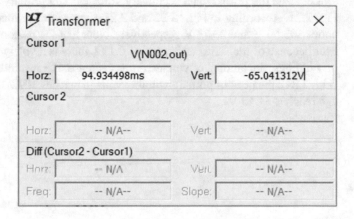

The positive region of the diode voltage graph in Fig. 2.18 gives the forward voltage drop of diode. According to Fig. 2.20, the forward voltage drop of the diode is about 0.8 V.

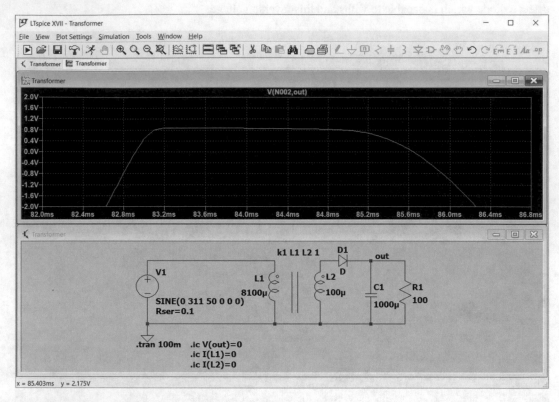

Fig. 2.20   Forward voltage drop of the diode is around 0.8 V

Let's see the primary and secondary voltages. The primary and secondary voltages are shown in Fig. 2.21. According to Figs. 2.22 and 2.23, the peak of primary and secondary winding voltages is about 303.66 V and 33.73 V, respectively. Note that the peak of input voltage source is 311 V; however because of the series resistance of 0.1 Ω, there is a voltage drop. If you increase the inductance of windings, the amount of voltage drops decrease. For instance, for L_p = 16200 μH and L_s = 200 μH (Fig. 2.24), the peak of primary winding voltage reaches 307.2 V and peak of secondary winding voltage reaches 34.13 V.

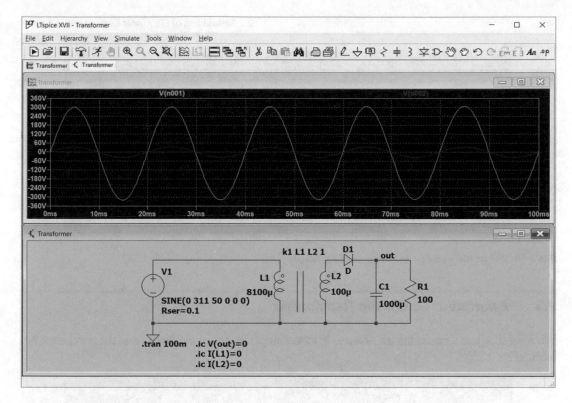

Fig. 2.21 Primary and secondary waveforms

Fig. 2.22 Peak of
primary voltage is
around 303.65 V

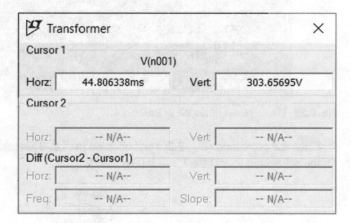

Fig. 2.23 Peak of
secondary voltage is
around 33.73 V

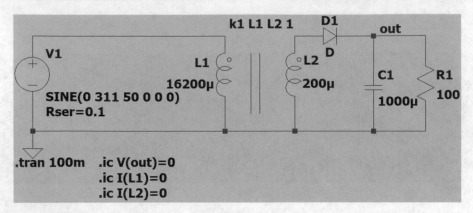

Fig. 2.24 Schematic with $L_p = 16200\ \mu H$ and $L_s = 200\ \mu H$

2.3   Example 2: Center Tap Transformer

We want simulate a center tap transformer in this example. Remove the diode from the previous schematic (Fig. 2.25).

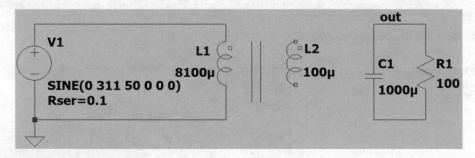

Fig. 2.25 Uncompleted schematic of Example 2

Click the copy icon (Fig. 2.26) and click on the L2. This makes a copy of L2 with name L3. Click on the schematic to add the L3 to it (Fig. 2.27).

Fig. 2.26 Copy icon

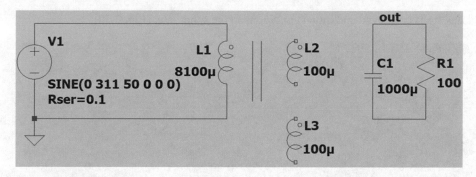

Fig. 2.27 L3 is a copy of inductor L2

Add two diodes to the schematic and convert the schematic to what is shown in Fig. 2.28. This circuit simulates a full wave rectifier.

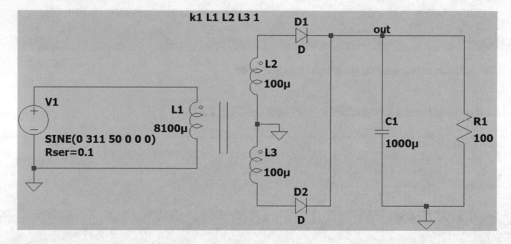

Fig. 2.28 Completed schematic of Example 2

The schematic shown in Fig. 2.29 is equivalent to schematic shown in Fig. 2.28.

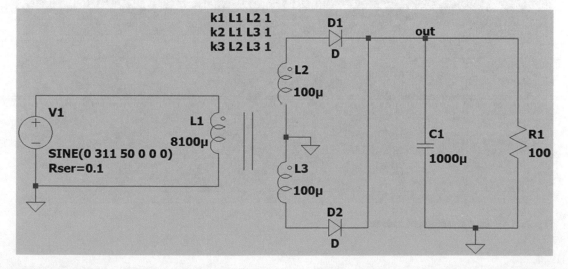

Fig. 2.29 Equivalent of schematic shown in Fig. 2.28

Add the .trans command and initial conditions to the schematic (Fig. 2.30).

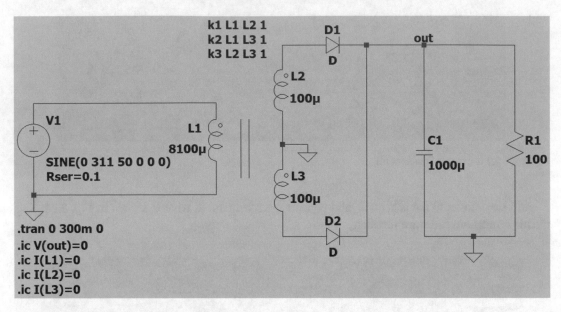

Fig. 2.30 Addition of .trans and .ic commands to the schematic

Run the simulation. Draw the node "out" voltage (Fig. 2.31).

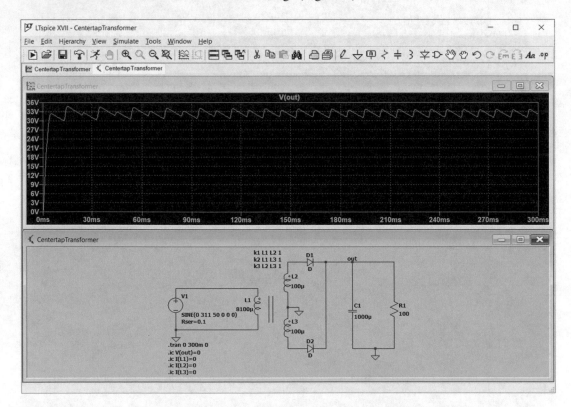

Fig. 2.31 Voltage waveform of node out

Zoom in the waveforms of node "out" and measure the output voltage ripple (Fig. 2.32). According to Fig. 2.33, the output voltage ripple is 2.46 V.

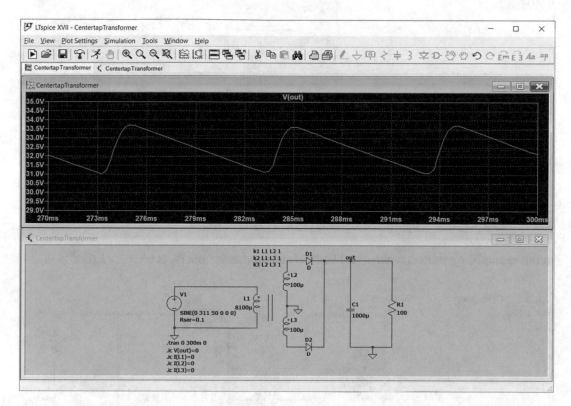

Fig. 2.32   Measurement of output voltage ripple

Fig. 2.33   Peak-peak of output voltage ripple is around 2.46 V

The frequency of output voltage ripple is 100 Hz (Fig. 2.34). The frequency of output voltage ripple is two times the frequency of input source. This is expected since the circuit is full wave rectifier.

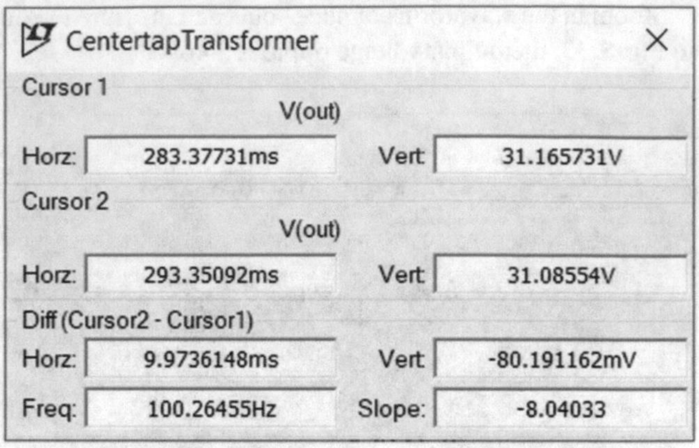

Fig. 2.34 Frequency of
output voltage ripple is
100 Hz

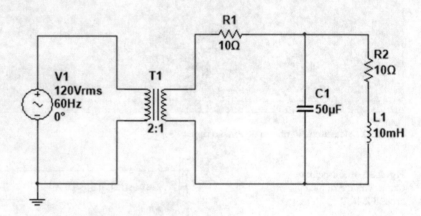

2.4   Example 3: Impedance Seen from Transformer

In this example, we want to measure the RMS of current drawn from input source V1 (Fig. 2.35).

Fig. 2.35 Circuit for
Example 3

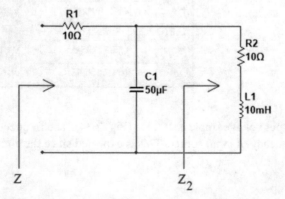

Let's use MATLAB to analyze the circuit. The impedance of the load (Fig. 2.36) is calculated with
the aid of commands shown in Fig. 2.37.

Fig. 2.36 Measurement of the input impedance

Fig. 2.37 MATLAB code

```
Command Window                                          ⊙
>> f=60;w=2*pi*f;R1=10;C1=50e-6;R2=10;L1=10e-3;
>> Xc=-j/(w*C1);
>> XL=j*w*L1;
>> Z2=R2+XL;
>> Z=R1+Xc*Z2/(Xc+Z2)

Z =

   21.1302 + 1.7998i

fx >>
```

The commands shown in Fig. 2.38 calculates the RMS of current drawn from source. According to the calculations shown in Fig. 2.38, the RMS of current drawn from input source is 1.4146 A.

```
Command Window                                          ⊙
>> n=2;Z=21.1302 + 1.7998i;
>> I=120/(n^2*Z)

I =

   1.4095 - 0.1201i

>> abs(I)

ans =

   1.4146

fx >>
```

Fig. 2.38 MATLAB code

The required LTspice schematic is shown in Fig. 2.39. The ratio of $\dfrac{L_2}{L_3}$ must be 4 to simulate a 2:1 transformer. Bigger values of L2 and L3 make the transformer closer to ideal case.

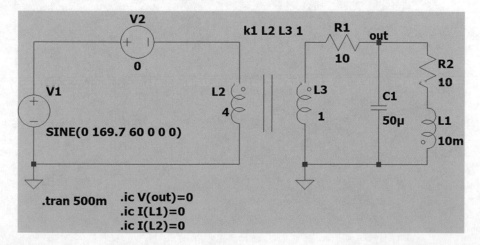

Fig. 2.39   LTspice equivalent of circuit shown in Fig. 2.35

Run the simulation (Fig. 2.40).

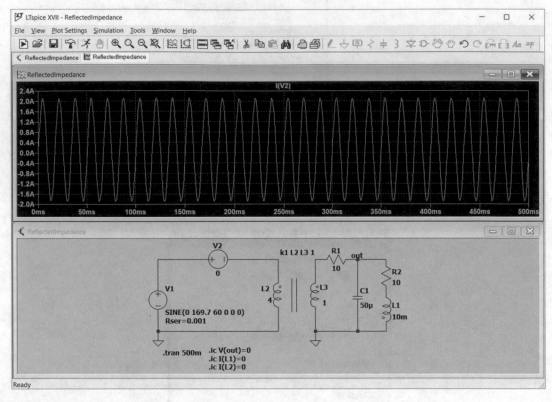

Fig. 2.40 Waveform of current drawn from input source V1

Right click on the time axis and fill the Left and Right boxes with 483.33 ms and 500 ms, respectively. This shows the graph for [483.33 ms, 500 ms] time interval (Fig. 2.41).

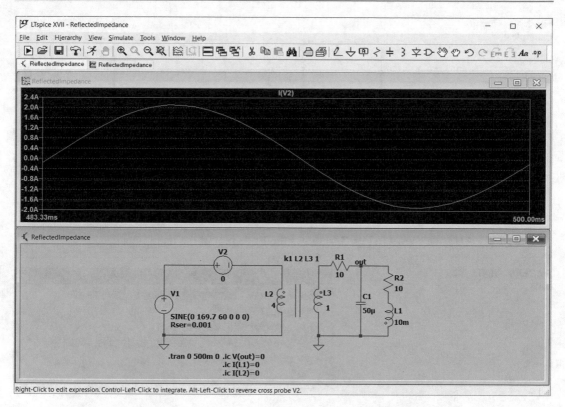

Fig. 2.41   One cycle of I(V2)

Hold down the Ctrl key and click on the I(V2). This shows the RMS of current drawn from the input AC source (Fig. 2.42). The obtained result is quite close to MATLAB result.

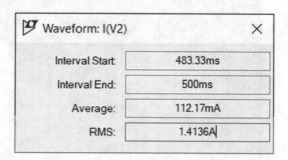

Fig. 2.42   Average and RMS values of waveform shown in Fig. 2.41

2.5   Example 4: Input Impedance of Electric Circuits

In this example, we want to measure the input impedance ($Z_{in}(s) = \dfrac{V_1(s)}{I_1(s)}$) of the circuit shown in Fig. 2.43 as a function of frequency.

Fig. 2.43 Circuit of
Example 4

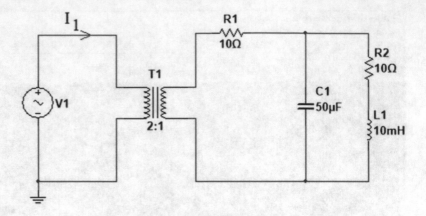

The MATLAB code shown in Fig. 2.44 draws the Bode plot of the input impedance.

Fig. 2.44 MATLAB
code

```
Command Window                                              ⊙
 >> n=2;R1=10;R2=10;L1=10e-3;C1=50e-6;
 >> s=tf('s');
 >> ZC1=1/C1/s;ZL1=L1*s;
 >> Z=n^2*(R1+(ZC1*(R2+ZL1)/(ZC1+R2+ZL1)));
 >> bode(Z),grid on
fx >> |
```

Output of the code in Fig. 2.44 is shown in Fig. 2.45.

Fig. 2.45 Output of
MATLAB code

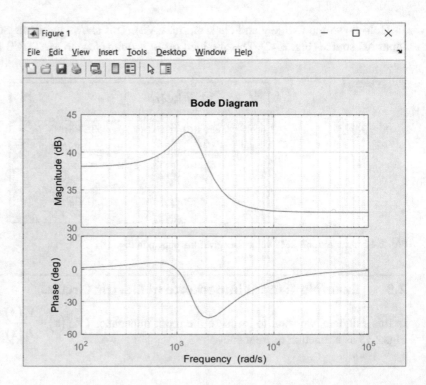

Note that unit of the horizontal axis is $\dfrac{\text{Rad}}{\text{s}}$. You can change it into Hz easily. In order to do this, right click on the graph and click the properties (Fig. 2.46).

Fig. 2.46   Properties permits you to change the unit of horizontal axis

After clicking the Properties, the Property Editor window is opened. Open the Units tab and select Hz for Frequency drop down list (Fig. 2.47). Select absolute for Magnitude drop down list as well (we don't want to use dB for input impedance graph). After applying the changes, click the Close button. Now, the horizontal axis has the unit of Hz (Fig. 2.48) and the vertical axis shows the magnate of impedance in Ohms.

Fig. 2.47   Property
Editor window

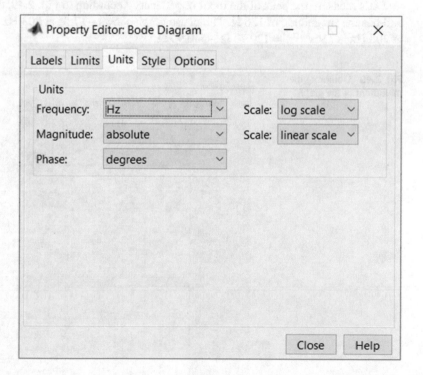

Fig. 2.48 The
horizontal axis has the
unit of Hz

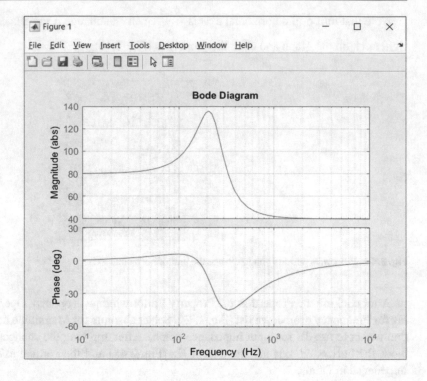

Let's measure the peak of the input impedance. According to Fig. 2.49, the peak occurs at 208 Hz and has the magnitude of 136 Ω. The phase graph shows −17.5° at 208 Hz. So, the input impedance at 208 Hz is $136e^{-j17.5°} = 129.70\ \Omega - j40.90\ \Omega$.

Fig. 2.49 Obtaining the
maximum of the graph

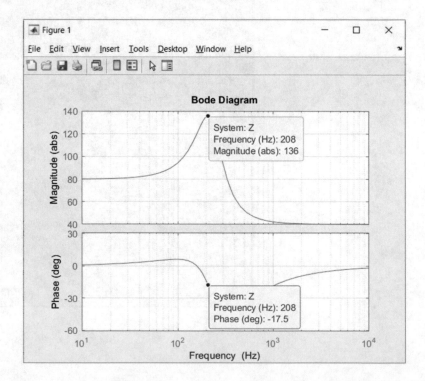

Let's check our calculations with LTspice. Open the schematic of Example 3 and remove the .trans and .ic commands from it (Fig. 2.50).

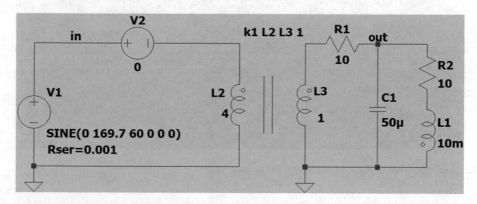

Fig. 2.50   Schematic of Example 3 without .trans and .ic commands

Right click on the V1 and do the settings similar to Fig. 2.51. After clicking the OK button, the schematic changes to what is shown in Fig. 2.52.

Independent Voltage Source - V1

Functions

- (none)
- PULSE(V1 V2 Tdelay Trise Tfall Ton Period Ncycles)
- SINE(Voffset Vamp Freq Td Theta Phi Ncycles)
- EXP(V1 V2 Td1 Tau1 Td2 Tau2)
- SFFM(Voff Vamp Fcar MDI Fsig)
- PWL(t1 v1 t2 v2...)
- PWL FILE: Browse

Additional PWL Points

Make this information visible on schematic: ☑

DC Value

DC value:
Make this information visible on schematic: ☑

Small signal AC analysis(.AC)

AC Amplitude: 1
AC Phase: 0
Make this information visible on schematic: ☑

Parasitic Properties

Series Resistance[Ω]:   0.001
Parallel Capacitance[F]:
Make this information visible on schematic: ☑

Cancel   OK

Fig. 2.51   Settings of voltage source V1

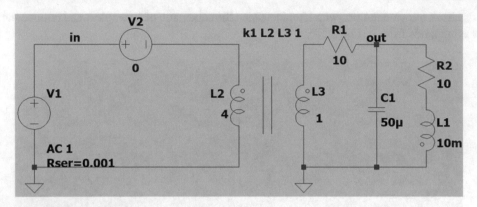

Fig. 2.52 Settings shown in Fig. 2.51 are shown on the schematic

Click the run button. After clicking the Run button, the Edit Simulation Command window appears. Go to the AC Analysis tab. Do the settings similar to Fig. 2.53 and click the OK button. These settings ask the LTspice to calculate the desired quantity (in this example input impedance) on the [10 Hz, 10 kHz] frequency range. Increasing the number entered to the Number of pointed per decade box increases the smoothness of output; however, the required time for simulation increases as well. 100 pointes per decade is good enough for many problems.

Fig. 2.53 Simulation settings

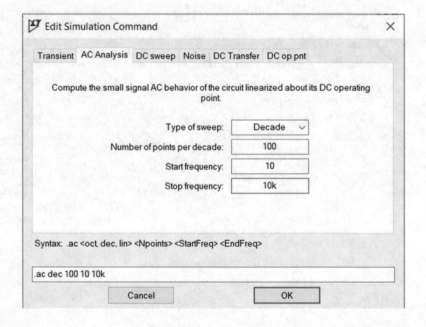

After clicking the OK button in Fig. 2.53, the schematic changes to what is shown in Fig. 2.54, and the simulation is done (Fig. 2.55).

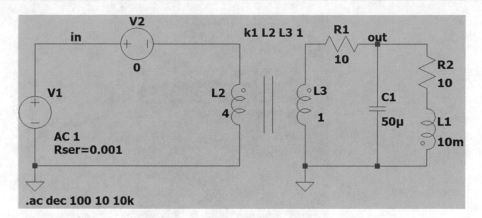

Fig. 2.54 Settings shown in Fig. 2.53 are added to the schematic

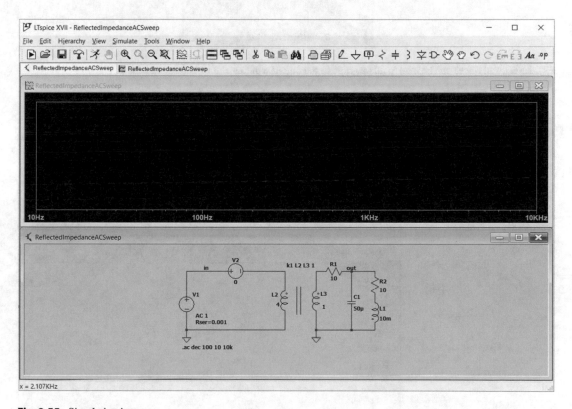

Fig. 2.55 Simulation is run

Right click on the black area of Fig. 2.55 and select the Add Traces (Fig. 2.56).

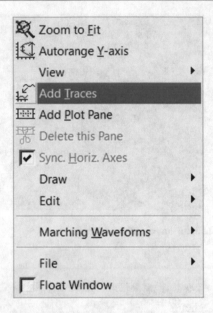

Fig. 2.56 Add traces can be used to add a graph to the output window

After clicking the Add Traces, the Add Traces to Plot window appears and permits you to determine what you want. Enter V(in)/I(V2) to the Expression(s) to add box and click the OK button (Fig. 2.57). Result is shown in Fig. 2.58. The solid line shows the magnitude graph and dotted line shows the phase graph.

Fig. 2.57 Add Traces to Plot window permits you to select the waveform that is shown on the screen

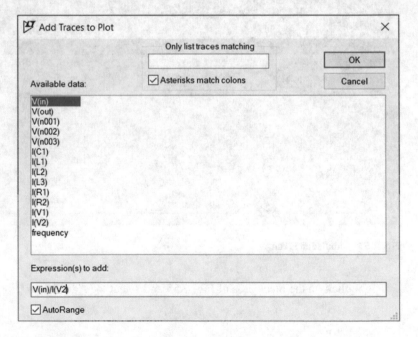

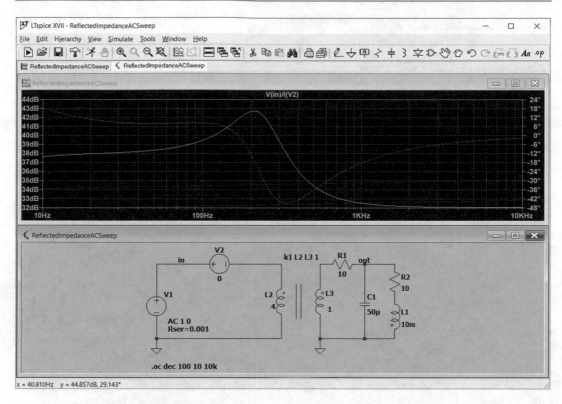

Fig. 2.58 Graph of input impedance

The vertical axis of Fig. 2.58 has the unit of dB. However, we want the absolute magnitude. In order to obtain the absolute magnitude, right click on the vertical axis of the graph and select Linear and click the OK button (Fig. 2.59). Now, the vertical axis has the unit of Ohms (Fig. 2.60).

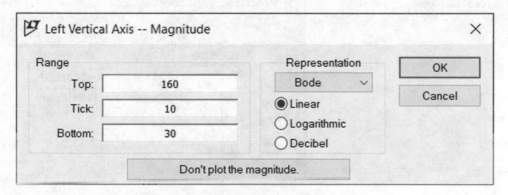

Fig. 2.59 Left Vertical Axis window

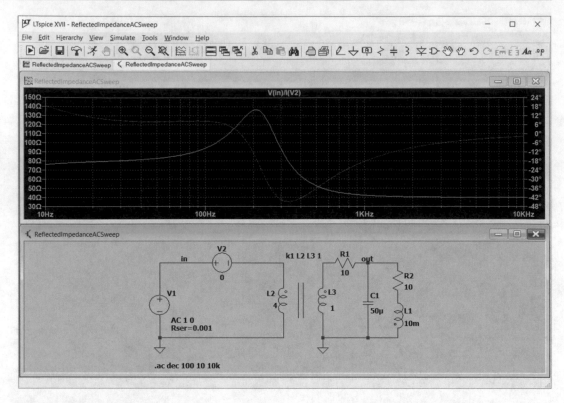

Fig. 2.60 Simulation result

Let's measure the peak of the input impedance graph shown in Fig. 2.60. According to Fig. 2.61, the maximum input impedance occurs at 208.1 Hz and equals to $136.65e^{j-16.34°}$ Ω. Obtained result is quite close to the MATLAB result.

Fig. 2.61 Reading the
coordinate of maximum
point

Note that the MATLAB result is calculated for ideal transformer case. If you increase the value of inductor L2 and L3, the transformer become closer to the ideal case and the obtained result become closer to the MATLAB result. For instance, for L3 = 10 H and L2 = 40 H, the result shown in Fig. 2.62 is obtained. According to this figure, the maximum input frequency is obtained at 208 Hz and its value is $135.723e^{j-17.58°}$ Ω.

Fig. 2.62 Obtaining the magnitude and phase for given frequency

⚡ ReflectedImpedanceACSweep		✕
Cursor 1		
	V(in)/I(V2)	
Freq: 208.00922Hz	Mag:	135.72289Ω ◉
	Phase:	-17.581652° ○
	Group Delay:	1.2382058ms ○
Cursor 2		
Freq: -- N/A--	Mag:	-- N/A-- ○
	Phase:	-- N/A-- ○
	Group Delay:	-- N/A-- ○
	Ratio (Cursor2 / Cursor1)	
Freq: -- N/A--	Mag:	-- N/A--
	Phase:	-- N/A--
	Group Delay:	-- N/A--

2.6   Example 5: Transfer Function of Linear Circuits

In the previous example we learned how to use AC Sweep analysis to obtain the input impedance of a circuit. In this example, we want to use AC sweep to obtain the frequency response of other quantities. Let's draw the frequency response of $\dfrac{V_{out}(j\omega)}{V_{in}(j\omega)}$ for circuit of previous example. Run the schematic of previous example (Fig. 2.63).

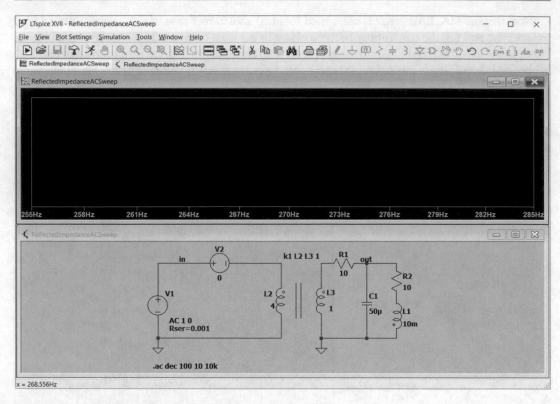

Fig. 2.63 Simulation is run

Right click on the black graph window and select Add Traces (Fig. 2.64).

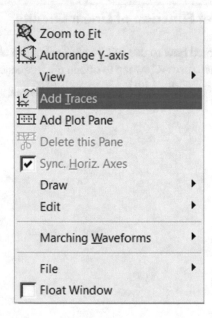

Fig. 2.64 Add Traces

Enter V(out)/V(in) to the Expression(s) to add box and click the OK button (Fig. 2.65).

Fig. 2.65 Defining the
V(out)/V(in)

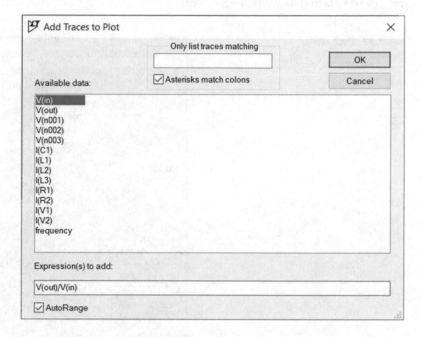

After clicking the OK button, graph of frequency response of $\dfrac{V_{\text{out}}(j\omega)}{V_{\text{in}}(j\omega)}$ appears on the screen(Fig. 2.66). The solid line shows the magnitude graph and dotted line shows the phase graph.

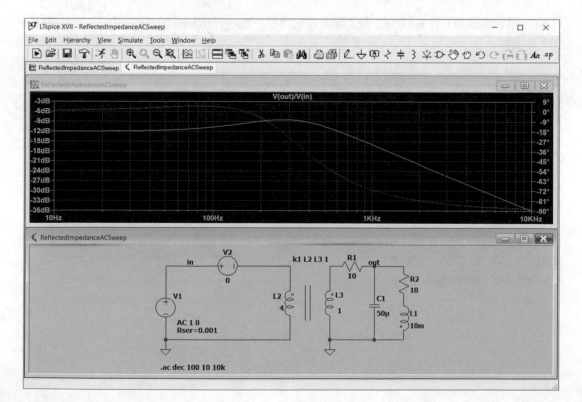

Fig. 2.66 Simulation result

You can use cursors to read the peak of the graph. According to Fig. 2.67, peak of $\dfrac{V_{out}(j\omega)}{V_{in}(j\omega)}$ occurs at 279.043 Hz and its value is −8.453 dB.

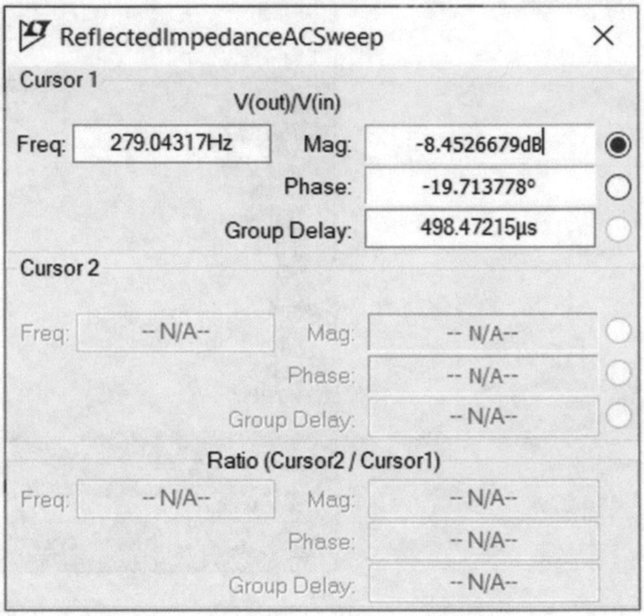

Fig. 2.67 Peak of the
graph shown in Fig. 2.66

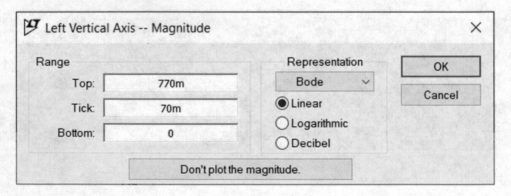

The unit of vertical axis in Fig. 2.66 is dB. If you like to see the absolute value of $\dfrac{V_{out}(j\omega)}{V_{in}(j\omega)}$, right-click on the vertical axis and select the Linear (Fig. 2.68). The graph with linear vertical axis is shown in Fig. 2.69.

Fig. 2.68 Left Vertical Axis window

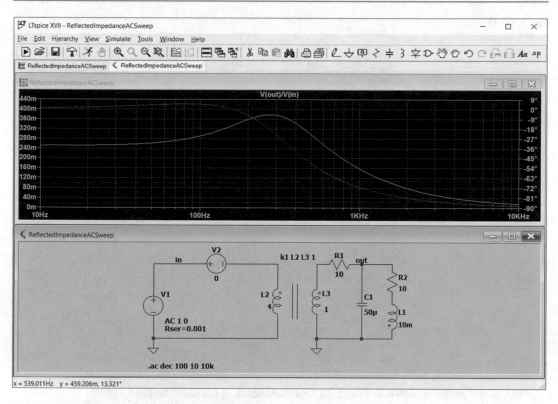

Fig. 2.69 Absolute value of gain is shown on the screen

Let's check the obtained result. The MATLAB code shown in Fig. 2.70 draws the frequencyresponse of $\dfrac{V_{out}(j\omega)}{V_{in}(j\omega)}$ for [10 Hz, 10 kHz] interval. Output of this code is shown in Fig. 2.71.

Fig. 2.70 MATLAB code

```
Command Window

>> n=2;R1=10;R2=10;L1=10e-3;C1=50e-6;
>> s=tf('s');
>> ZC1=1/C1/s;ZL1=L1*s;
>> Z2=(ZC1*(R2+ZL1)/(ZC1+R2+ZL1));
>> fmin=10;wmin=2*pi*fmin;
>> fmax=10000;wmax=2*pi*fmax;
>> w=logspace(log10(wmin),log10(wmax),500);
>> bode(Z2/(R1+Z2)*(1/n),w),grid on
fx >>
```

Fig. 2.71 Output of
MATLAB code

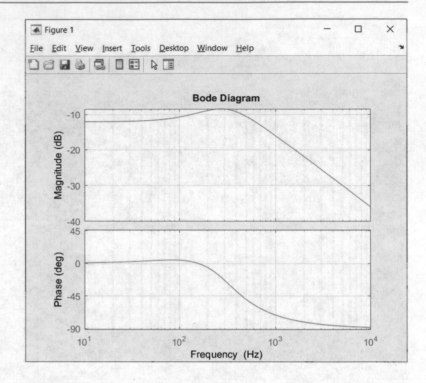

Peak of the graph is shown in Fig. 2.72. Obtained result is the same as the LTspice.

Fig. 2.72 Output of
MATLAB code

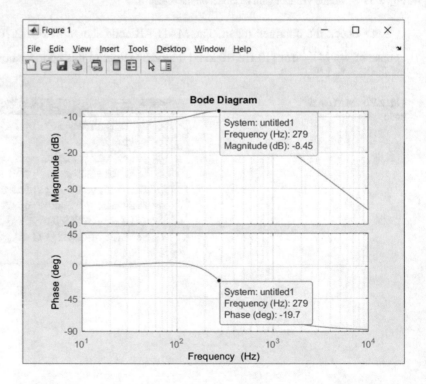

2.7   Example 6: DC Sweep Analysis

In this example, we want to obtain the I–V characteristic of a diode. DC sweep analysis is used for this purpose. Draw the schematic shown in Fig. 2.73.

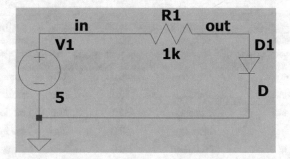

Fig. 2.73   Schematic of Example 6

Right click on the diode D1 and click the Pick New Diode button (Fig. 2.74).

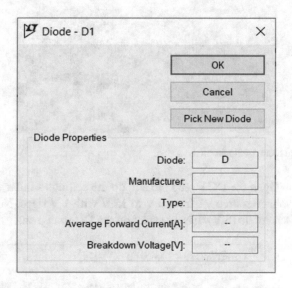

Fig. 2.74   Diode window

Click on the 1N4148 and click the OK button (Fig. 2.75). The schematic changes to what is shown in Fig. 2.76.

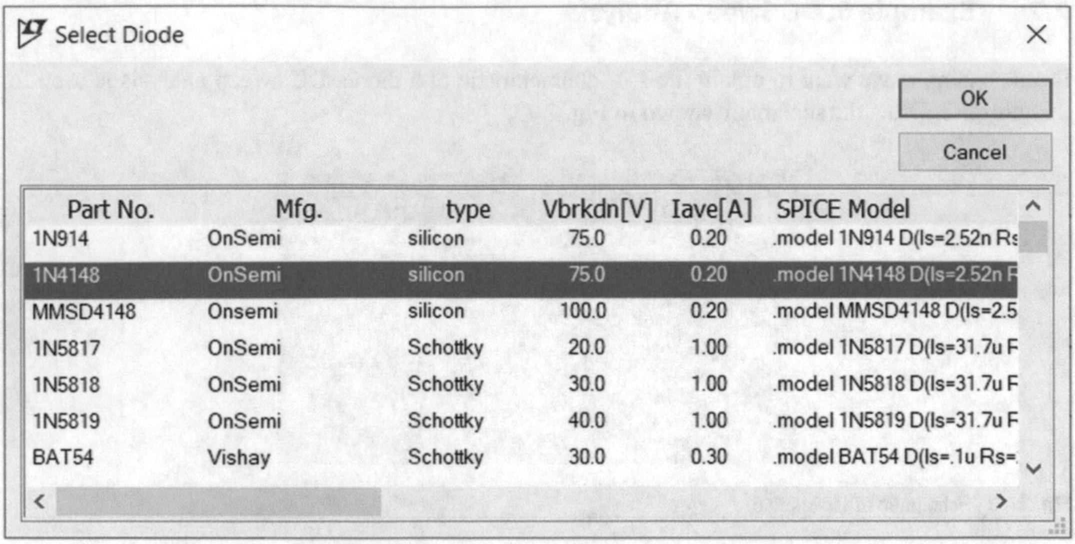

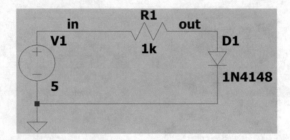

Fig. 2.75 1N4148 is selected

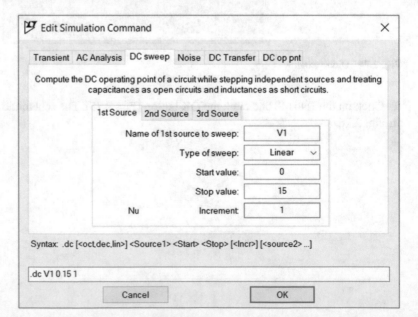

Fig. 2.76 Name of selected diode is shown on the screen

Click the Run button. Open the DC sweep tab and do the settings similar to Fig. 2.77. These settings change the input voltage source V1 form 0 V to 15 V with 1 V steps. Note that the value of V1 which is shown on the schematic (5 V), has no effect on the DC sweep analysis.

Fig. 2.77 Simulation settings

After clicking the OK button, window shown in Fig. 2.78 appears.

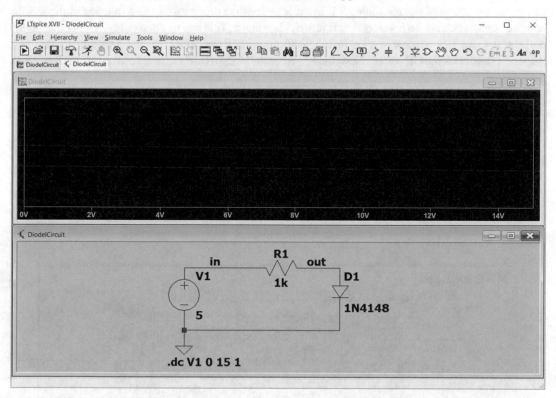

Fig. 2.78 Simulation is run

Put the mouse cursor on the diode D1 and click it. This draws the graph of diode current as a function of voltage source V1 (Fig. 2.79).

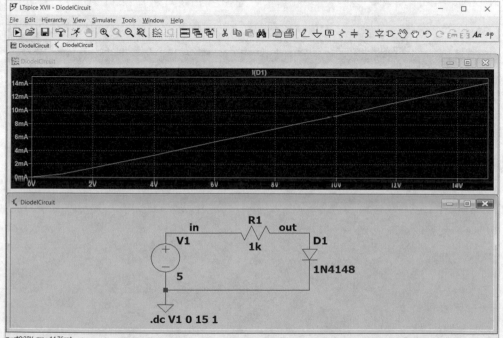

Fig. 2.79 Closer look at the diode D1 current

We want to see the current of diode as a function of its voltage. However, Fig. 2.79 shows the diode current as a function of voltage source V1. In order to draw the diode current as a function of diode voltage, we need to change the horizontal axis into diode voltage. This can be done easily. Right click on the horizontal axis. The window shown in Fig. 2.80 appears. Enter V(out) to the Quantity Plotted box (Fig. 2.81) and click the OK button. Now, the I-V characteristics of diode appears on the screen (Fig. 2.82).

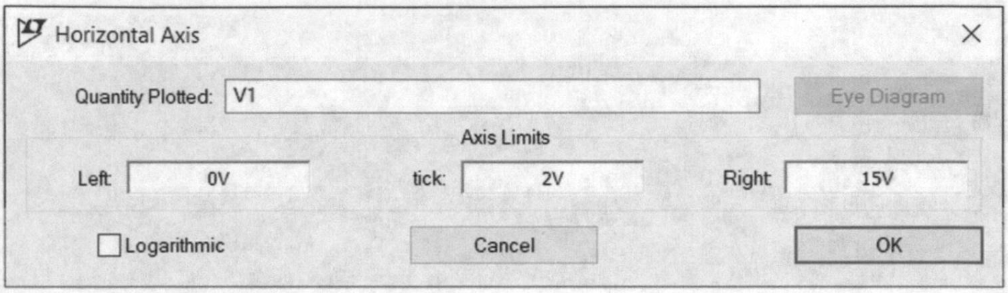

Fig. 2.80   Horizontal Axis window

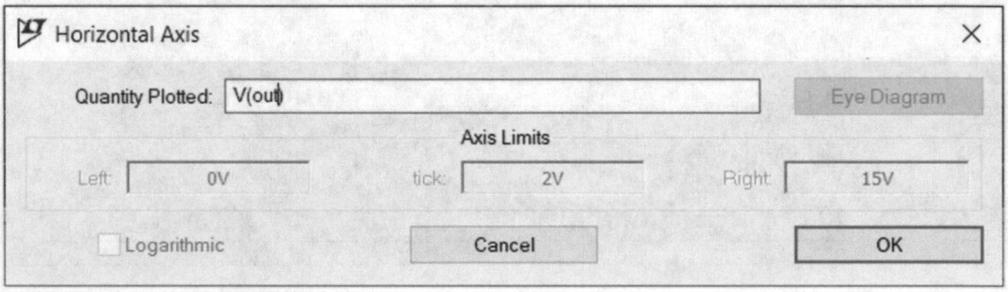

Fig. 2.81   V(out) is entered to the box

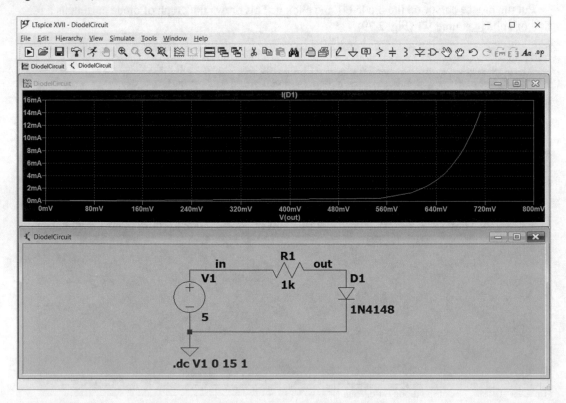

Fig. 2.82   Simulation result (IV characteristic)

2.8   Example 7: I–V Characteristics of Zener Diode

In this example, we want to draw the I–V characteristics of a zener diode. We use the DC sweep analysis again. Draw the schematic shown in Fig. 2.83.

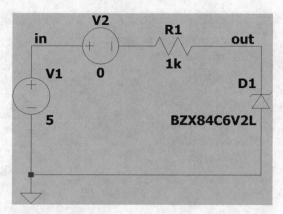

Fig. 2.83 Schematic of Example 7

Click the Run icon and open the DC sweep tab. Do the settings similar to Fig. 2.84. After clicking the OK button, the graph window appears (Fig. 2.85).

Fig. 2.84 Simulation settings

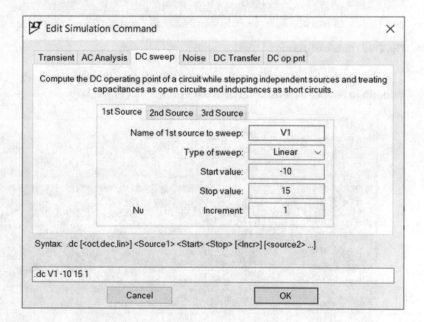

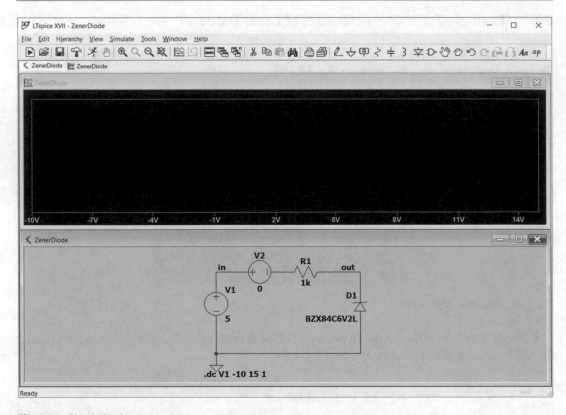

Fig. 2.85 Simulation is run

Click on the V2. This draws the current which pass through the diode (i.e., current that goes from cathode to anode) as a function of input voltage V1 (Fig. 2.86).

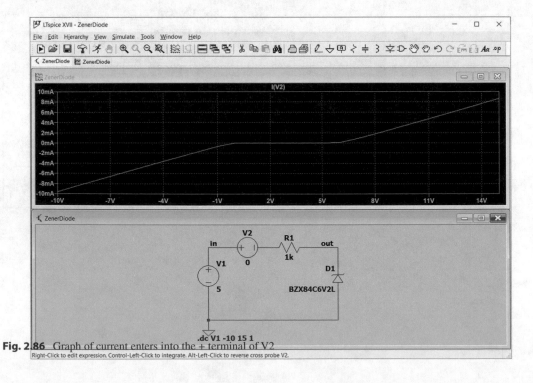

Fig. 2.86 Graph of current enters into the + terminal of V2

We want to see the current of zener diode as a function of its voltage. In order to draw the diode current as a function of diode voltage, we need to change the horizontal axis into diode voltage. Right click on the horizontal axis and enter V(out) to the Quantity Plotted box (Fig. 2.87). After clicking the OK button, the I–V characteristics of the zener diode appears on the screen (Fig. 2.88).

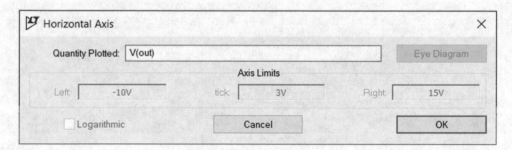

Fig. 2.87 Horizontal Axis window

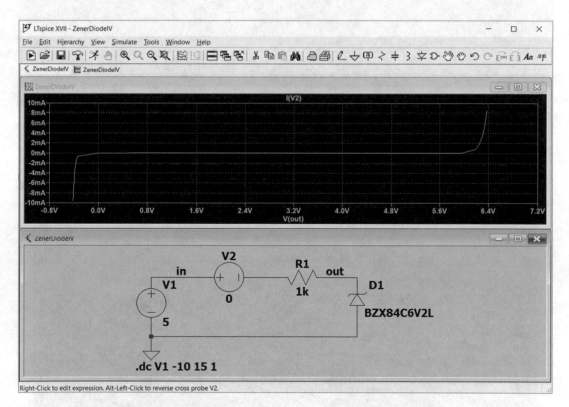

Fig. 2.88 I–V characteristic of the zener diode

If you decrease the increase of voltage in each step, the curve become smoother. The curve in Fig. 2.89 is drawn with 0.1 V steps and it is smoother than the curve of Fig. 2.88.

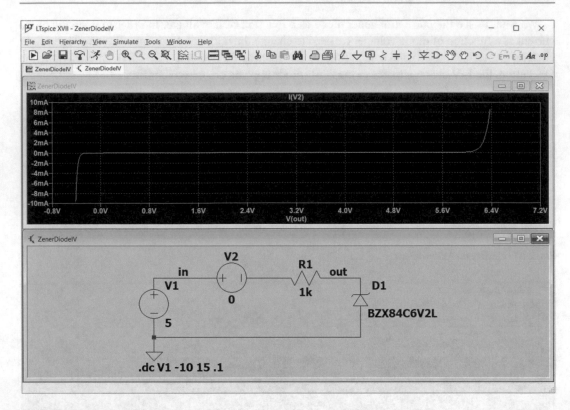

Fig. 2.89 Obtained graph is more smooth in comparison with the graph shown in Fig. 2.88

You can measure the knee of the graph with a cursor (Fig. 2.90). According to Fig. 2.91, the knee of the graph is at 6.17 V.

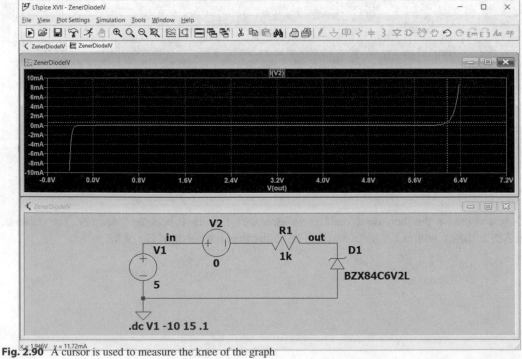

Fig. 2.90 A cursor is used to measure the knee of the graph

Fig. 2.91 Knee of the graph is around 6.17 V

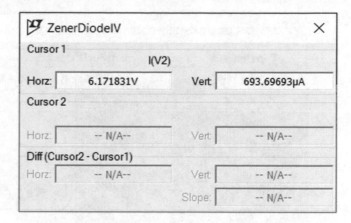

2.9 Example 8: Common Emitter Amplifier

In this example, we want to analyze a common emitter amplifier. The NPN and PNP transistors are simulated with the aid of blocks shown in Figs. 2.92 and 2.93, respectively.

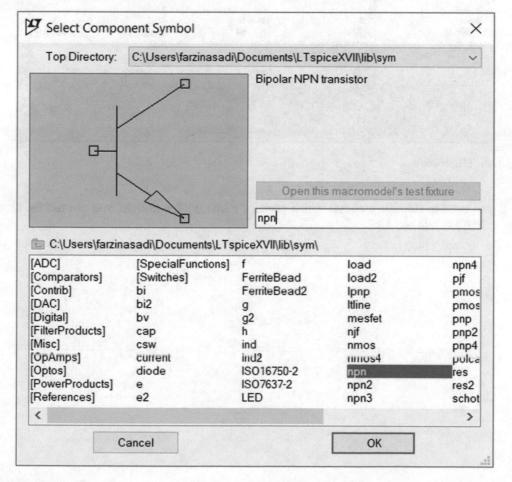

Fig. 2.92 NPN block

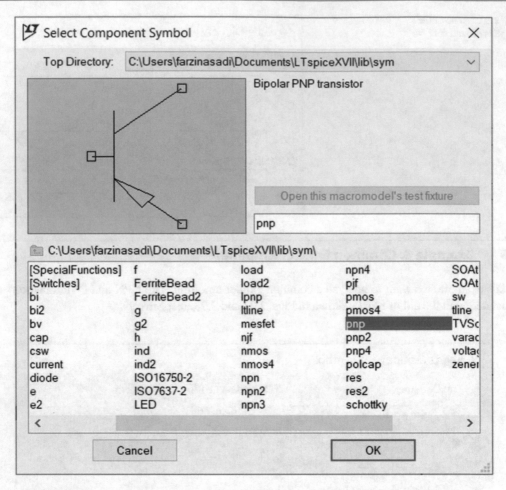

Fig. 2.93 PNP block

If you need to simulate a circuit which contains a MOSFET transistor, you can use the blocks shown in Figs. 2.94 and 2.95.

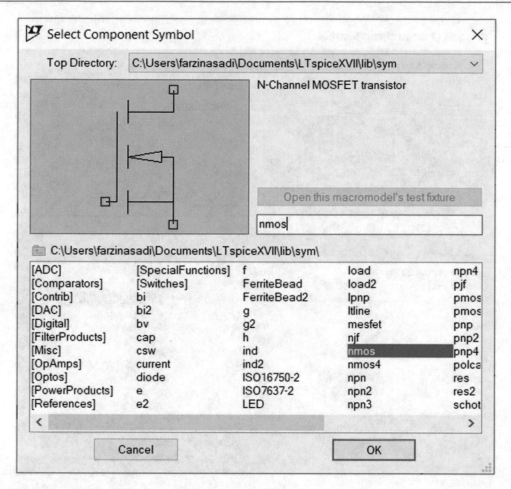

Fig. 2.94 NMOS block

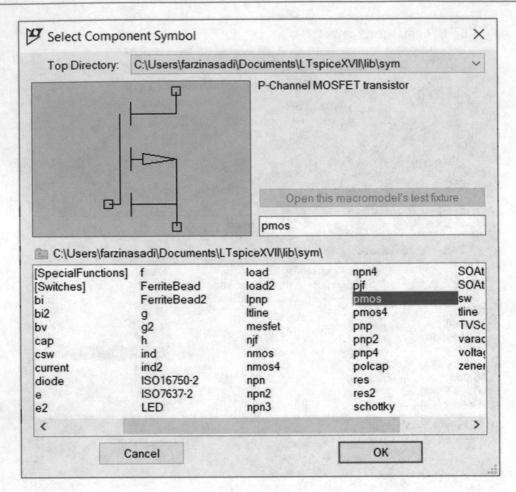

Fig. 2.95 PMOS block

Draw the schematic shown in Fig. 2.96.

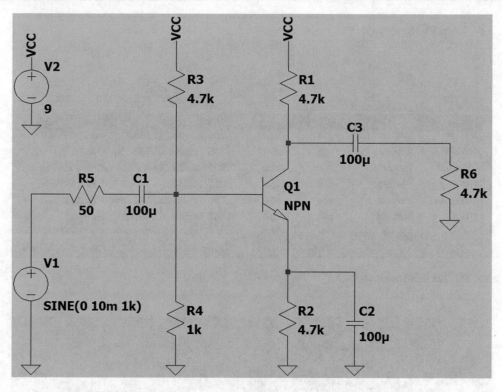

Fig. 2.96 Schematic of Example 8

Right click on the transistor Q1 and click the Pick New Transistor button (Fig. 2.97). Then select 2N2222 and click the OK button (Fig. 2.98). After clicking the OK button, the schematic changes to what is shown in Fig. 2.99.

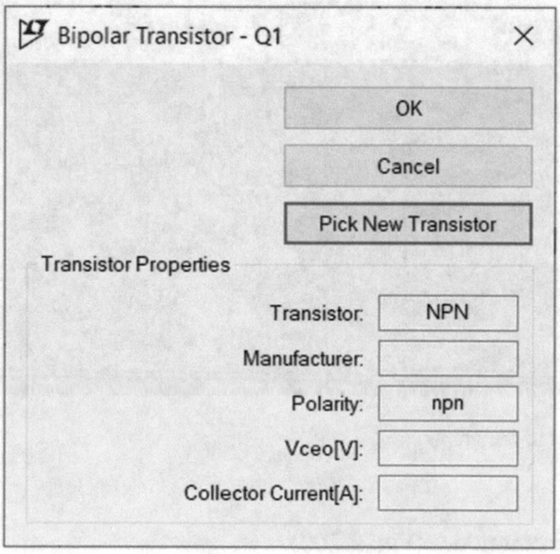

Fig. 2.97 Bipolar Transistor window

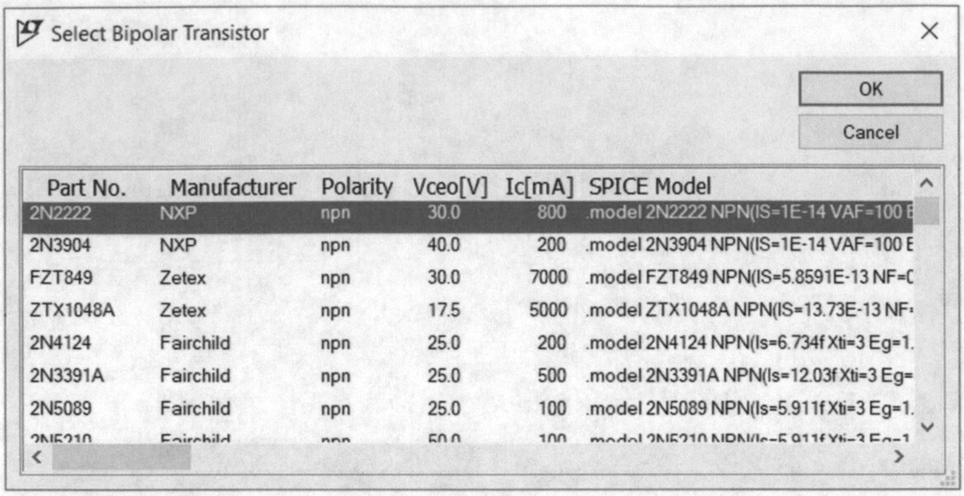

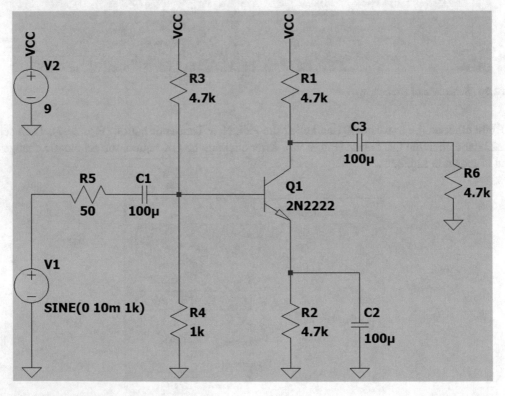

Fig. 2.98 2N2222 transistor is selected

Fig. 2.99 Changes are applied to the graph

Give the name to the circuit nodes (Fig. 2.100).

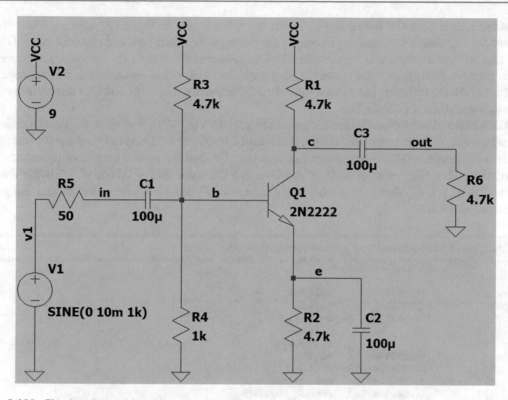

Fig. 2.100   Circuit nodes are labeled

Click the Run icon and open the DC op pnt tab (Fig. 2.101). Then click the OK button.

Fig. 2.101   Simulation
settings

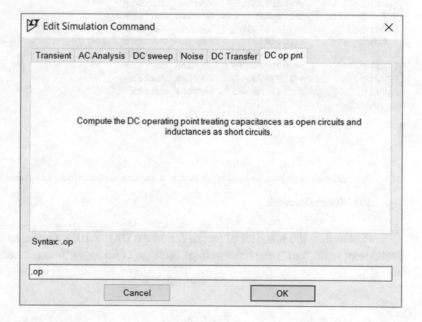

After clicking the OK button in Fig. 2.101, the result shown in Fig. 2.102 appears. This result shows the DC steady state values of voltages and currents. Note that base and collector currents are positive and emitter current is negative. In LTspice, the currents that enter the transistor are assumed to be positive. Since the emitter current in NPN transistors goes out of the transistor, it is negative. If you apply the DC operating point to a circuit with PNP transistor, base and collector currents are negative and emitter current is positive.

Let's take a closer look to the result shown in Fig. 2.102. V(c), V(b), V(e) show the voltage of node c, voltage of node b and voltage of node e, respectively. Ic(Q1), Ib(Q1), and Ie(Q1) show the collector current of transistor Q1, the base current of transistor Q1, and the emitter current of transistor Q1, respectively. So, the operating point of transistor is VCE = 8.03888 − 0.965636 = 7.0732 V and IC = 0.204 mA. I(V2) shows the current drawn from source V2 and I(R3) shows the current that pass through resistor R3.

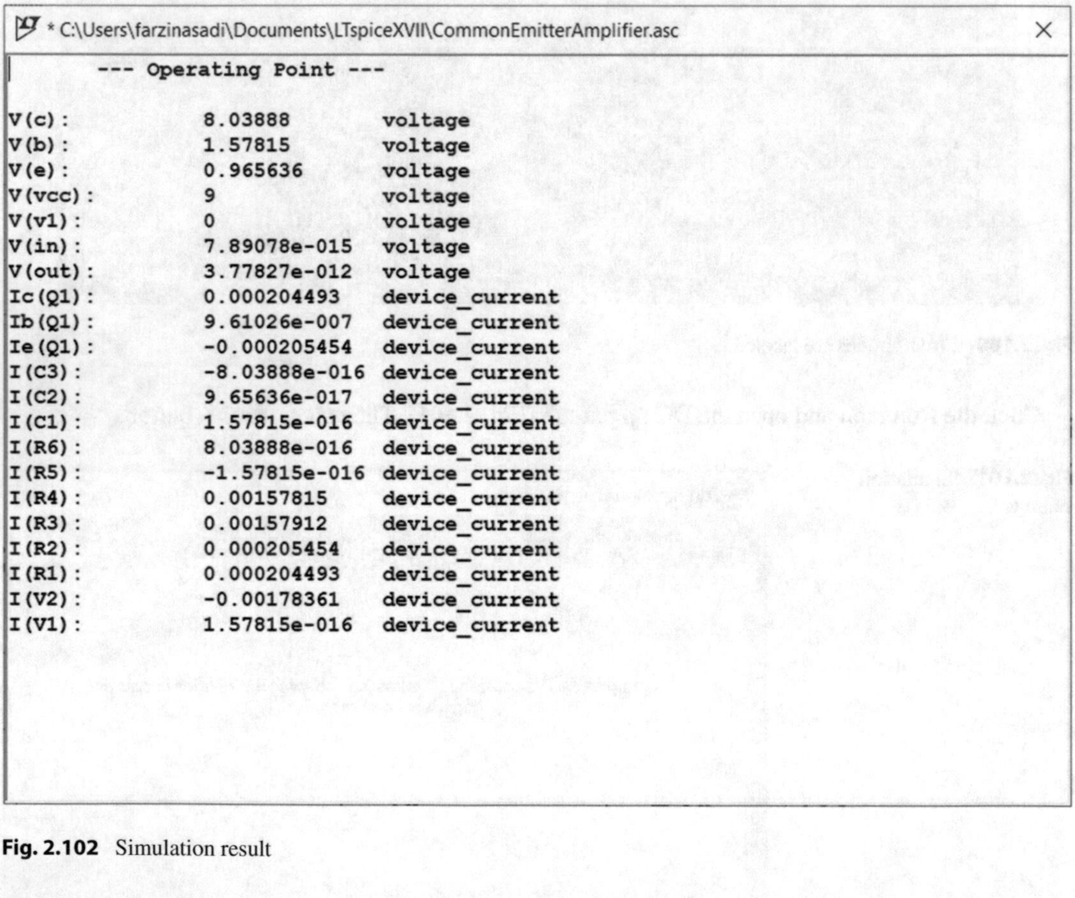

📄 * C:\Users\farzinasadi\Documents\LTspiceXVII\CommonEmitterAmplifier.asc		✕
--- Operating Point ---		
V(c):	8.03888	voltage
V(b):	1.57815	voltage
V(e):	0.965636	voltage
V(vcc):	9	voltage
V(v1):	0	voltage
V(in):	7.89078e-015	voltage
V(out):	3.77827e-012	voltage
Ic(Q1):	0.000204493	device_current
Ib(Q1):	9.61026e-007	device_current
Ie(Q1):	-0.000205454	device_current
I(C3):	-8.03888e-016	device_current
I(C2):	9.65636e-017	device_current
I(C1):	1.57815e-016	device_current
I(R6):	8.03888e-016	device_current
I(R5):	-1.57815e-016	device_current
I(R4):	0.00157815	device_current
I(R3):	0.00157912	device_current
I(R2):	0.000205454	device_current
I(R1):	0.000204493	device_current
I(V2):	-0.00178361	device_current
I(V1):	1.57815e-016	device_current

Fig. 2.102 Simulation result

Let's simulate the behavior of circuit for 50 ms (Fig. 2.103). Figure 2.104 shows the load voltage (voltage of node "out") and input voltage (voltage of node "in") simultaneously.

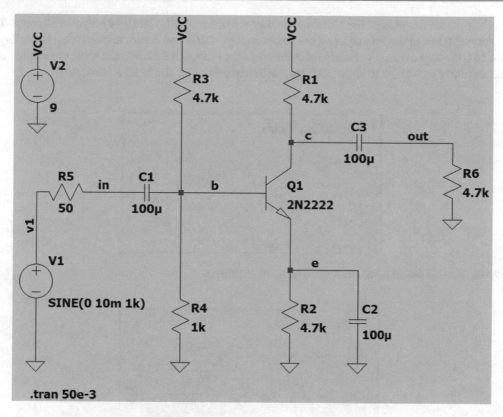

Fig. 2.103 .trans 50e-3 command is added to the schematic

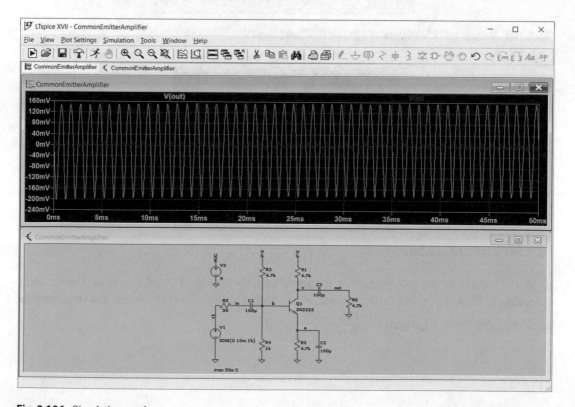

Fig. 2.104 Simulation result

Let's measure the gain of the system. One way it to compare the RMS of output with input. The approximate RMS of output and input are shown in Figs. 2.105 and 2.106, respectively. According to Fig. 2.107, the approximate voltage gain of the amplifier is about 18.37. We can write the voltage gain of the amplifier as −18.37 to show 180 phase difference between the input and output.

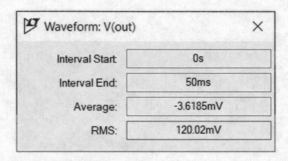

Fig. 2.105 Average and RMS values of V(out) for [0, 50 ms] interval

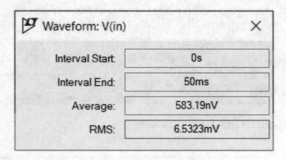

Fig. 2.106 Average and RMS values of V(in) for [0, 50 ms] interval

```
Command Window                          ⊙
   >> 120.02e-3/6.5323e-3


   ans =


   18.3733

fx >> |
```

Fig. 2.107 MATLAB calculations

You can measure the voltage gain of the amplifier with the aid of cursors as well. For instance, according to Figs. 2.108 and 2.109, at $t = 2.24$ ms, output voltage is −186.34 mV and input voltage is 9.36 mV. So, the voltage gain is $\dfrac{-186.34\,\text{mV}}{9.36\,\text{mV}} = 19.91$.

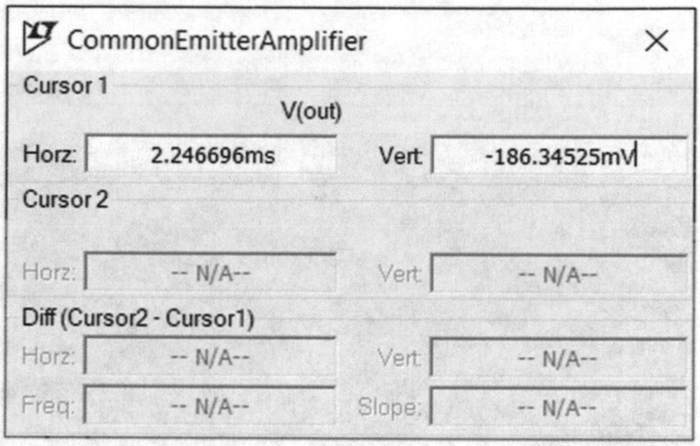

Fig. 2.108 Value of
V(out) at t = 2.24 ms

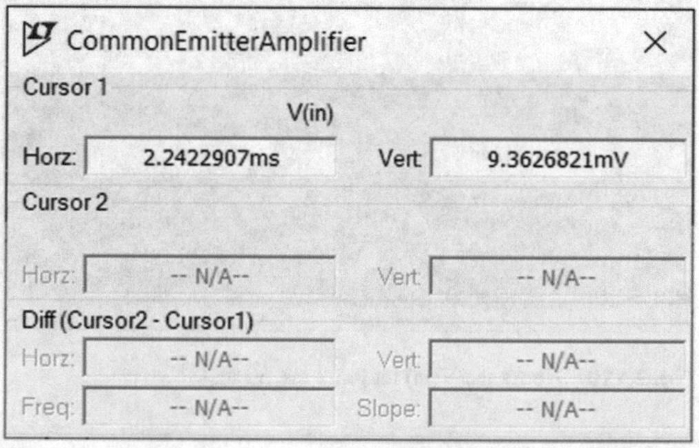

Fig. 2.109 Value of
V(in) at t = 2.24 ms

Figure 2.110 shows the output voltage and input voltage for [45.2 ms, 50 ms] time interval. You can easily see that the output voltage is distorted.

Let's use cursors to ensure that output voltage is distorted (Fig. 2.111). Value of peak voltages are shown in Fig. 2.112. According to Fig. 2.112, positive peak value is 155.89 mV and negative peak value is −186.23 mV. So, you can obviously see that the positive and negative half cycles are not amplified with the same gain. Remember that transistor is a nonlinear component so such a behavior is expected. You can obtain a good linear amplifier with the aid of feedback.

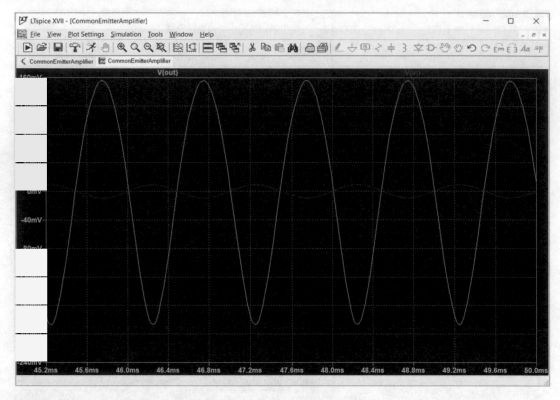

Fig. 2.110 V(out) and V(in) for [45.2 ms, 50 ms]

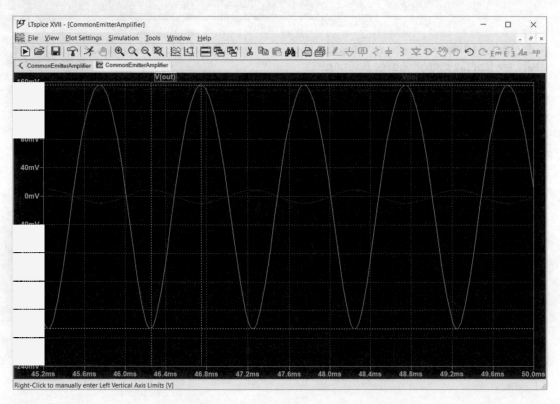

Fig. 2.111 Addition of cursors to the graph

Fig. 2.112 Coordinates
read by the cursors

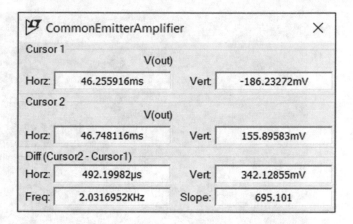

2.10   Example 9: FFT Analysis

In the previous example, we saw that positive and negative half cycles of the input are not amplified with the same gain. This shows that out amplifier is nonlinear. If you stimulate a linear amplifier with a sinusoidal input with frequency of f_0, the output contains a component with frequency f_0 only. However, if you stimulate an amplifier with a sinusoidal with frequency f_0, the output contains f_0 and its harmonics, i.e., kf_0 where $k \in N$. Let's see the harmonic contents of the common emitter amplifier of previous example. Open the schematic of previous example and click the Simulate> Edit Simulation Cmd. Enter 100n to the Maximum Timestep box and click the OK button (Fig. 2.113). This cause the simulation time step to be less than 100 ns which increases the accuracy of simulation. Small step size is preferred for FFT and Fourier (Example 10) simulations. After clicking the OK button, the schematic changes to what is shown in Fig. 2.114.

Fig. 2.113 Simulation
settings

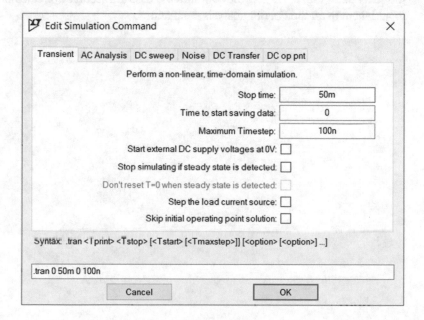

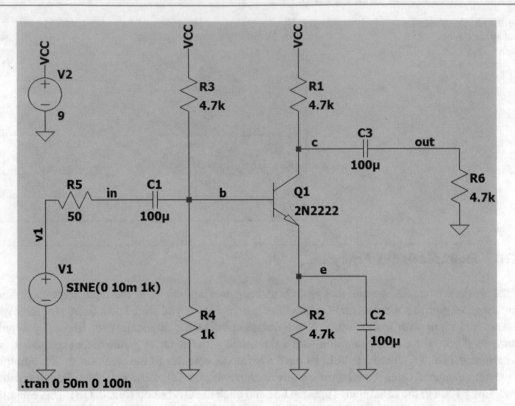

Fig. 2.114 Simulation command is added to the schematic

Click the SPICE Directive button and add the .options numdgt = 7 and .options plotwinsize = 0 to the schematic (Fig. 2.115). These two lines increase the accuracy of calculations. So, it is a good idea to add them to all the Fourier analysis simulations.

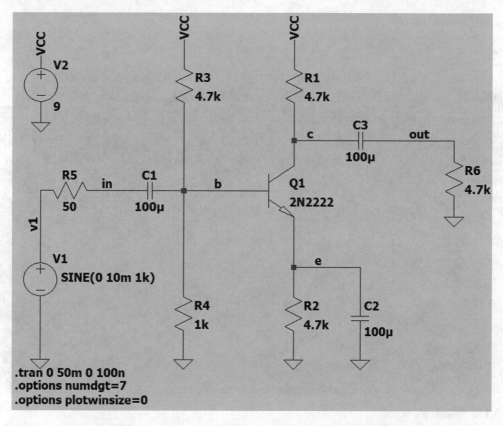

Fig. 2.115   .options numdgt = 7 and .options plotwinsize = 0 commands are added to the schematic

Run the simulation (Fig. 2.116).

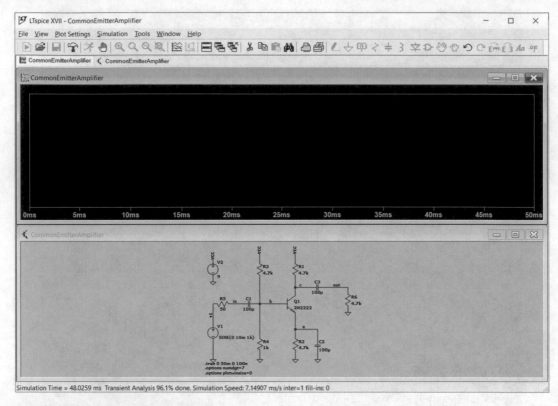

Fig. 2.116 Simulation is run

Right click on the black area and click the View> FFT (Fig. 2.117).

Fig. 2.117 View> FFT

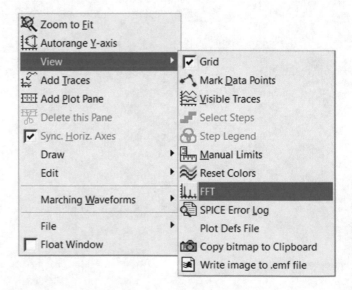

After clicking the View> FFT, the window shown in Fig. 2.118 appears. Select V(out) since we want to study the harmonic content of voltage of node "out" and click the OK button.

Fig. 2.118   V(out) is selected

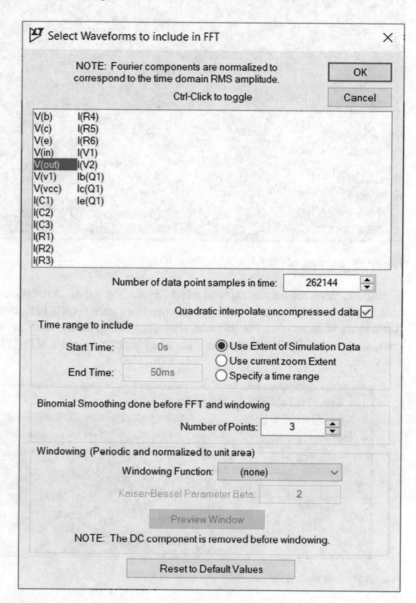

After clicking the OK button, the spectrum (frequency content) of the V(out) appears on the screen (Fig. 2.119). The peaks show presence of harmonics.

Fig. 2.119 Spectrum of V(out)

Right click on the vertical axis and select Linear (Fig. 2.120). After selecting the Linear, the amplitude (peak value) of the harmonics appears one the screen (Fig. 2.121). Note that the FFT shows the RMS value of harmonics. For instance, according to Fig. 2.121, RMS value of fundamental (1 kHz) component is 121.7 mV. So, the amplitude of 1 kHz component is $\sqrt{2} \times 121.7\,\text{mV} = 172.108\,\text{mV}$.

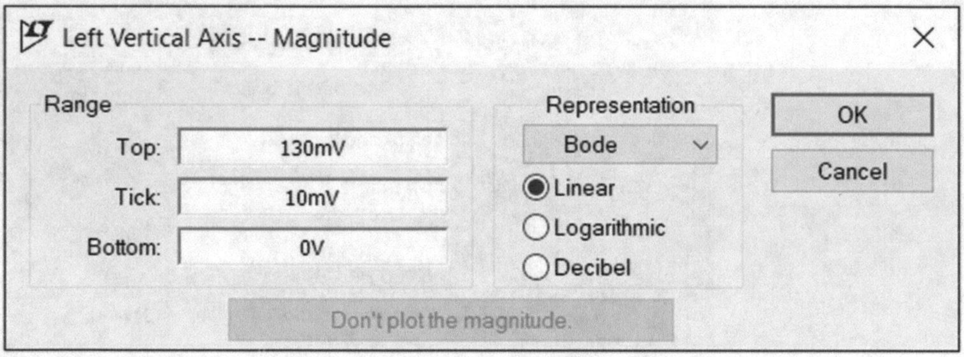

Fig. 2.120 Left Vertical Axis window

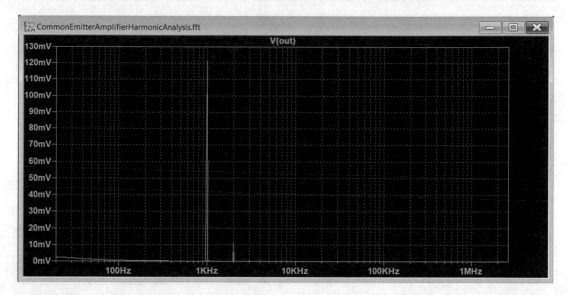

Fig. 2.121 Amplitude of harmonics

You can see the harmonic content in the desired range by right clicking on the horizontal axis and entering the desired range. For instance, if you want to see the harmonic content in the [500 Hz, 10 kHz] range, right click on the horizontal axis and enter 500 to the Left box and 10 kHz to the Right box (Fig. 2.122). After clicking the OK button, [500 Hz, 10 kHz] interval will be shown (Fig. 2.123). Figure 2.123 shows that the output has a component at 2 kHz. This shows that out amplifier is not linear. Note that the input signal has only one component at 1 kHz, so the 2 kHz component is generated during the amplification process. If you zoom in the graph, you can see the presence of third harmonics as well (Fig. 2.124). If you right click on the vertical axis and select the Decibel, you can see the presence of harmonics more easily (Fig. 2.125).

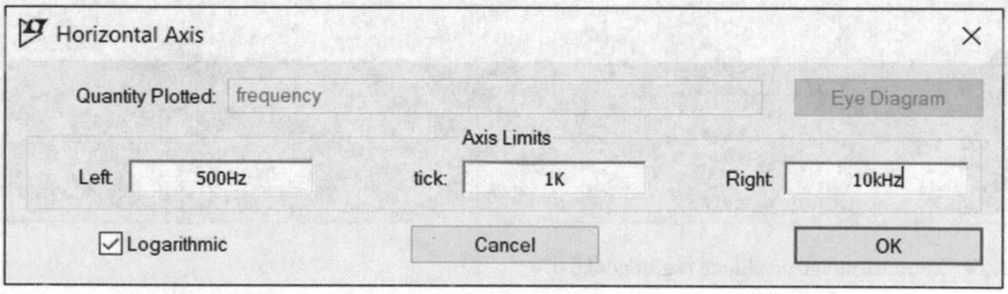

Fig. 2.122 Horizontal Axis window

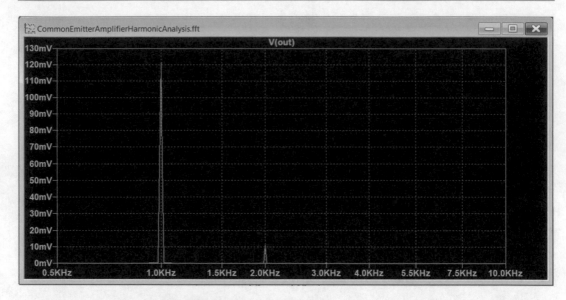

Fig. 2.123   Spectrum for [500 Hz, 10 kHz] interval

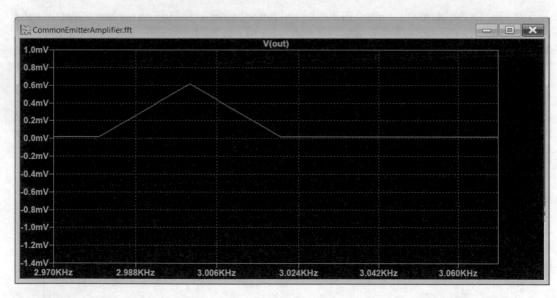

Fig. 2.124   Third harmonic amplitude is around 0.6 mV

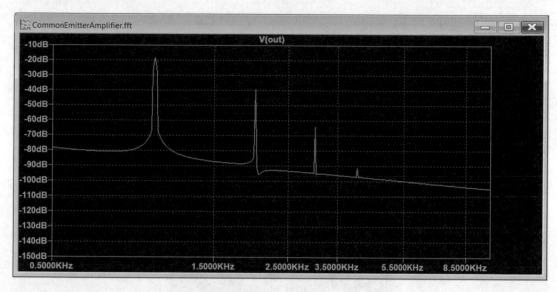

Fig. 2.125 Decibel mode permits you to see the harmonics easily

2.11   Example 10: .Four Command

In the previous example we used the FFT capability built into the waveform viewer to see the harmonic contents of the output voltage. In this example, we will introduce the Fourier analysis (.four command) which is used to measure the amplitude harmonics. The .four command calculated the Total Harmonic Distortion (THD) as well. The help page of .Four command is shown in Fig. 2.126.

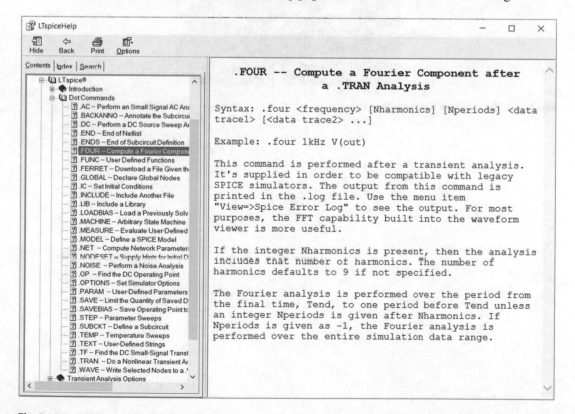

Fig. 2.126 .FOUR section of LTspice Help

Let's study a simple example. Draw the schematic shown in Fig. 2.127. Settings of the V1, V2, and V3 sources are shown in Figs. 2.128, 2.129, and 2.130, respectively. The voltage of node "a" can be written as $V_a = 10 \sin (2\pi \times 50 \times t + 30°) + 7 \sin (2\pi \times 150 \times t) + 5 \sin (2\pi \times 250 \times t + 45°)$. Settings of the used transient analysis is shown in Fig. 2.131. The .four 50 V(a) asks the LTspice to calculate the harmonics of node "a" voltage. It tells the LTspice that the fundamental frequency of V(a) is 50 Hz. So, amplitudes of components at 50 Hz, 100 Hz, 150 Hz, … are calculated. By default, .four command calculates the harmonics up to ninth harmonic. If you need more harmonics, you can write it after the fundamental frequency. For instance, .four 50 15 V(a) analyze the voltage of node "a" up to the 15th harmonic.

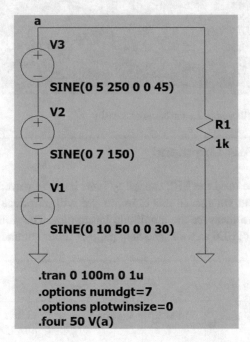

Fig. 2.127 Schematic of Example 10

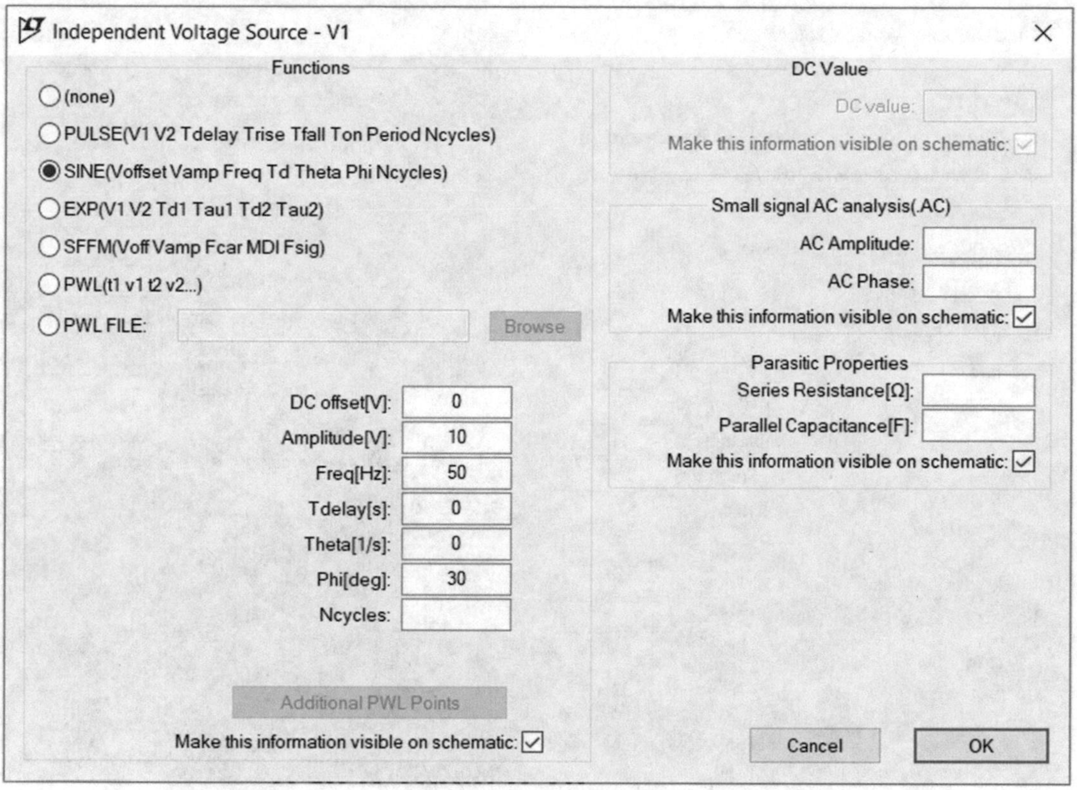

Fig. 2.128 Settings of voltage source V1

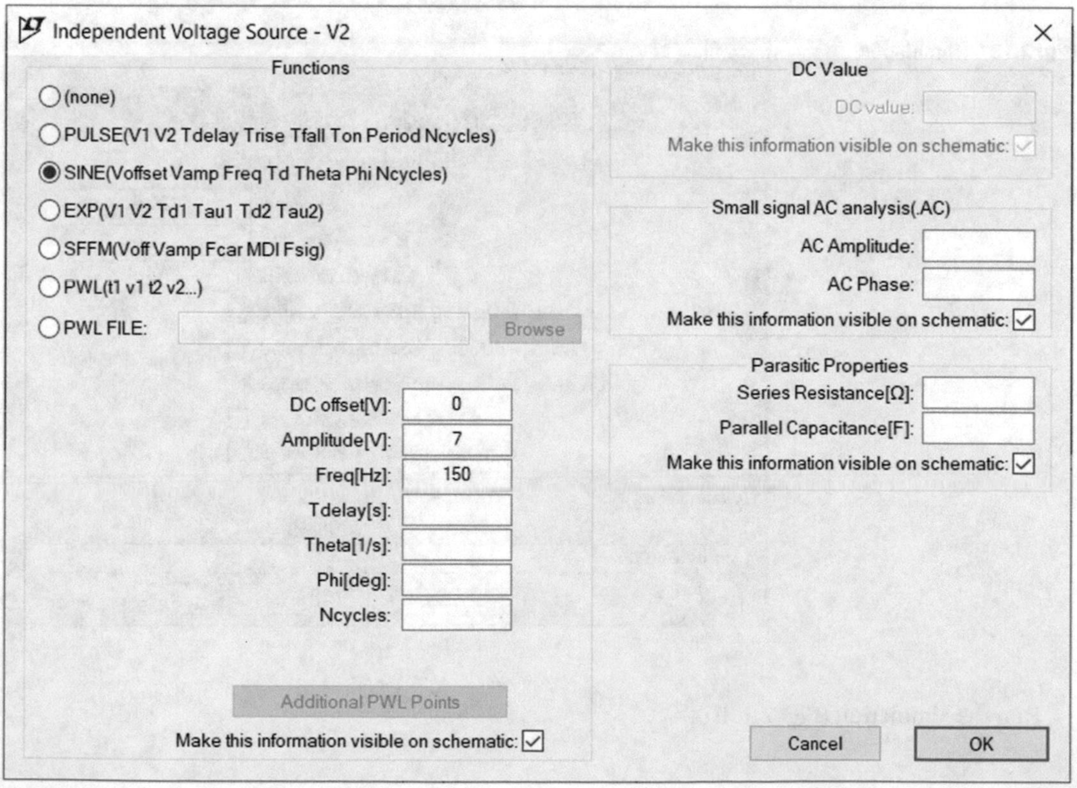

Fig. 2.129 Settings of voltage source V2

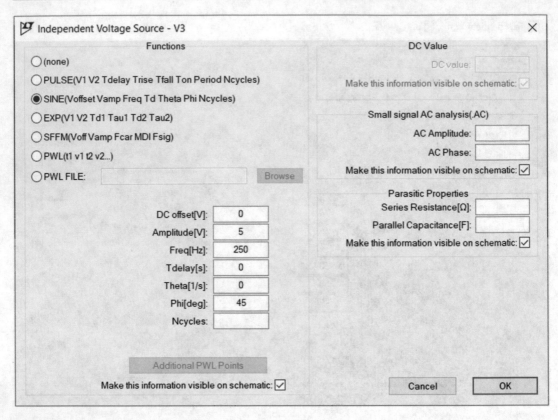

Fig. 2.130 Settings of voltage source V3

Fig. 2.131 Simulation settings

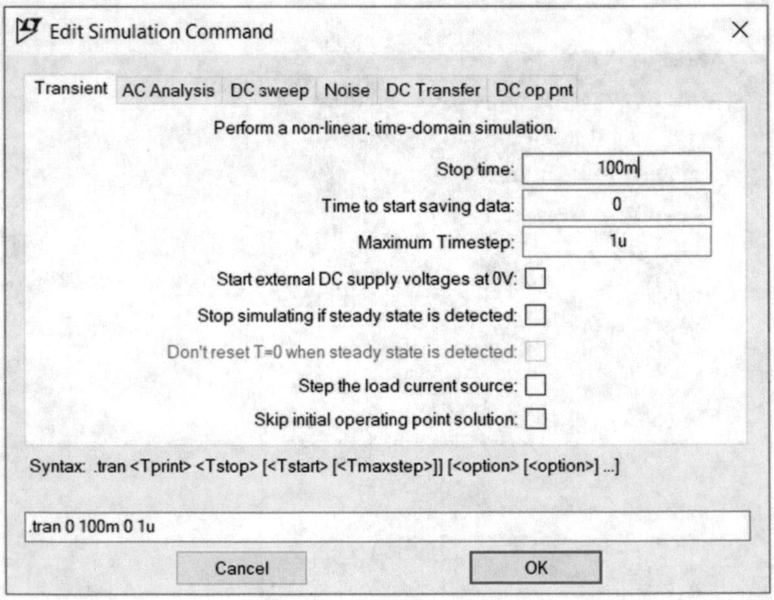

Run the simulation (Fig. 2.132).

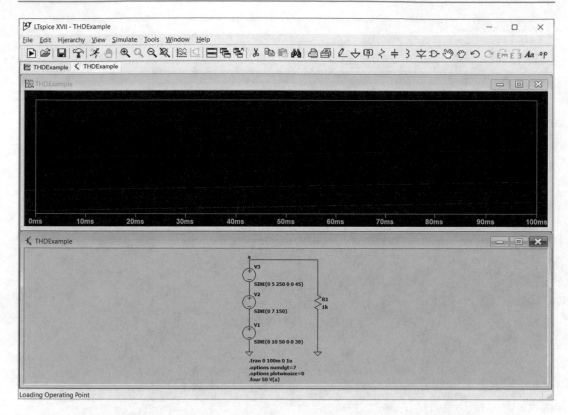

Fig. 2.132 Simulation is run

Press the Ctrl+L or click the View> SPICE Error Log (Fig. 2.133). This opens the SPICE Error Log window (Fig. 2.134).

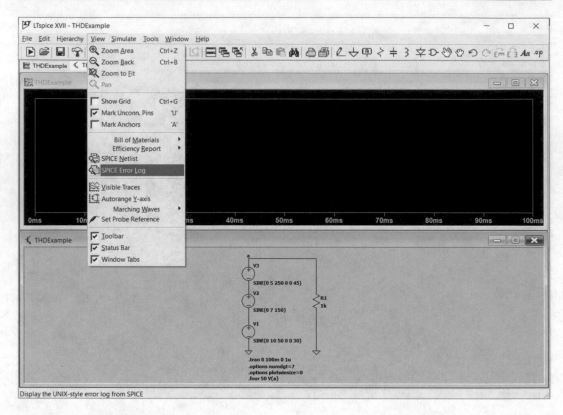

Fig. 2.133 View> SPICE Error Log

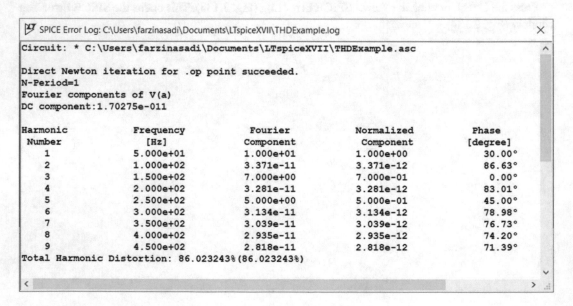

Fig. 2.134 Simulation result

Let's study the obtained results more carefully. According to the "Fourier Component" column of Fig. 2.134, the amplitude of components at 50 Hz, 150 Hz and 250 Hz are 10 V, 7 V, and 5 V, respectively. Amplitude of other harmonics are very small and can be ignored. According to "Phase [degree]" column of Fig. 2.134, the phase of components at 50 Hz, 150 Hz and 250 Hz are 30 ° , 0 ° , and 45°, respectively.

The DC component of the voltage of node "a" is 1.70275×10^{-11} V which can be considered as zero. We have no DC source in the circuit, so the DC component must be zero. If you put a DC source in series with the sources, then its value appears in the DC component line.

The Total Harmonic Distortion line displays the THD. Let's calculate the THD for voltage of node "a": $THD = \dfrac{\sqrt{\sum_{h=2}^{N} V_{h,RMS}^2}}{V_{1,RMS}} = \dfrac{\sqrt{\left(\dfrac{7}{\sqrt{2}}\right)^2 + \left(\dfrac{5}{\sqrt{2}}\right)^2}}{\dfrac{10}{\sqrt{2}}} = 0.8602$ or 86.02%. So, LTspice result is correct.

2.12   Example 11: THD of Common Emitter Amplifier

THD is a figure of merit specially in power amplifiers and inverters. Lower THD is preferred. In this example, we use the .four command to measure the THD of the common emitter amplifier of Example 8.

Draw the schematic shown in Fig. 2.135. The .four 1 kHz 10 25 V(out) command asks LTspice to do a Fourier analysis on the node "out" voltage and move ahead up to tenth harmonic. It tells the LTspice that fundamental frequency of node "out" voltage is 1 kHz and asks it to do the Fourier analysis on the last 25 cycles. Since, 1 cycle of a wave with frequency of 1 kHz takes 1 ms, 25 cycle takes 25 ms. The stop time of transient analysis is 50 ms. So, the Fourier analysis is applied to the [25 ms, 50 ms] of voltage of node "out." If you write a command like, .four 1 kHz 10 V(out), only the last cycle of node "out" voltage, i.e., portion of waveform between 49 ms and 50 ms, is used for calculation of harmonics. Increasing the number of cycles which are used in calculations increases the accuracy of result. However, the simulation takes more time to be done.

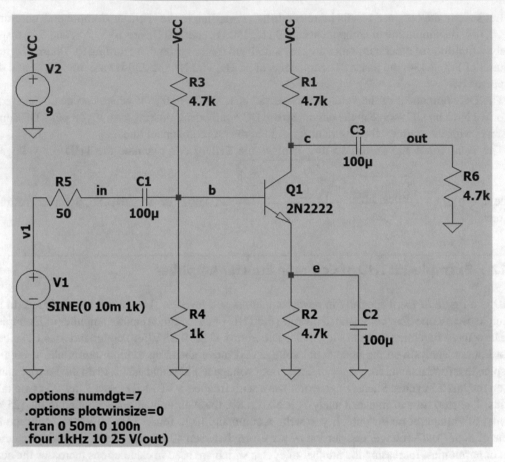

Fig. 2.135 Schematic of Example 11

Run the simulation. Click the View> SPICE Error Log (Fig. 2.136) or press the Ctrl+L to see the result.

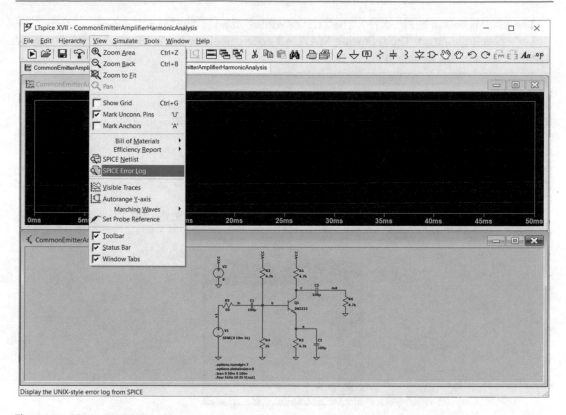

Fig. 2.136 View> SPICE Error Log

SPICE Error Log: C:\Users\farzinasadi\Documents\LTspiceXVII\CommonEmitterAmplifierHarmonicAnalysis.log

```
Direct Newton iteration for .op point succeeded.
N-Period=25
Fourier components of V(out)
DC component:-0.00094146

Harmonic    Frequency       Fourier       Normalized     Phase         Normalized
 Number       [Hz]         Component      Component     [degree]       Phase [deg]
   1        1.000e+03      1.712e-01      1.000e+00     -179.17°         0.00°
   2        2.000e+03      1.524e-02      8.903e-02      92.02°         271.19°
   3        3.000e+03      8.853e-04      5.170e-03       3.82°         183.00°
   4        4.000e+03      3.705e-05      2.164e-04     -90.19°          88.98°
   5        5.000e+03      4.750e-06      2.774e-05     179.87°         359.04°
   6        6.000e+03      3.020e-06      1.764e-05     176.24°         355.42°
   7        7.000e+03      2.580e-06      1.507e-05     177.15°         356.33°
   8        8.000e+03      2.258e-06      1.319e-05     177.53°         356.70°
   9        9.000e+03      2.007e-06      1.172e-05     177.81°         356.99°
  10        1.000e+04      1.807e-06      1.055e-05     178.03°         357.21°
Total Harmonic Distortion: 8.918202%(8.924962%)


Date: Sat Jun 06 11.09.00 2021
```

Fig. 2.137 Simulation result

The result is shown in Fig. 2.137. The THD of voltage of node "out" is 8.924962%.

Let's check the result. The following MATLAB code calculates the THD of node "out" voltage based on the amplitudes shown in Fig. 2.137.

```
f=1e3;w=2*pi*f;
A1=1.712e-1;A2=1.524e-2;A3=8.853e-4;A4=3.705e-5;A5=4.75e-6;
A6=3.020e-6;A7=2.58e-6;A8=2.258e-6;A9=2.007e-6;A10=1.807e-6;

Num=sqrt(.5*(A2^2+A3^2+A4^2+A5^2+A6^2+A7^2+A8^2+A9^2+A10^2));
Den=A1/sqrt(2);
THD=Num/Den*100
```

Output of the code is shown in Fig. 2.138. According to Fig. 2.138, the THD is 8.916% which is very close to the LTspice value.

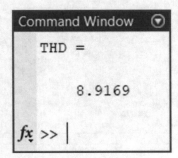

Fig. 2.138 Output of MATLAB code

Let's draw the graph of node "out" voltage. Following MATLAB code draws the graph of node "out" voltage based on the result shown in Fig. 2.137.

```
f=1e3;w=2*pi*f;
V0=-.00094146;
A1=1.712e-1;A2=1.524e-2;A3=8.853e-4;A4=3.705e-5;A5=4.75e-6;
A6=3.020e-6;A7=2.58e-6;A8=2.258e-6;A9=2.007e-6;A10=1.807e-6;
phi1=-179.17;phi2=92.02;phi3=3.82;phi4=-90.19;phi5=179.87;
phi6=176.24;phi7=177.15;phi8=177.53;phi9=177.81;phi10=178.03
;
DR=pi/180; %Degree to Radian Conversion Coefficient

syms t
y=V0+A1*sin(w*t+phi1*DR)+A2*sin(2*w*t+phi2*DR)+A3*sin(3*w*t+
phi3*DR)+...
A4*sin(4*w*t+phi4*DR)+A5*sin(5*w*t+phi5*DR)+A6*sin(6*w*t+phi
6*DR)+...
A7*sin(7*w*t+phi7*DR)+A8*sin(8*w*t+phi8*DR)+A9*sin(9*w*t+phi
9*DR)+...
A10*sin(10*w*t+phi10*DR);
ezplot(y,[0,1e-3])
grid on
```

After running the code, the result shown in Fig. 2.139 is obtained. You can compare it with the [0, 100 ms] portion of V(out) (Fig. 2.140). The two waveforms are not exactly the same; however, they are very close. Note that the MATLAB graph considers the first 10 harmonics and ignores the higher order harmonics. The small difference between the graphs come from this. If you increase the number harmonics in the MATLAB code, the difference between the two curve becomes smaller.

Fig. 2.139 Output of MATLAB code

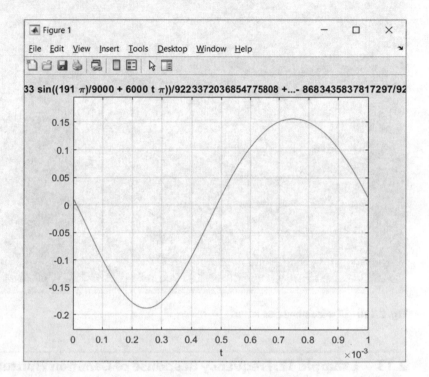

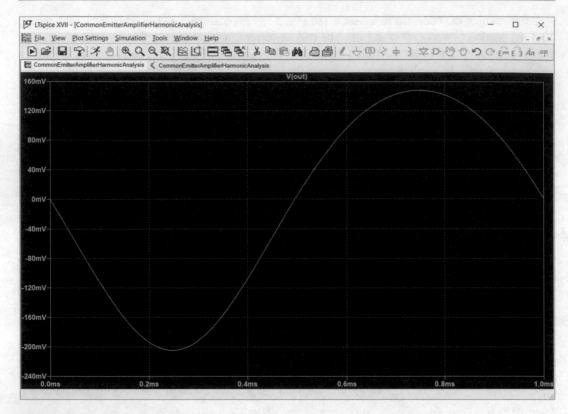

Fig. 2.140 LTspice simulation result

2.13   Example 12: Frequency Response of Common Emitter Amplifier

We want to obtain the frequency response of the common emitter amplifier of Example 8. Open the schematic of Example 8 and remove the .trans command from it (Fig. 2.141).

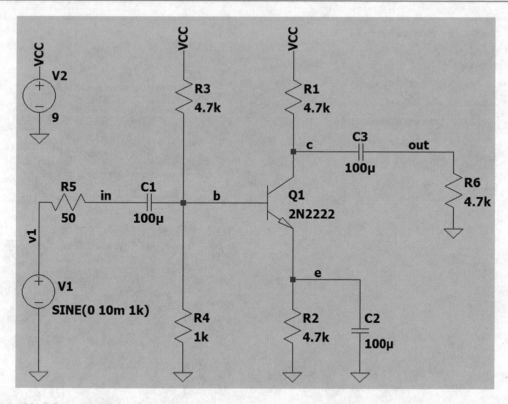

Fig. 2.141 Schematic of Example 12

Right click on the V1 and do the settings similar to Fig. 2.142. After applying the changes, click the OK button. After clicking the OK button, the schematic changes to what is shown in Fig. 2.143.

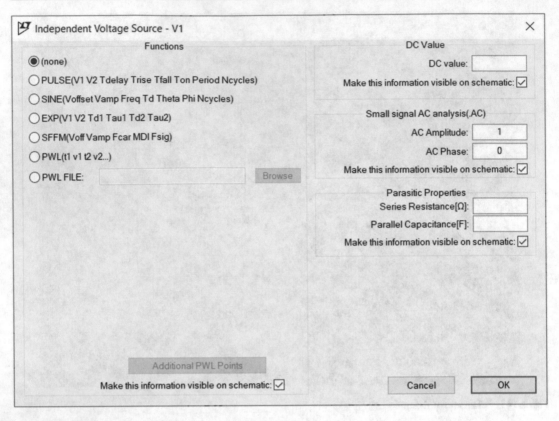

Fig. 2.142 Settings of voltage source V1

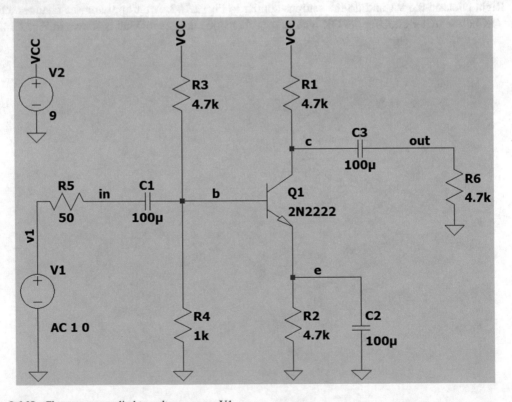

Fig. 2.143 Changes are applied to voltage source V1

Now, click the Run icon. Open the AC sweep tab and do the settings similar to Fig. 2.144 and click the OK button. After clicking the OK button, the simulation is done (Fig. 2.145).

Fig. 2.144 Simulation settings

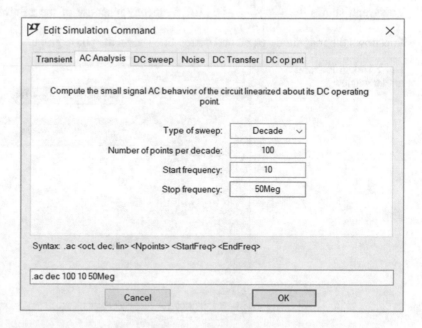

Fig. 2.145 Simulation is run

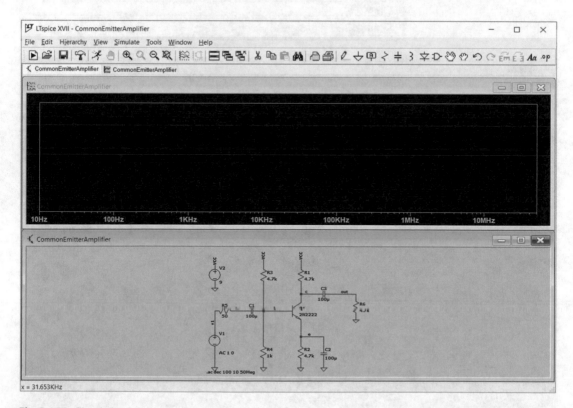

Right click on the black area and enter V(out)/V(in) to the Expression(s) to add box (Fig. 2.146) and click the OK button. After clicking the OK button, the result shown in Fig. 2.147 appears. Thisgraph shows the $\dfrac{V_{out}(j\omega)}{V_{in}(j\omega)}$, i.e., the frequency response of the gain of the amplifier. The solid lineshows the magnitude graph and dotted line shows the phase graph.

Fig. 2.146 Add Traces to Plot window

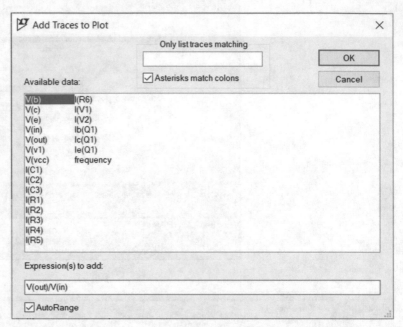

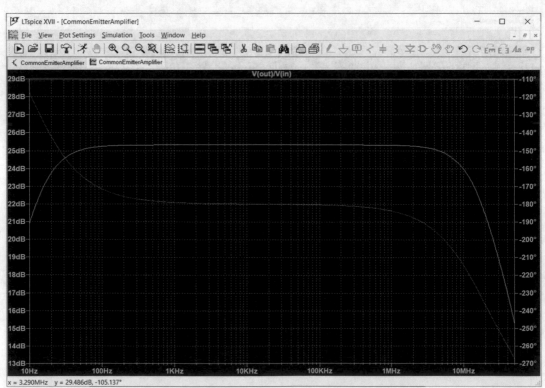

Fig. 2.147 Bode diagram of $\dfrac{V_{out}(s)}{V_{in}(s)}$

Let's measure the midband gain of the amplifier. According to Fig. 2.148, the midband gain is about 25.318 dB which shows the gain of $10^{\frac{25.318}{20}} = 18.446$.

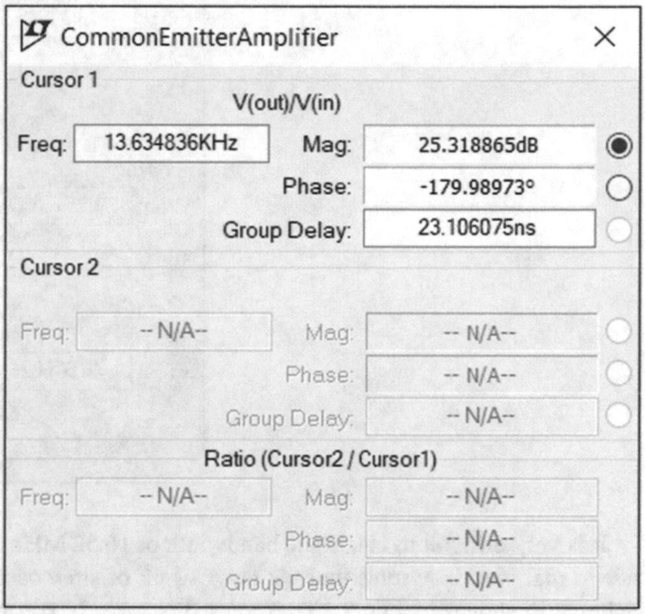

Fig. 2.148 Midband gain is around 25 dB

Let's measure the cut off frequency of the amplifier. We need to search for points which gain decreased to $25.318 - 3 = 22.318$ dB (Fig. 2.149). According to Fig. 2.150, the midband gain decreased by 3 dB at $f_L = 13.33$ Hz and $f_H = 16.52$ MHz. So, the bandwidth of the amplifier is $f_H - f_L \approx f_H \approx 16.52$ MHz.

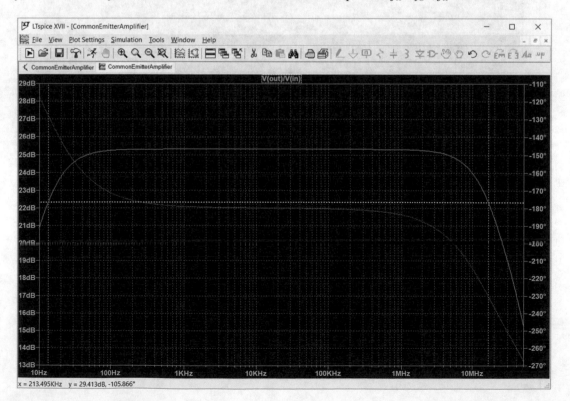

Fig. 2.149 Measurement of cutoff frequency

Fig. 2.150 Coordinates
of cursors in Fig. 2.149

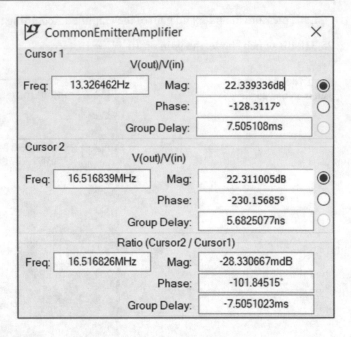

It is very difficult to obtain the bandwidth of 16.52 MHz for a single stage common emitter ampli-
fier in real world. Assume that we have 10 pF of stray capacitance between the base and collector.
Schematic shown in Fig. 2.151, shows this case. If you run this simulation, the result shown in
Fig. 2.152 appears.

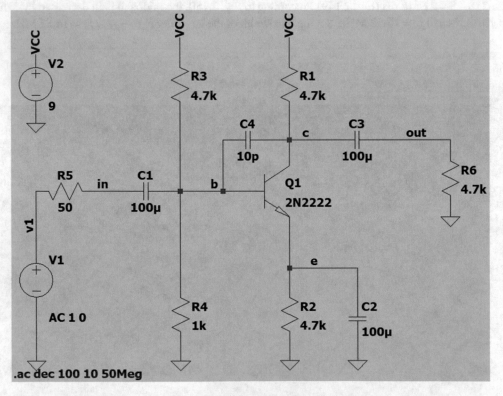

Fig. 2.151 10 pF stray capacitor is added between collector and emitter

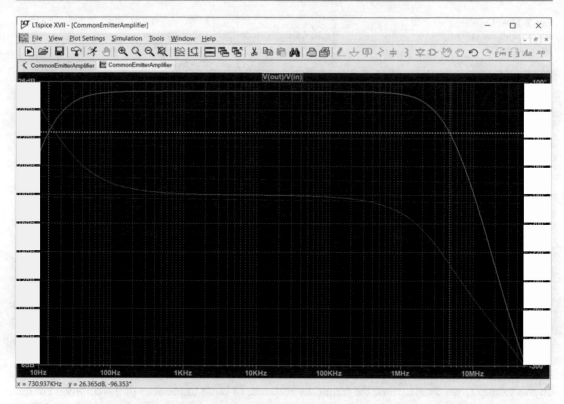

Fig. 2.152   Bode diagram in presence of stray capacitor

According to Fig. 2.153, the midband gain and low cut off frequency (f_L) does not change. However, the high cut off frequency decreased to $f_H = 4.76$ MHz. For 100 pf capacitor, the high cut off frequency decreases to 646 kHz (Fig. 2.154).

Fig. 2.153   Cutoff frequencies for 10 pF stray capacitor

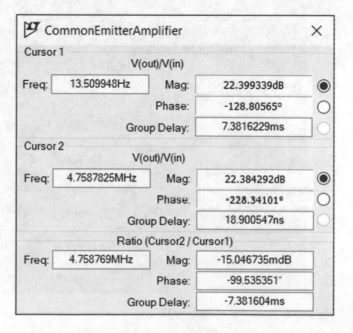

Fig. 2.154 Cutoff
frequencies for 100 pF
stray capacitor

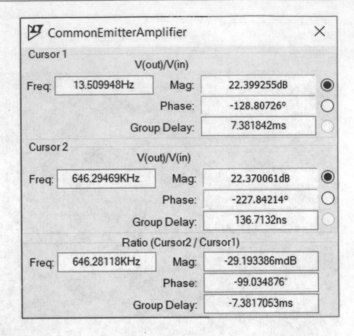

2.14   Example 13: Input Impedance of Common Emitter Amplifier

In this example, we want to draw the frequency response of input impedance of the common emitter of Example 8. We use the AC sweep for this purpose. Consider the schematic shown in Fig. 2.155.

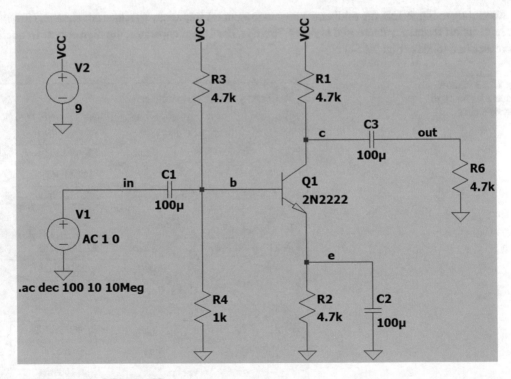

Fig. 2.155 Schematic of Example 13

Run the simulation (Fig. 2.156).

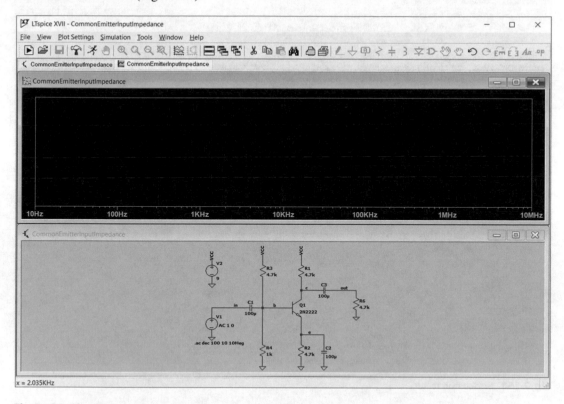

Fig. 2.156   Simulation is run

Right click on the black area and enter –V(in)/I(in) to the Expression(s) to add box (Fig. 2.157) and click the OK button. After clicking the OK button, the result shown in Fig. 2.158 appears. The solid line shows the magnitude graph and dotted line shows the phase graph.

Fig. 2.157   Add Traces to Plot window

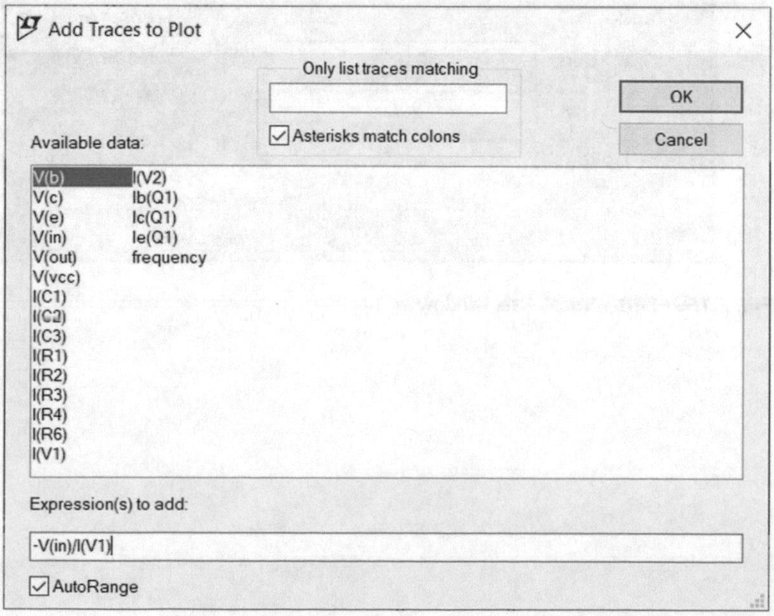

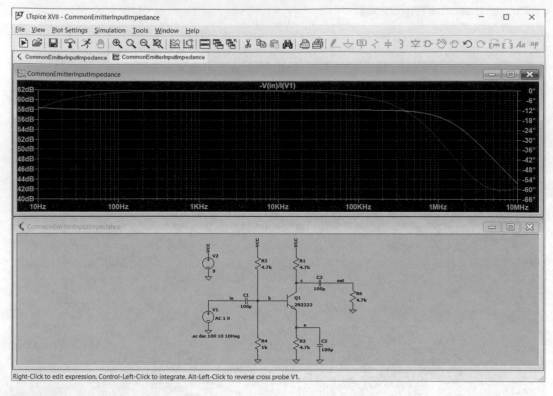

Fig. 2.158 Simulation result

Right click on the vertical axis and select the Linear (Fig. 2.159). Now the vertical axis has the unit of Ohms and you can read it easily (Fig. 2.160).

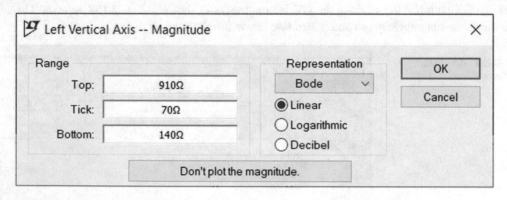

Fig. 2.159 Left Vertical Axis window

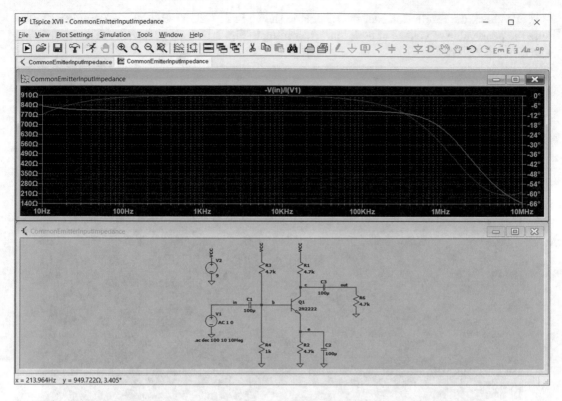

Fig. 2.160 Simulation result

Let's check the obtained result. According to Fig. 2.161, the input impedance at 1 kHz is $800.10e^{-j0.167534°} \ \Omega \approx 800.10 \ \Omega$.

Fig. 2.161 Input
impedance at 1 kHz

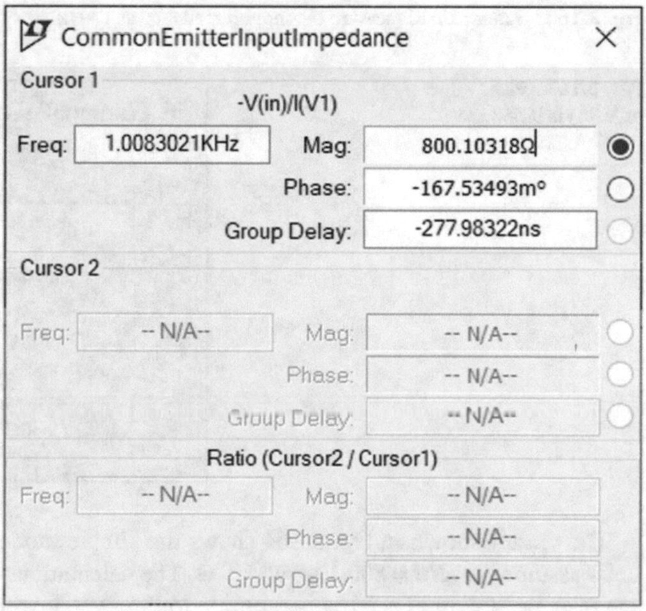

Let's check the result. Consider the schematic shown in Fig. 2.162. Run this schematic and measure the peak value of voltage of node "in." According to Fig. 2.163, the peak value of node "in" voltage is about 9.39 mV.

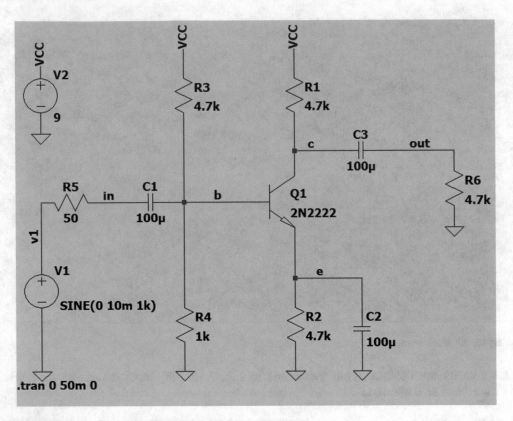

Fig. 2.162 Schematic to measure the input impedance at 1 kHz

Fig. 2.163 Peak value of V(in) is 9.388 mV

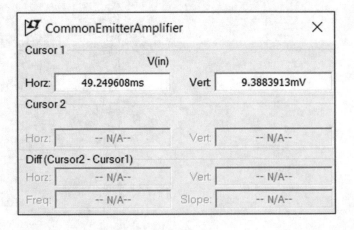

The circuit shown in Fig. 2.164 shows the simple model of the schematic shown in Fig. 2.162. Let's assume that $RB = 800.10e^{-j0.167534°}$ Ω. The calculations shown in Fig. 2.165, calculated the voltage of node "in" in Fig. 2.164. According to Fig. 2.165, with $RB = 800.10e^{-j0.167534°}$ Ω, the voltage of node "in" is 9.4118 mV which is quite close to the value we measured in Fig. 2.163. So, the assumption that $RB = 800.10e^{-j0.167534°}$ Ω is correct. You can use the same technique for other frequencies to ensure that input impedance graph given by LTspice is correct.

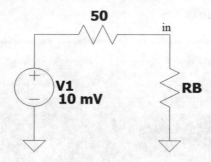

Fig. 2.164 Simple model for Fig. 2.162

Fig. 2.165 MATLAB calculations

```
Command Window
>> Rin=800.10318*exp(j*-0.16753493/180*pi);Rsource=50;
>> abs(Rin/(Rin+Rsource)*10)

ans =

    9.4118

fx >>
```

2.15   Example 14: Output Impedance of Common Emitter Amplifier

In this example, we want to draw the output impedance of the common emitter amplifier of Example 8.

Draw the schematic shown in Fig. 2.166. The output impedance of the amplifier is the impedance seen from the output node of the amplifier (in this example, the impedance seen from the node "out"). Input of the amplifier must be zero. That is why the node "in" is connected to ground.

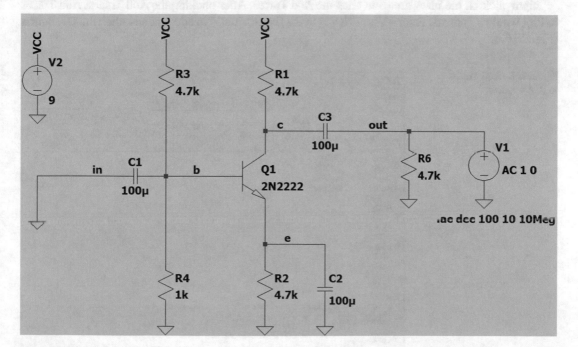

Fig. 2.166 Schematic of Example 14

Run the simulation (Fig. 2.167).

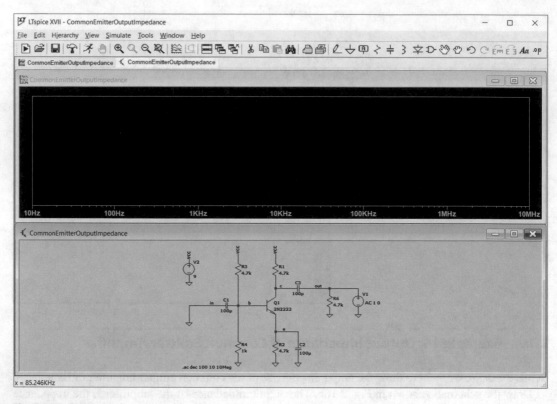

Fig. 2.167 Simulation is run

Right click on the black area and click the Add Traces. After clicking the Add Traces, Add Traces to Plot window appears. Enter –V(out)/I(V1) to the Expression(s) to add box and click the OK button (Fig. 2.168).

Fig. 2.168 Add Traces
to Plot window

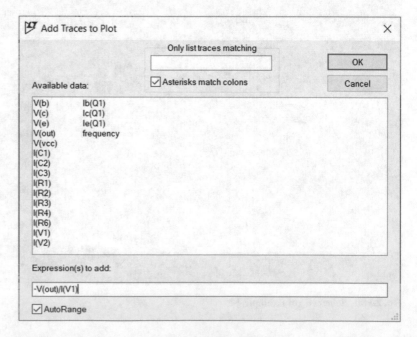

After clicking the OK button, the graph of output impedance of the amplifier is shown on the screen (Fig. 2.169). The solid line shows the magnitude graph and dotted line shows the phase graph.

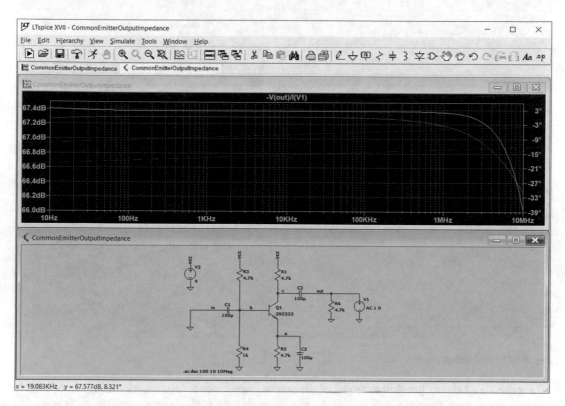

Fig. 2.169 Simulation result

Right click on the vertical axis and select the Linear (Fig. 2.170). Now, the vertical axis has the unit of Ohms (Fig. 2.171).

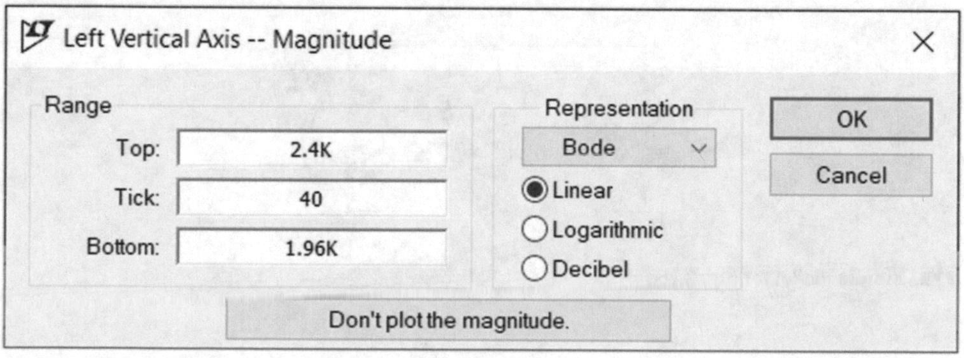

Fig. 2.170 Left Vertical Axis window

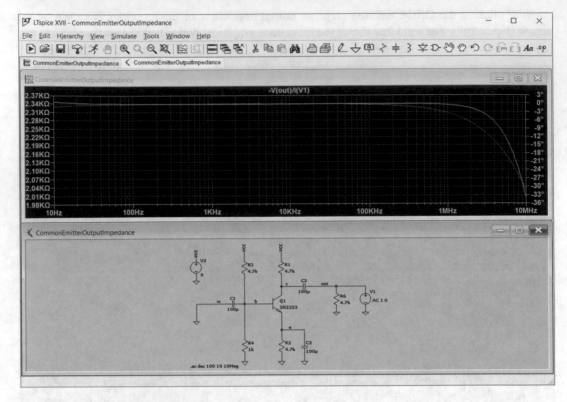

Fig. 2.171 Simulation result

Figure 2.172 shows the simple model of the output impedance measurement for Fig. 2.166. RC is the resistance seen from the collector of the transistor and R6 is the output load. So, what we measured in Fig. 2.166 is the parallel combination of the RC and R6. In other words, the output load resistor is considered in the calculation of output impedance. You can draw the graph of RC only by entering a big value for resistor R6 (Fig. 2.173). After running the schematic shown in Fig. 2.174, the graph shown in Fig. 2.175 is obtained. According to Fig. 2.175, the resistance seen from the collector node of transistor Q1 is about 4.7 kΩ.

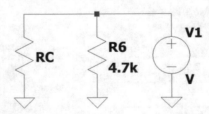

Fig. 2.172 Simple model for Fig. 2.166

Fig. 2.173 R6 value is
changed to 4.7 MΩ

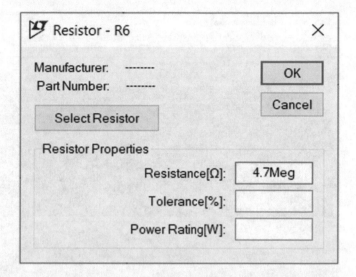

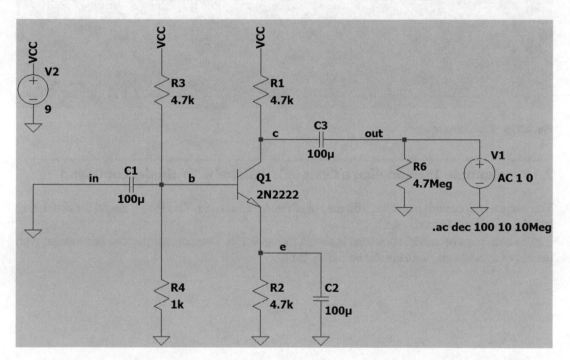

Fig. 2.174   Schematic with R6 = 4.7 MΩ

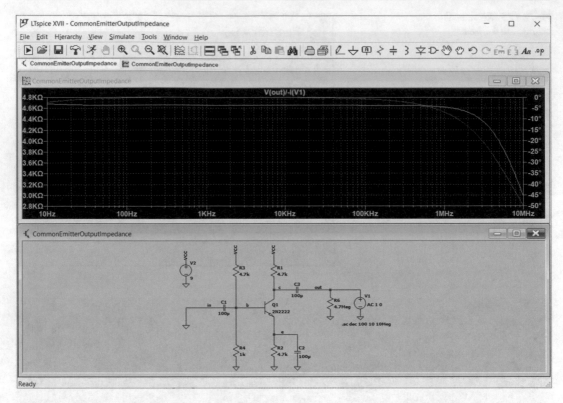

Fig. 2.175 Simulation result

2.16   Example 15: Modeling a Custom Transistor with .model Command

You can make a custom transistor with the aid of .model command. This is very useful for simulating the circuits shown in textbooks.

The help page of .model command is shown in Fig. 2.176. Components that can be modeled with the aid of .model command are shown in Fig. 2.177.

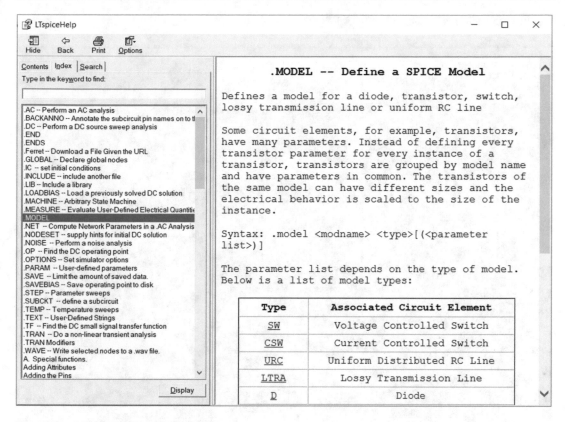

Fig. 2.176 .MODEL section of LTspice Help

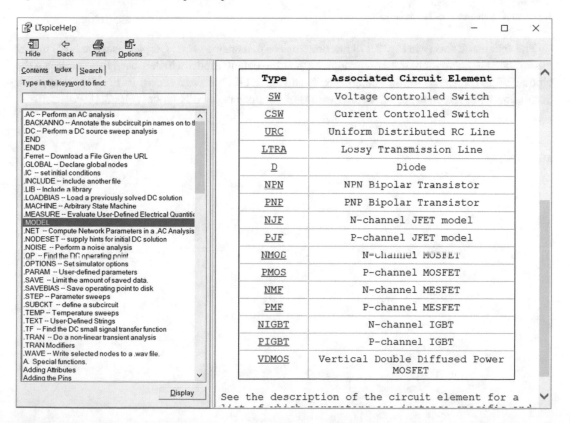

Fig. 2.177 .MODEL section of LTspice Help

Draw the schematic shown in Fig. 2.178.

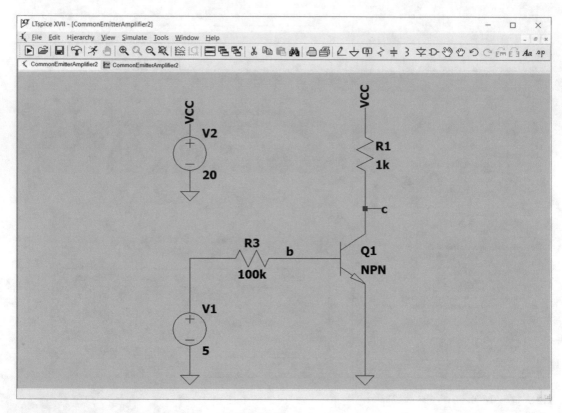

Fig. 2.178 Schematic of Example 15

The schematic shown in Fig. 2.178 has one transistor. Its name is Q1. LTspice uses the NPN model to simulate the behavior of this component. NPN is the default model which is used for simulating the bipolar NPN transistor. If you click the NPN in Fig. 2.177, the page shown in Fig. 2.179 appears. Scroll down the screen to see the default values of parameters of NPN model (Fig. 2.180).

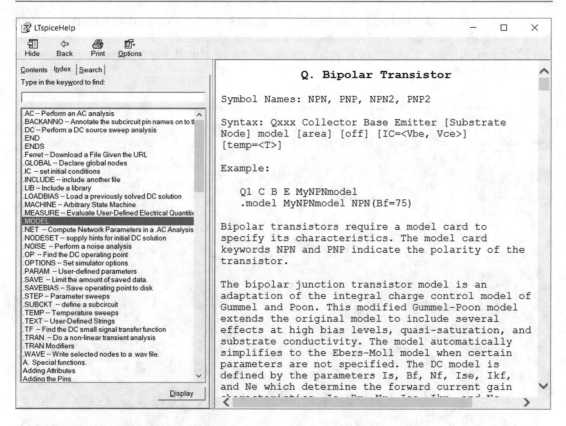

Fig. 2.179 Bipolar section of LTspice Help

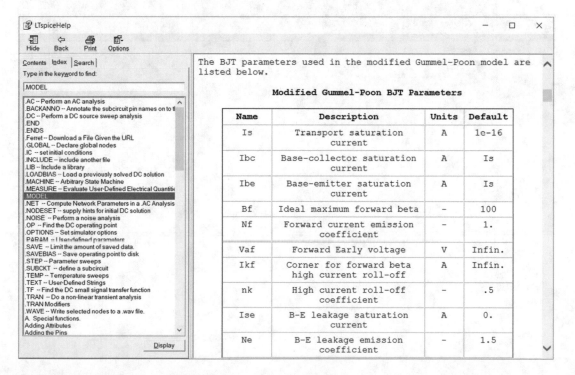

Fig. 2.180 Parameters of BJT transistor model

Right click on the NPN text behind the transistor (Fig. 2.181) and change it to MYNPN (Fig. 2.182). Now, the LTspice uses the MYNPN model to simulate the behavior of this transistor. However, the MYNPN model is not defined yet. If you try to simulate the circuit, the error message shown in Fig. 2.183 appears.

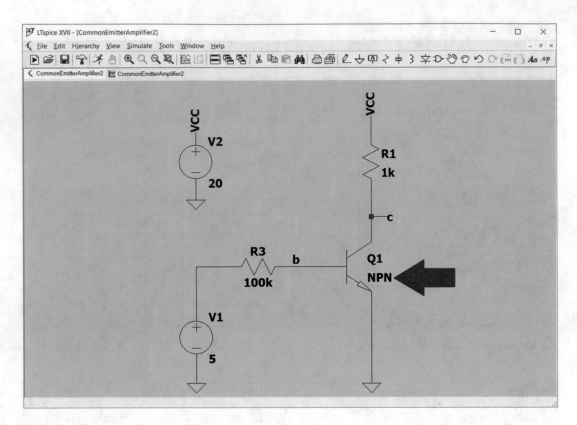

Fig. 2.181 Schematic of Example 15

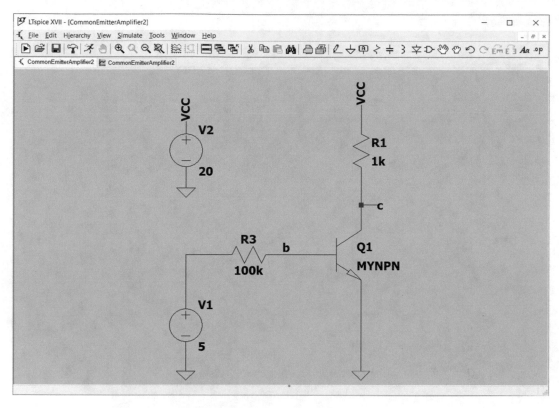

Fig. 2.182   Transistor Q1 will use the MYNPN model

Fig. 2.183   Error message

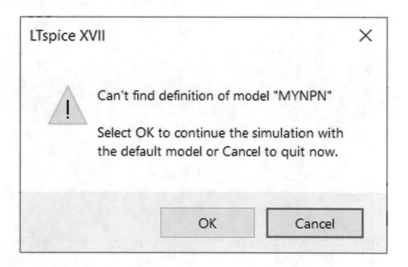

Add the .model command shown in Fig. 2.184. This command makes the current gain of transistor Q1 to be 175. Default values are used for other parameters of the transistor Q1 (Default values are shown in Fig. 2.180).

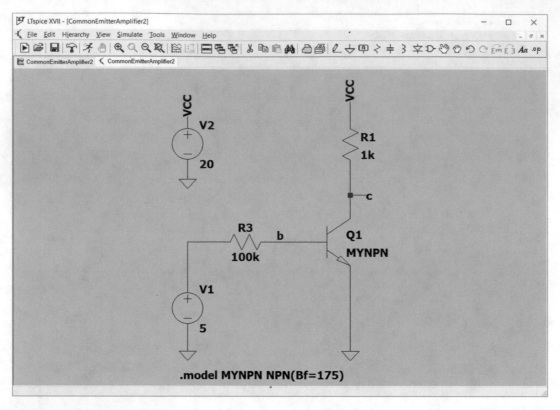

Fig. 2.184 Current gain of Q1 is set to 175

You can use the schematic shown in Fig. 2.185 as well. This schematic is equivalent to Fig. 2.184.

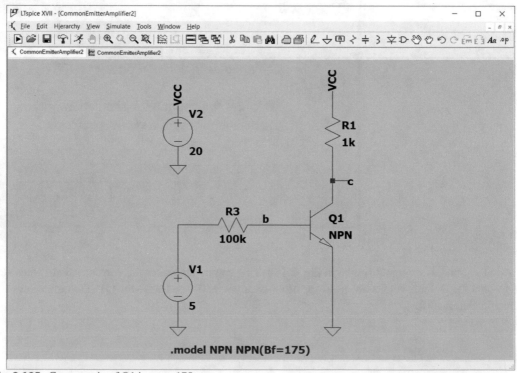

Fig. 2.185 Current gain of Q1 is set to 175

Let's test the circuit and ensure that the current gain of transistor is 175. Add the .dc V1 0 20 0.2 command to the schematic (Fig. 2.186). This command changes the value of V1 from 0 V to 20 V with 0.2 V steps.

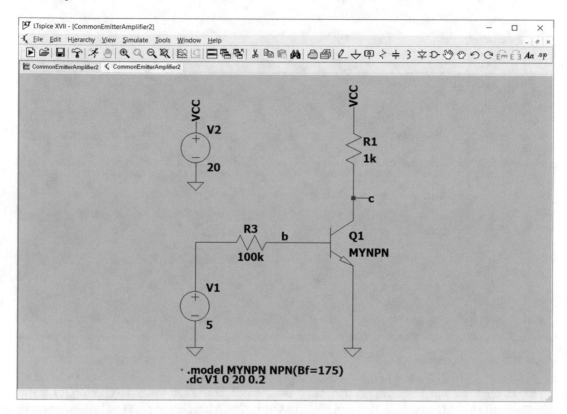

Fig. 2.186 Addition of .dc V1 0 20 0.2 command to the schematic

Run the simulation (Fig. 2.187).

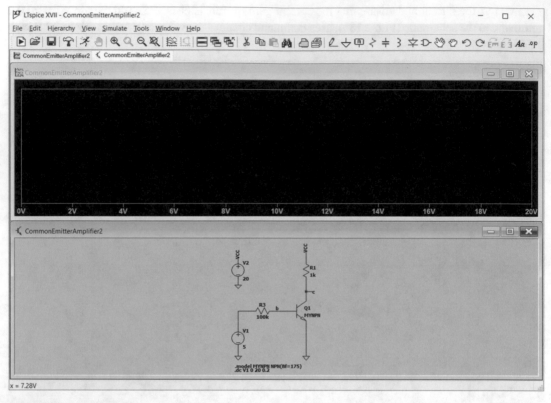

Fig. 2.187 Simulation is run

Right click on the black region and click the Add Traces. Then enter Ic(Q1)/Ib(Q1) to the Expression(s) to add box (Fig. 2.188) and click the OK button. After clicking the OK button, the graph shown in Fig. 2.189 appears. According to Fig. 2.189, the current gain is 175 for the active forward region. When V1 > 12.15 V, saturation occurs and the current gain decreases.

Fig. 2.188 Add Traces to Plot window

Add Traces to Plot

Only list traces matching

☑ Asterisks match colons

OK

Cancel

Available data:

V(b)
V(c)
V(vcc)
V(n001)
I(R1)
I(R3)
I(V1)
I(V2)
Ib(Q1)
Ic(Q1)
Ie(Q1)
v1

Expression(s) to add:

Ic(Q1)/Ib(Q1)

☑ AutoRange

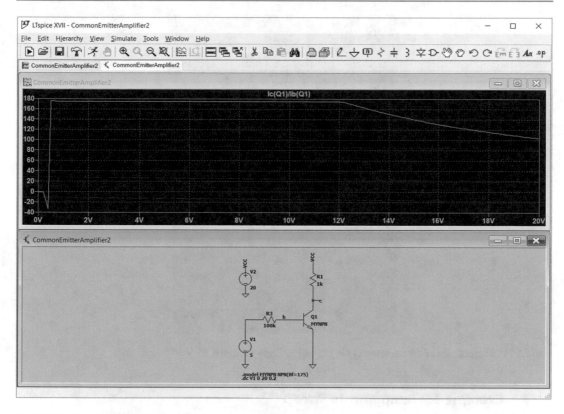

Fig. 2.189 Simulation result

Let's assume that we want to set the Early voltage of transistor Q1 to be 75 V. In order to do this, right click on the .model command and change it to what is shown in Fig. 2.190 and click the OK button. Now you have a transistor with current gain 175 and Early voltage of 75 V (Fig. 2.191).

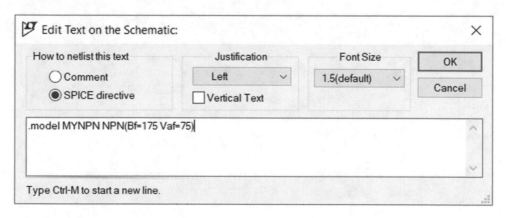

Fig. 2.190 Defining the transistor parameters

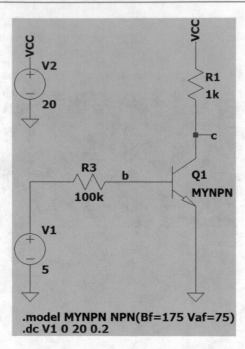

Fig. 2.191 Current gain of the transistor Q1 is 175 and its Early voltage is 75 V

2.17   Example 16: Temperature Sweep

You can use the .temp command to simulate the behavior of a circuit under different temperatures. In this example, we draw the gain of common emitter amplifier circuit for different temperatures.

The schematic shown in Figs. 2.192 and 2.193 draw the frequency response of the amplifier in the [100 Hz, 1 MHz] frequency range for three different temperatures: $-40\,°\,C,\ 25\,°\,C$ and $100\,°\,C$. Both of the schematics are equivalent.

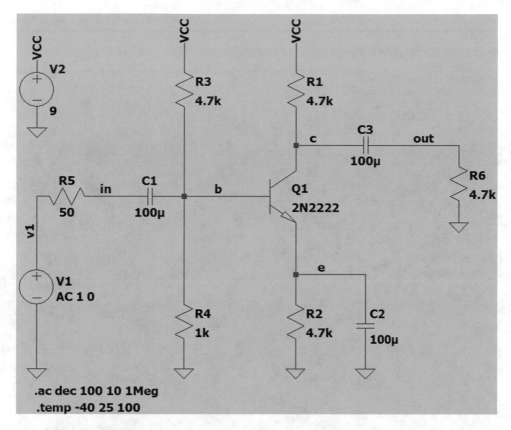

Fig. 2.192 Schematic of Example 16

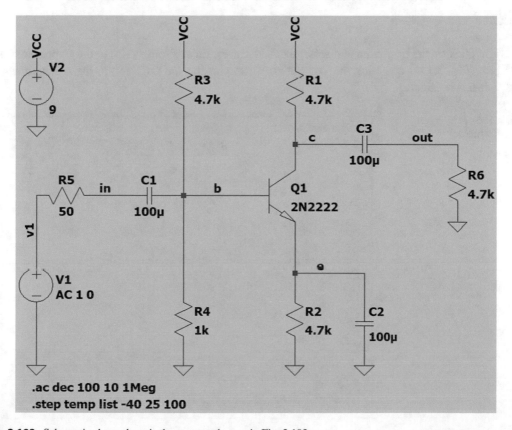

Fig. 2.193 Schematic shown here is the same as the one in Fig. 2.192

Run the simulation (Fig. 2.194).

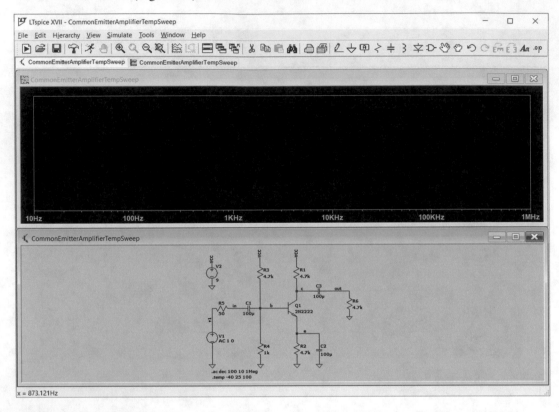

Fig. 2.194 Simulation is run

Right click on the black area and select the Add Traces. Enter V(out)/V(in) to the Expression(s) to add box (Fig. 2.195) and click the OK button. After clicking the OK button, the result shown in Fig. 2.196 is obtained.

Fig. 2.195 Add Traces to Plot window

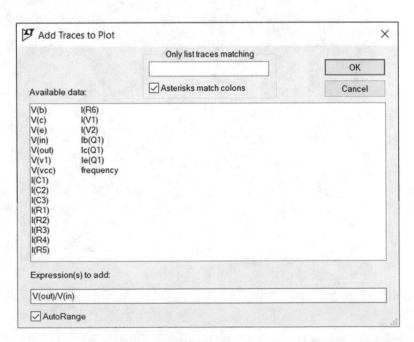

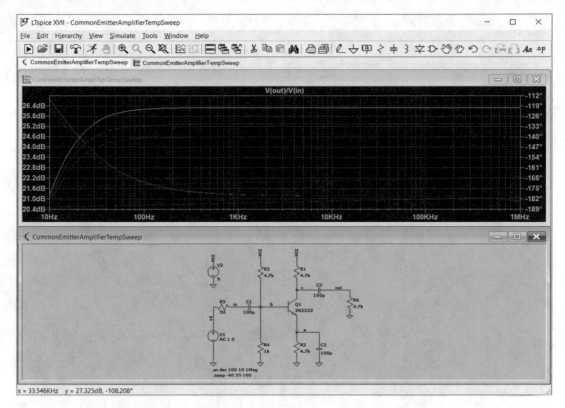

Fig. 2.196 Simulation result

We are not interested in the phase graph. We want to study the effect of temperature on the magnitude graph. You can remove the phase graph by right clicking on the right axis and clicking the Don't plot phase button (Fig. 2.197). After clicking the Don't plot phase button, the phase graph is removed from the screen (Fig. 2.198).

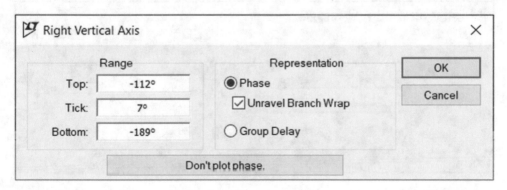

Fig. 2.197 Right Vertical Axis window

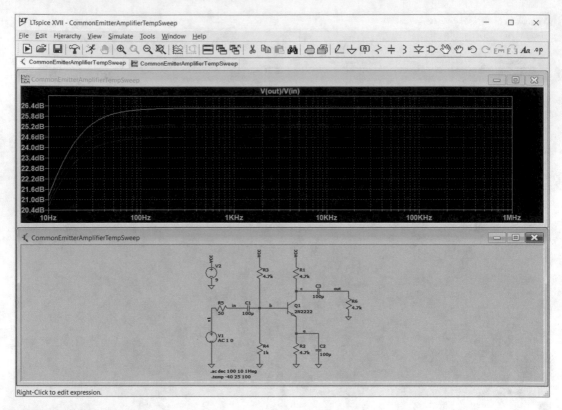

Fig. 2.198 Simulation result

Right click on the vertical axis and select the Linear (Fig. 2.199). Now, the gain of the amplifier is shown on the vertical axis (Fig. 2.200). The gain changes from about 17 to about 21 as temperature changes.

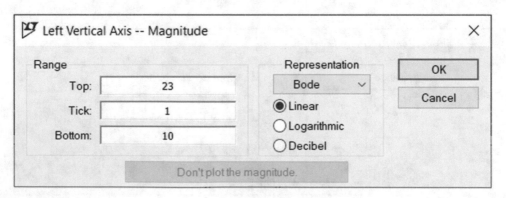

Fig. 2.199 Left Vertical Axis window

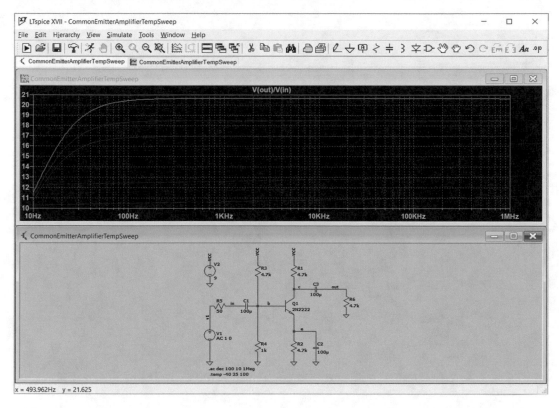

Fig. 2.200 Simulation result

Let's see which graph belongs to which temperature. Click on the V(out)/V(in) to add a cursor to the graph (Fig. 2.201). Use the keyboard's up and down arrow keys (Fig. 2.202) to select the desired graph. If you right click on the cursor, the temperature associated with the selected graph is shown. In this example, increase in temperature leads to decrease in gain.

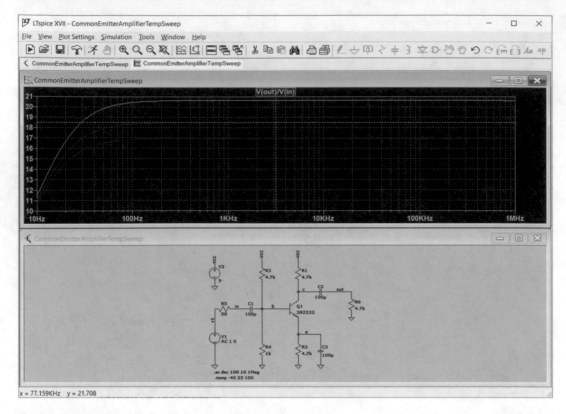

Fig. 2.201 A cursor is added to the output window

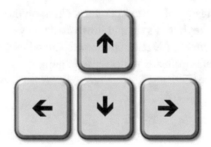

Fig. 2.202 Selection of desired graph with the aid of up and down arrow keys

2.18   Example 17: Effect of Temperature on the Forward Voltage Drop of Diode

In this example, we want to study the effect of temperature on the forward voltage drop of a diode. Consider the schematic shown in Fig. 2.203. The .step temp -20 100 5 line changes the temperature from −20 ° C to 100 ° C with 5 ° C steps. After simulating the circuit (Fig. 2.204), you can see the voltage of node "out" decreases as the temperature increases.

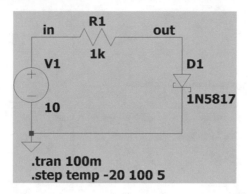

Fig. 2.203   Schematic of Example 17

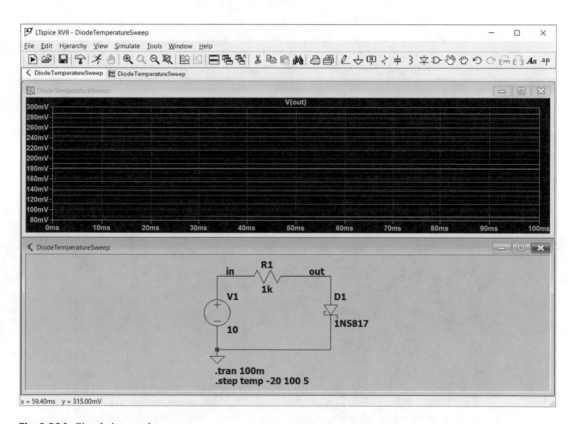

Fig. 2.204   Simulation result

2.19   Example 18: Noninverting op amp Amplifier

In this example, we want to simulate the circuit shown in Fig. 2.205. The gain of this amplifier is $1 + \dfrac{R_1}{R_2} = 1 + 9 = 10$. So, we expect to observe 100 mV at output of the amplifier.

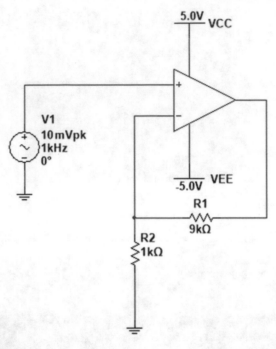

Fig. 2.205 Circuit for Example 18

The op amp in Fig. 2.205 is LT 6016 (Fig. 2.206). The op amps available in LTspice can be found in [OpAmps] section (Fig. 2.207).

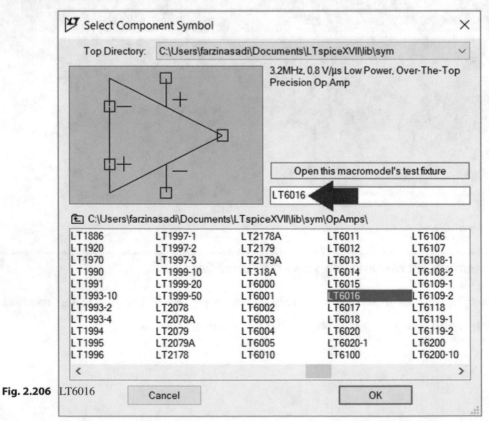

Fig. 2.206 LT6016

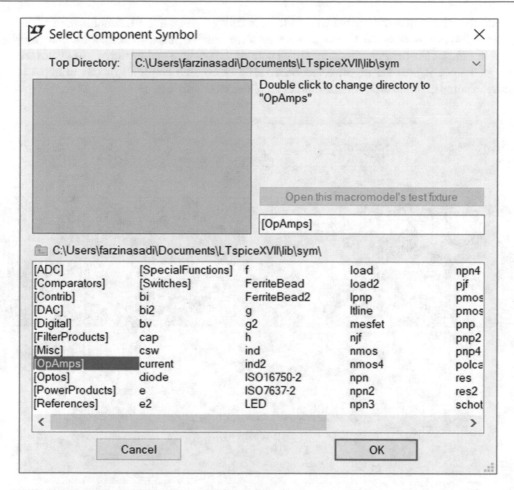

Fig. 2.207 OpAmps section of Select Component Symbol window

Draw the schematic shown in Fig. 2.208.

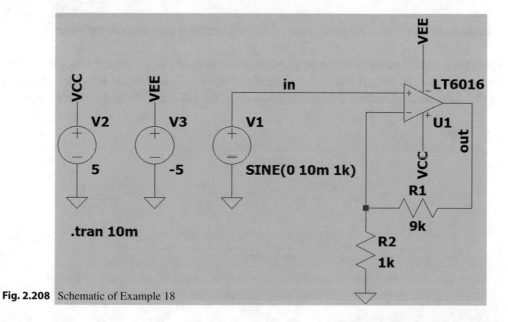

Fig. 2.208 Schematic of Example 18

Run the simulation and draw the waveform of node "out" (Fig. 2.209). Amplitude of node "out" voltage is 100 mV as expected. Compare this waveform with the output of common emitter amplifier shown in Fig. 2.111. Op amp circuit amplifies both of the positive and negative half cycles with the same gain since it uses the negative feedback. THD of op amp amplifier of this example is very smaller in comparison to the common emitter amplifier of Example 8.

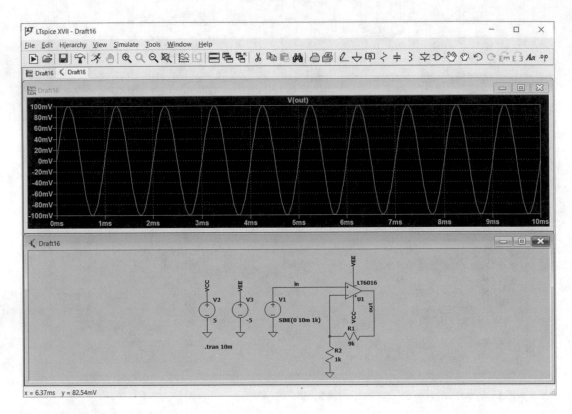

Fig. 2.209 Simulation result

2.20   Example 19: Input Impedance of Noninverting op amp Amplifier

Let's measure the input impedance of the amplifier of previous example. We expect a very high input impedance since the input voltage source is connected to the + terminal of the op amp. The schematic shown in Fig. 2.210 is used to draw the input impedance on the [10 Hz, 1 MHz] frequency range.

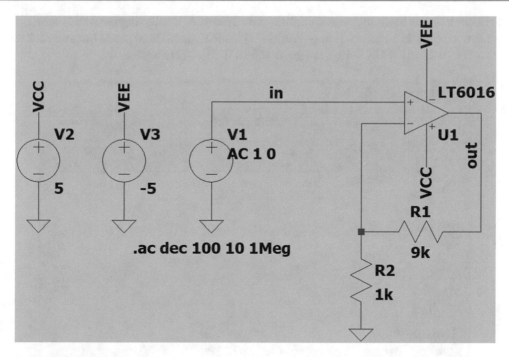

Fig. 2.210 Schematic of Example 19

Run the simulation (Fig. 2.211).

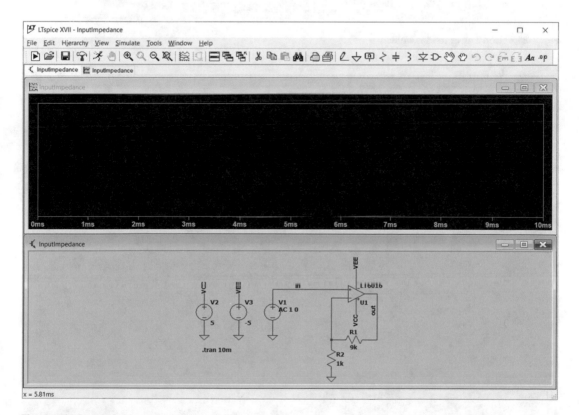

Fig. 2.211 Simulation is run

Right click on the black region and click the Add Traces. After clicking the Add Traces, the window shown in Fig. 2.212 appears. Enter –V(in)/I(V1) to the Expression(s) to add box and click the OK button. After clicking the OK button, the graph shown in Fig. 2.213 appears.

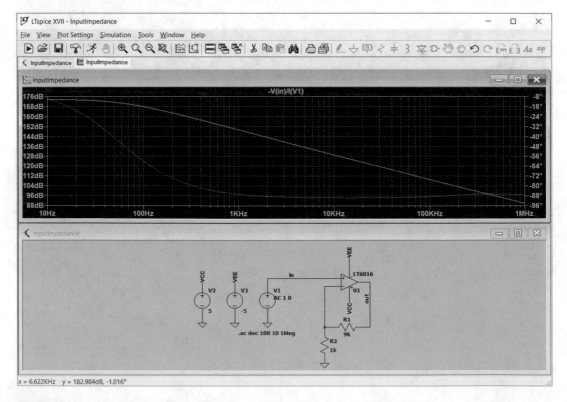

Fig. 2.212 Add Traces to Plot window

Fig. 2.213 Simulation result

Right click on the vertical axis and select the Linear (Fig. 2.214). Now the vertical axis has the unit of Ohms (Fig. 2.215).

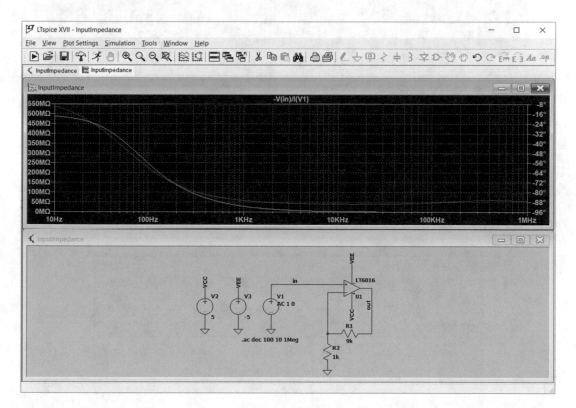

Fig. 2.214 Left Vertical Axis window

Fig. 2.215 Simulation result

Note that the input impedance is very large for small frequencies. As the frequency increases the input impedance decreases and reaches to about 31.55 kΩ at 1 MHz. The value of phase graph at 1 MHz is $-87°$. So, the input impedance is capacitive at high frequencies.

2.21   Example 20: Output Impedance of Noninverting op amp Amplifier

In this example, we want to measure the output impedance of the amplifier of previous example. The required schematic to measure the output impedance is shown in Fig. 2.216. In calculation of output impedance, the input of amplifier must be zero. That is why + terminal of the op amp is connected to the ground.

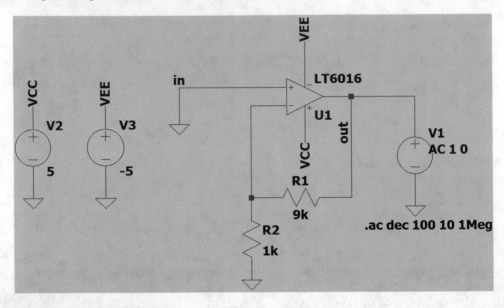

Fig. 2.216 Schematic of Example 20

Run the simulation. Right click on the black region and select the Add Traces. After clicking the Add Traces, the window shown in Fig. 2.217 appears. Enter –V(out)/I(V1) to the Expression(s) to add box (Fig. 2.217) and click the OK button. After clicking the OK button, the graph shown in Fig. 2.218 appears on the screen.

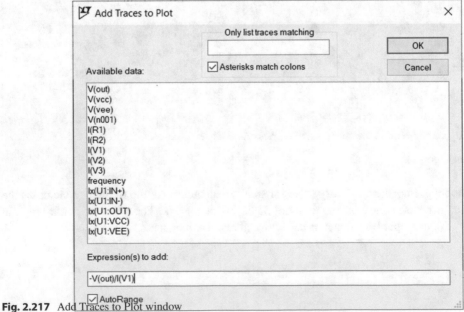

Fig. 2.217 Add Traces to Plot window

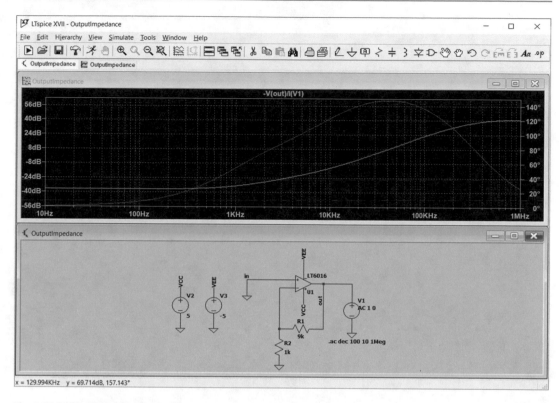

Fig. 2.218 Simulation result

Right click on the vertical axis and select the Linear (Fig. 2.219). After clicking the Linear, the vertical axis changes to Ohms (Fig. 2.220).

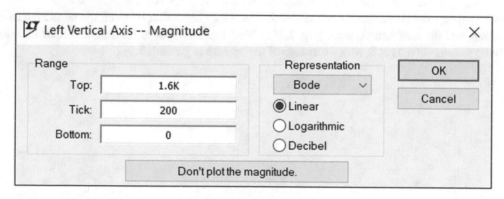

Fig. 2.219 Left Vertical Axis window

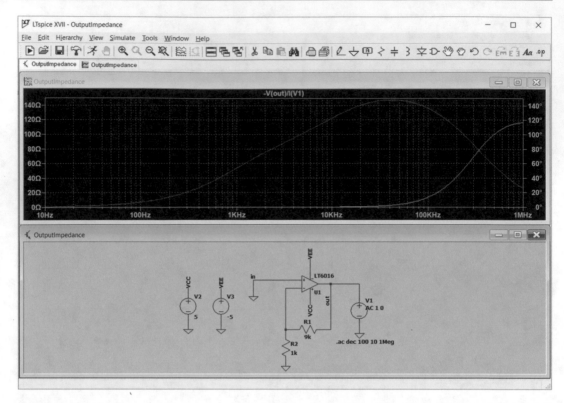

Fig. 2.220 Simulation result

2.22 Example 21: Stability of op amp Amplifiers

In this example, we want to study the stability of noninverting op amp amplifier of Example 18. We want to measure the phase margin of the circuit. Open the connection between the negative terminal of op amp and the feedback network (Fig. 2.221). Note that the input of the amplifier must be zero in studying the stability. That is why + terminal of the op amp is grounded.

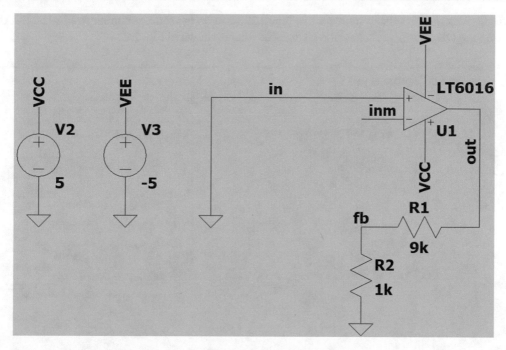

Fig. 2.221 Schematic of Example 21

Add a voltage source between the "fb" and "inm" nodes (Fig. 2.222).

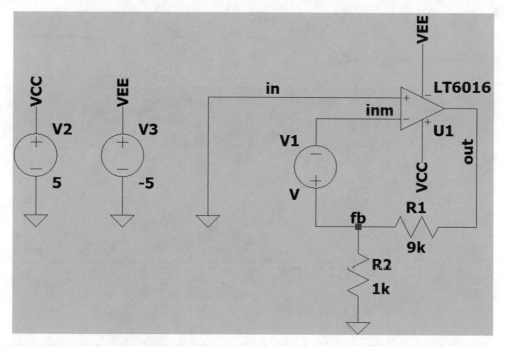

Fig. 2.222 Voltage source V1 is added to the circuit

Right click on the voltage source V1 in Fig. 2.222 and do the settings similar to Fig. 2.223. After clicking the OK button, the schematic changes to what is shown in Fig. 2.224.

Independent Voltage Source - V1   ×

Functions   DC Value

◉ (none)   DC value: 0

○ PULSE(V1 V2 Tdelay Trise Tfall Ton Period Ncycles)   Make this information visible on schematic: ☑

○ SINE(Voffset Vamp Freq Td Theta Phi Ncycles)

○ EXP(V1 V2 Td1 Tau1 Td2 Tau2)   Small signal AC analysis(.AC)

○ SFFM(Voff Vamp Fcar MDI Fsig)   AC Amplitude: 1

○ PWL(t1 v1 t2 v2...)   AC Phase: 0

○ PWL FILE: [   ]   Browse   Make this information visible on schematic: ☑

Parasitic Properties

Series Resistance[Ω]: [   ]

Parallel Capacitance[F]: [   ]

Make this information visible on schematic: ☑

Additional PWL Points

Make this information visible on schematic: ☑   Cancel   OK

Fig. 2.223 Settings of voltage source V1

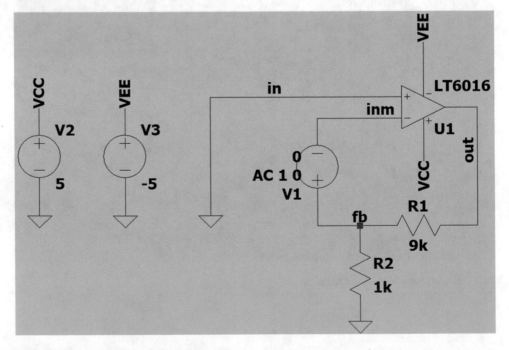

Fig. 2.224 Changes are applied to voltage source V1

Add the AC sweep command shown in Fig. 2.225 and run the simulation. The added AC sweep command calculated the frequency response on the [10 Hz, 1 MHz] range.

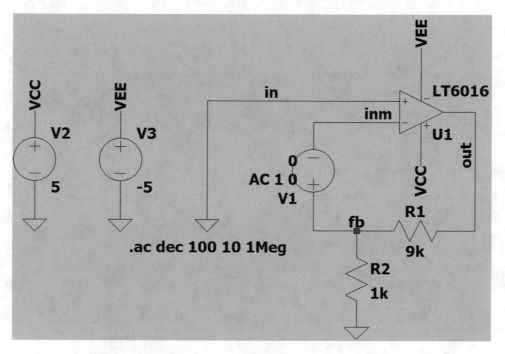

Fig. 2.225 Addition of AC sweep command to the schematic

After running the simulation, right click on the black region and select Add Traces. After clicking the Add Traces, the window shown in Fig. 2.226 appears. Enter V(fb)/V(inm) to the Expression(s) to add box and click the OK button. After clicking the OK button, the result shown in Fig. 2.227 appears.

Add Traces to Plot	✕

Only list traces matching

☑ Asterisks match colons

OK

Cancel

Available data:

```
V(fb)          Ix(U1:VEE)
V(inm)
V(out)
V(vcc)
V(vee)
I(R1)
I(R2)
I(V1)
I(V2)
I(V3)
frequency
Ix(U1:IN+)
Ix(U1:IN-)
Ix(U1:OUT)
Ix(U1:VCC)
```

Expression(s) to add:

```
V(fb)/V(inm)
```

☑ AutoRange

Fig. 2.226 Add Traces to Plot window

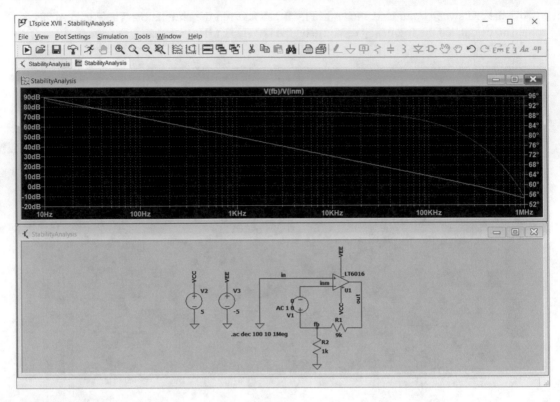

Fig. 2.227   Simulation result

Now, use a cursor to find the gain cross over frequency, i.e., the frequency which gain is 0 dB. According to Fig. 2.228, at 280.82 kHz, the gain is approximately 0 dB. The value of Phase box at this frequency, gives the phase margin to us. So, the phase margin of the studied amplifiers is about 78.305°.

Fig. 2.228   Measurement
of gain cross over
frequency and phase
margin

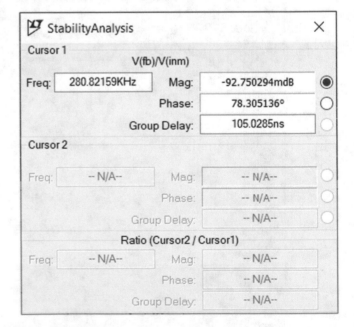

2.23   Example 22: Addition of LM 741 op amp to LTspice

The LM 741 op amp is not available in the LTspice. In this example, we see how to add LM 741 to LTspice.

Go to the Texas Instruments website and search for lm741 (Fig. 2.229).

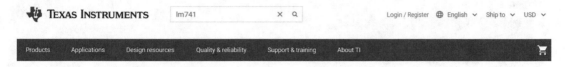

Fig. 2.229 Search for LM 741 SPICE model

Click on the first search result (Fig. 2.230) to open it (Fig. 2.231).

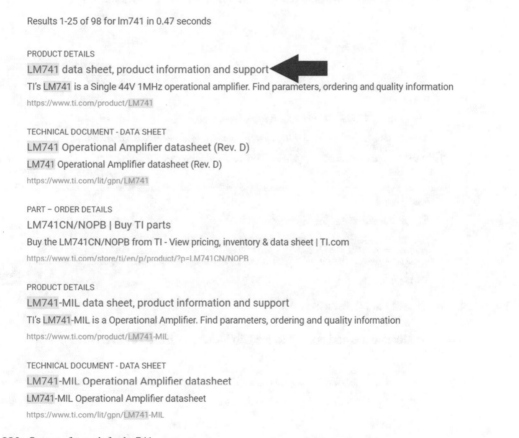

Fig. 2.230 Output of search for lm741

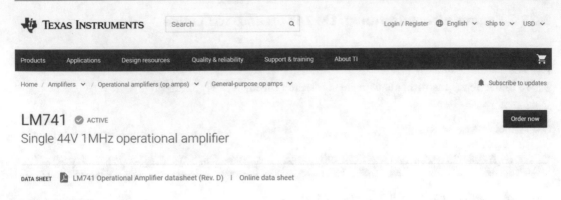

Fig. 2.231 LM741 page

Scroll down the page and download the LM 741 PSPICE Model (Fig. 2.232).

Fig. 2.232 Download of LM 741 PSPICE model

Unzip the downloaded file and copy it to the "lib" folder of LTspice (Fig. 2.233).

Fig. 2.233 Downloaded PSPICE model is copied into the "lib" folder of LTspice

Double click the snom211 folder to open it. It contains a file named LM741.mod. Click on the address bar and press the Ctrl+C to copy the address of this file (Fig. 2.234).

Fig. 2.234 Address of LM741.mod is copied into the clipboard

Open a new file in LTspice (Fig. 2.235). Click on the SPICE Directive and type .inc then press the Ctrl+V to paste the copied address (Fig. 2.236).

Fig. 2.235 Opening a new schematic file

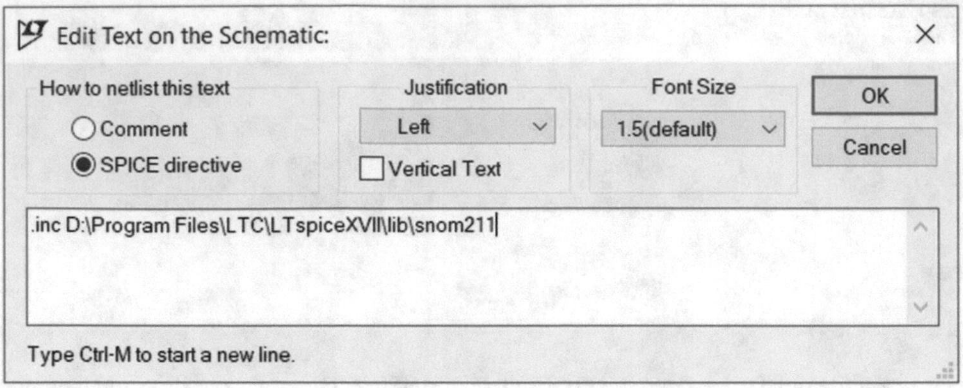

Fig. 2.236 Edit Text on the Schematic window

Add the model name to the .inc command (Fig. 2.237) and click the OK button. After clicking the OK button, click on the schematic to add the entered command to it (Fig. 2.238). The .inc command adds (includes) the prepared model to our schematic.

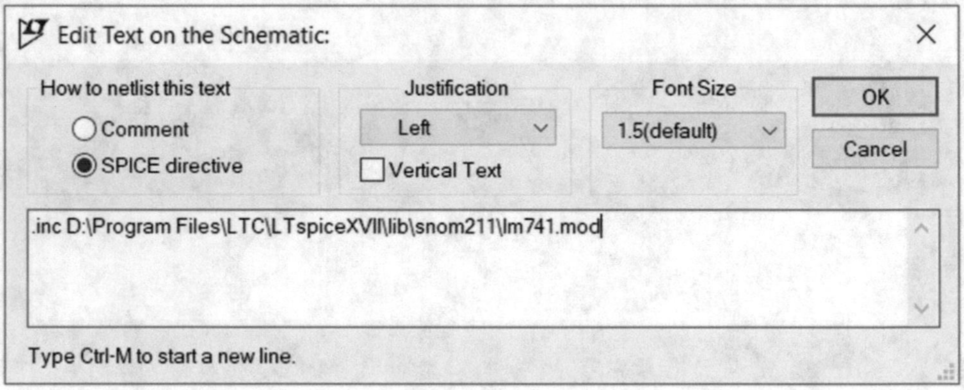

Fig. 2.237 Edit Text on the Schematic window

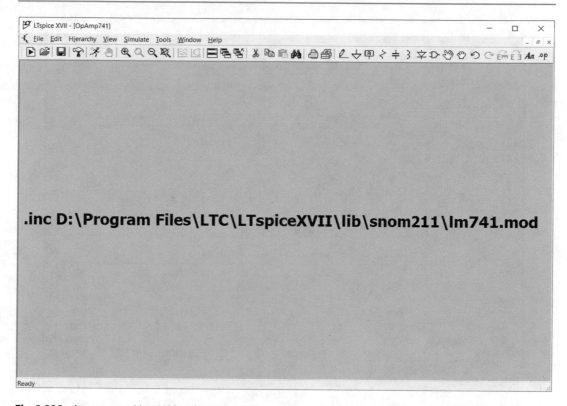

Fig. 2.238   .inc command is added to the schematic

Press F2 key. This opens the Select Component Symbol window (Fig. 2.239). Double click the OpAmps section to open it.

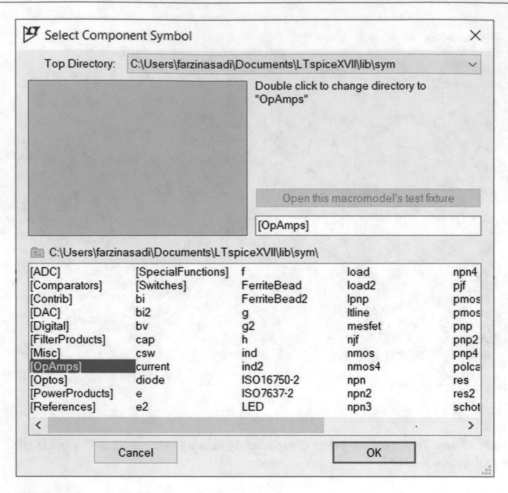

Fig. 2.239 OpAmps section of Select Component Symbol window

Select the opamp2 (Fig. 2.240) and add it to the schematic (Fig. 2.241).

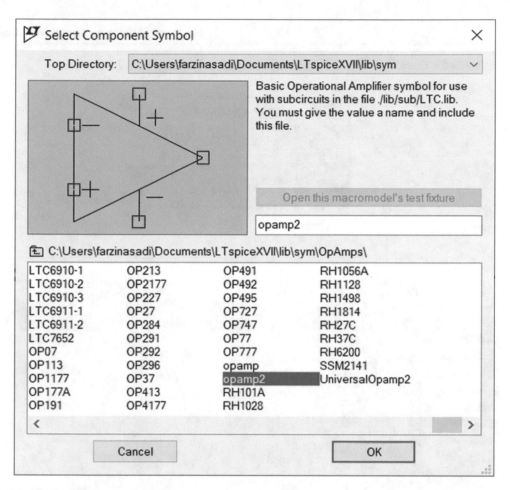

Fig. 2.240 opamp2 block

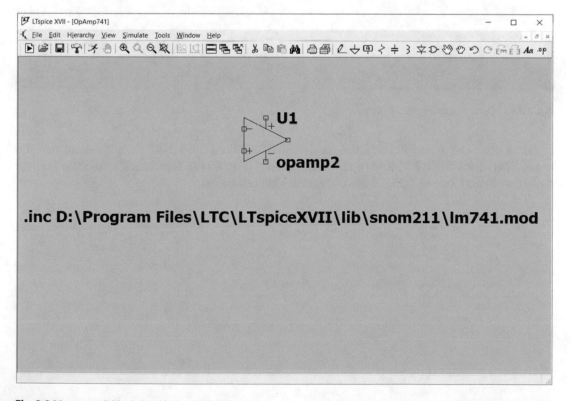

Fig. 2.241 opamp2 block is added to the schematic

Now click the Open icon (Fig. 2.242) and open the LM741.mod (Fig. 2.243).

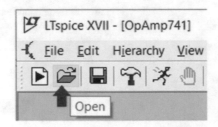

Fig. 2.242 Open icon

Fig. 2.243 Open an existing file window

Scroll down the opened file until you see the .SUBCKT line (Fig. 2.244). Copy the name that you see in front of the .SUBCKT. Right click on the opamp2 (Fig. 2.245). Double click the Value box and paste the copied name to it (Fig. 2.246). Then click the OK button.

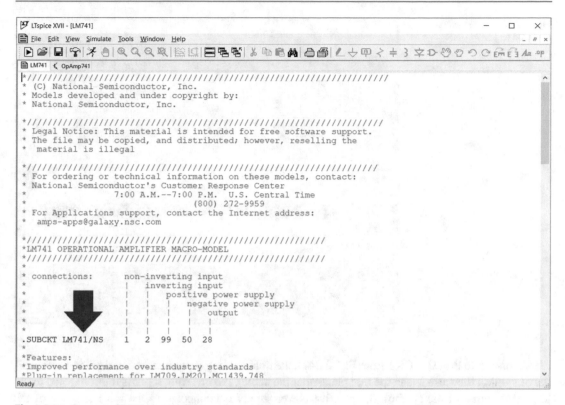

Fig. 2.244 Content of LM741.mod file

Fig. 2.245 Component
Attribute Editor window

Fig. 2.246 Value box is
changed to LM741/NS

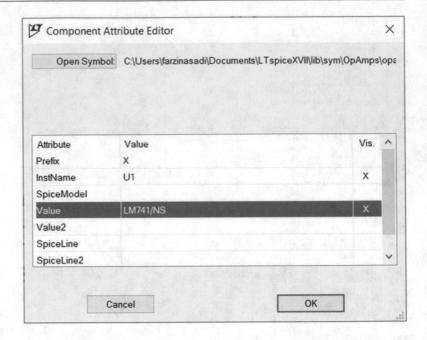

According to the .SUBCKT line (Fig. 2.247), the noninverting input is connected to the first terminal of symbol (because it is in the first place after the name), the inverting input is the connected to the
second terminal of the symbol, the positive power supply is connected to the third terminal of the
symbol, the negative power supply is connected to the fourth terminal of the symbol and the output is
connected to the fifth terminal of the symbol. We need to ensure that the symbol is compatible with
this order. In order to do this, right click on the op amp block and click the Open Symbol button
(Fig. 2.248). This opens the symbol (Fig. 2.249).

Fig. 2.247 .SUBCKT
line of the model

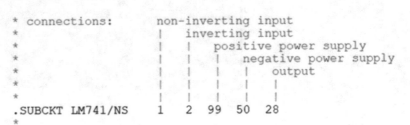

Fig. 2.248 Open Symbol button

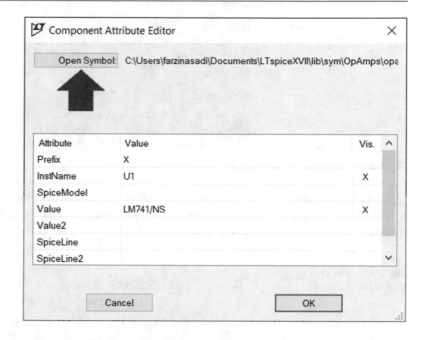

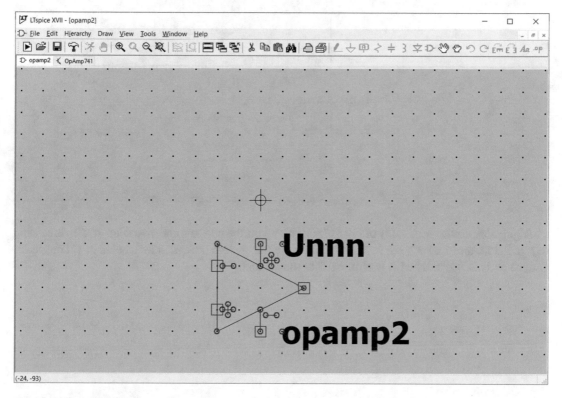

Fig. 2.249 Opened symbol

The op amp terminals are labeled in Fig. 2.250 for ease of reference.

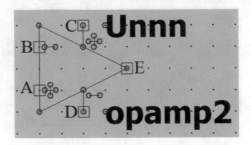

Fig. 2.250 Assigning the labels A, B, C, D and E to the symbol

Right click on the terminal A of Fig. 2.250. This opens the window shown in Fig. 2.251. According to Fig. 2.251, terminal A is the noninverting terminal of the op amp (note that its label is In+). The Netlist Order box is 1. So, terminal A is terminal 1 of the symbol.

Fig. 2.251 Pin/Port
Properties window

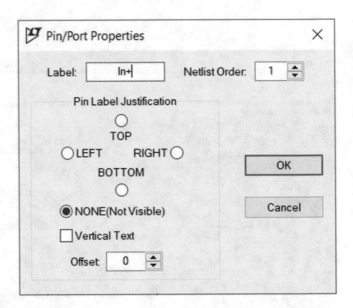

Right click on the terminal B of Fig. 2.250. This opens the window shown in Fig. 2.252. According to Fig. 2.252, terminal B is the inverting terminal of the op amp (note that its label is In-). The Netlist Order box is 2. So, terminal B is terminal 2 of the symbol.

Fig. 2.252 Pin/Port
Properties window

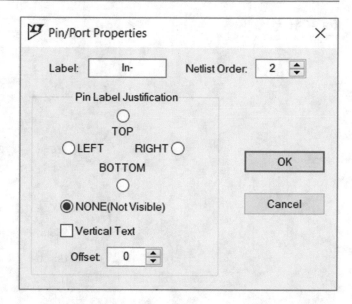

Right click on the terminal C of Fig. 2.250. This opens the window shown in Fig. 2.253. According to Fig. 2.253, terminal C is the positive power supply terminal of the op amp (note that its label is V+). The Netlist Order box is 3. So, terminal C is terminal 3 of the symbol.

Fig. 2.253 Pin/Port
Properties window

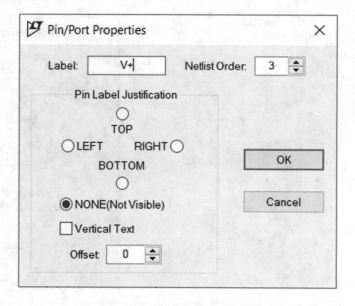

Right click on the terminal D of Fig. 2.250. This opens the window shown in Fig. 2.254. According to Fig. 2.254, terminal D is the negative power supply terminal of the op amp (note that its label is V-). The Netlist Order box is 4. So, terminal D is terminal 4 of the symbol.

Fig. 2.254 Pin/Port
Properties window

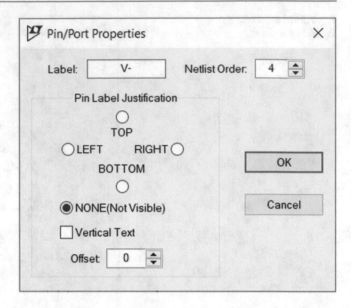

Right click on the terminal E of Fig. 2.250. This opens the window shown in Fig. 2.255. According to Fig. 2.255, terminal E is the output terminal of the op amp (note that its label is OUT). The Netlist Order box is 5. So, terminal E is terminal 5 of the symbol.

Fig. 2.255 Pin/Port
Properties window

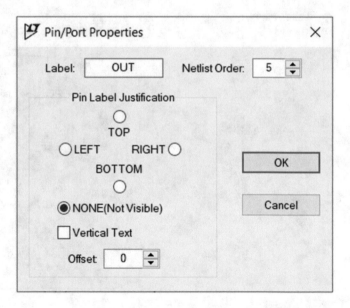

We are lucky since all the connections are compatible with Fig. 2.247. Sometimes the model is not compatible with the symbol and you need to change the Netlist Order box according to the order you see in the .SUBCKT line of the model (see Example 10 in Chap. 4).

Let's test the imported 741 op amp with a simple circuit. The schematic shown in Fig. 2.256 is a simple inverting amplifier with gain of −10. Run the simulation. The result is shown in Fig. 2.257.

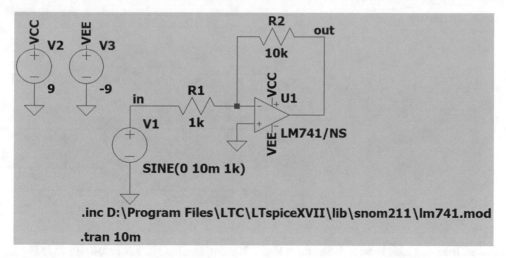

Fig. 2.256 Simple inverting amplifier

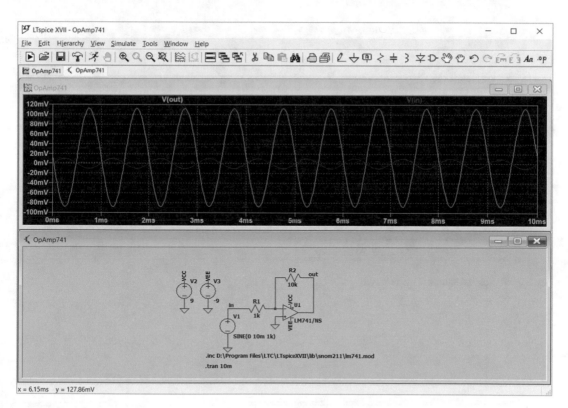

Fig. 2.257 Simulation result

It is obvious that the output has DC offset. Because the maximum of output is +111.51 mV and its minimum is −87.48 mV (Fig. 2.258). Peak-to-peak of output voltage is about 200 mV which gives the voltage gain of 10 as expected ($A_V = -\dfrac{V_{p-p,\text{out}}}{V_{p-p,\text{in}}} = -\dfrac{200\,\text{mV}}{20\,\text{mV}} = -10$).

Fig. 2.258 Readings of
the cursors

🏳 OpAmp741		✕
Cursor 1		
	V(out)	
Horz: 9.2524186ms	Vert:	-87.489024mV
Cursor 2		
	V(out)	
Horz: 9.7537379ms	Vert:	111.51454mV
Diff (Cursor2 - Cursor1)		
Horz: 501.31926µs	Vert:	199.00357mV
Freq: 1.9947368KHz	Slope:	396.96

Convert the schematic to what is shown in Fig. 2.259 and simulate it. The result shown in Fig. 2.260 shows that we have 12 mV output (offset) for zero input.

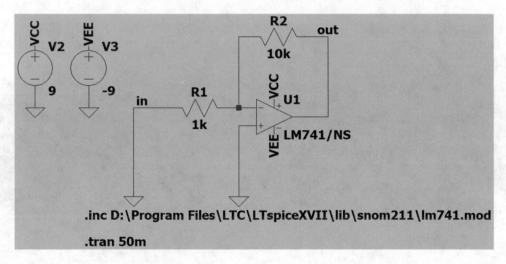

.inc D:\Program Files\LTC\LTspiceXVII\lib\snom211\lm741.mod

.tran 50m

Fig. 2.259 Voltage source V1 is removed

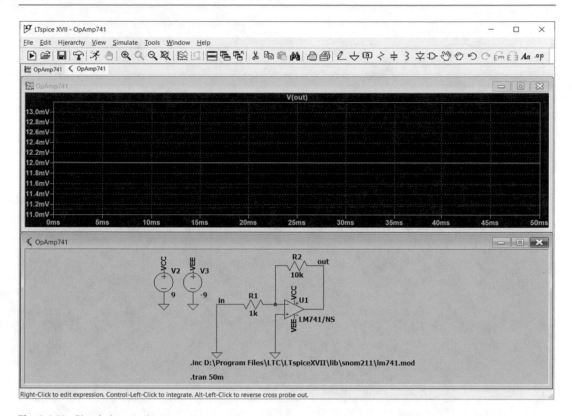

Fig. 2.260 Simulation result

We need to force the output to be around zero for zero input. In order to do this, add the resistor R3 to the schematic (Fig. 2.261). Value of R3 is set with trial and error. Enter different values for R3 and see the simulation result. R3 = 18.8 kΩ is a good value since it decreased the DC offset to about 50 μV (Fig. 2.262).

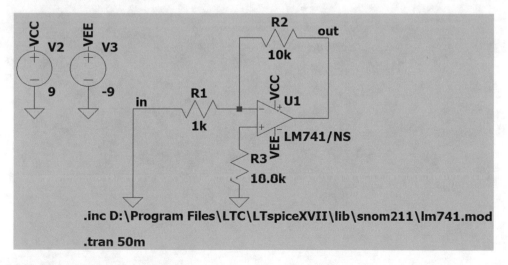

Fig. 2.261 Resistor R3 is added to the schematic

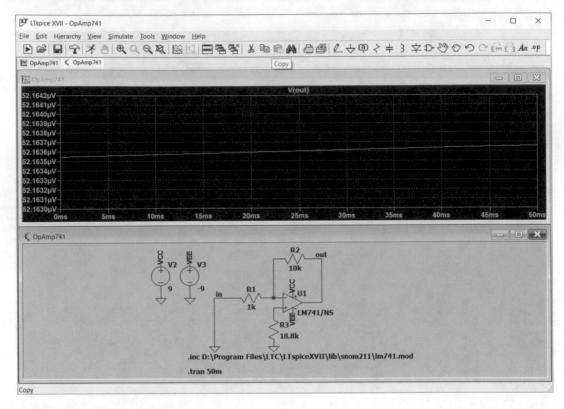

Fig. 2.262 Simulation result

Now add the input signal source to the circuit and simulate the circuit. According to the result shown in Fig. 2.263, the DC offset problem is solved.

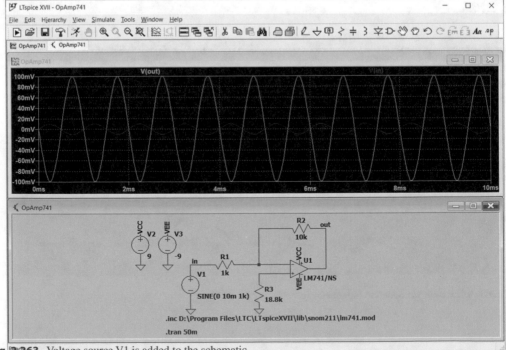

Fig. 2.263 Voltage source V1 is added to the schematic

2.24   Example 23: Measurement of Common Mode Rejection Ratio (CMRR) of an op amp Difference Amplifier

In this example, we want to measure the Common Mode Rejection Ratio (CMRR) of an op amp difference amplifier. Schematic of this example is shown in Fig. 2.264.

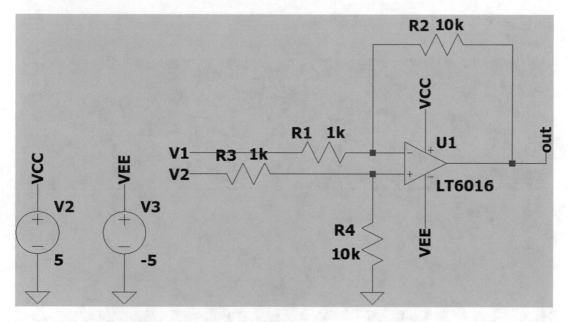

Fig. 2.264   Schematic of Example 23

Let's measure the common mode gain of the amplifier. Change the schematic to what is shown in Fig. 2.265.

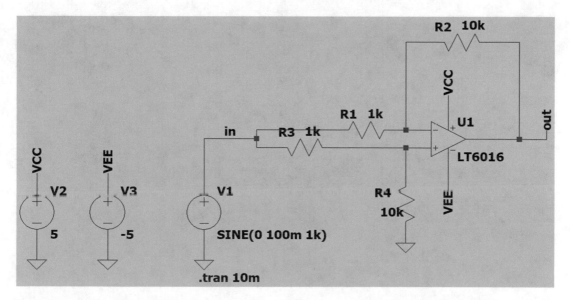

Fig. 2.265   Measurement of common mode gain of the amplifier

Run the simulation. The result is shown in Fig. 2.266. The output has DC offset. The schematic shown in Fig. 2.267 can measure the DC offset of output. According to the result shown in Fig. 2.268, the DC offset is −450.49 nV.

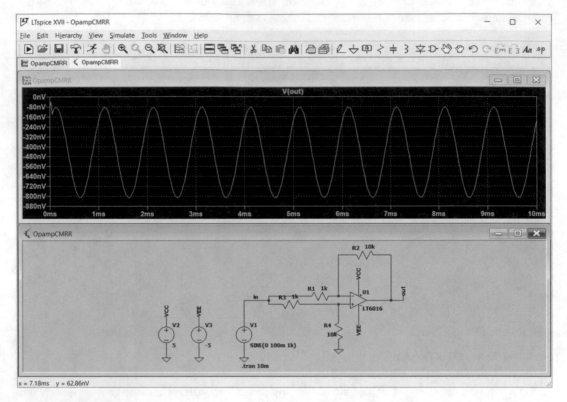

Fig. 2.266 Simulation result

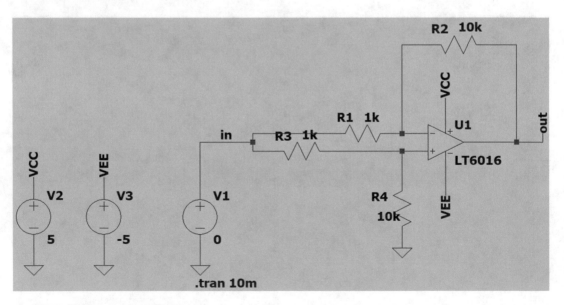

Fig. 2.267 Measurement of DC offset

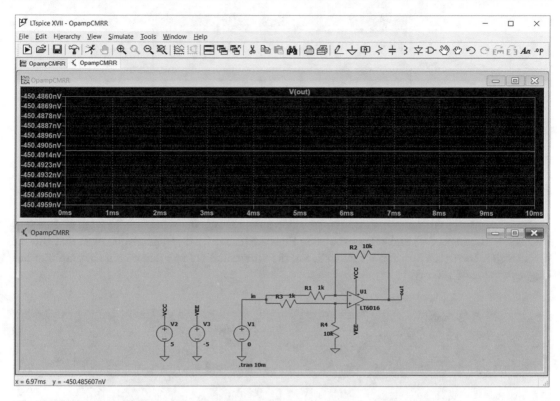

Fig. 2.268 Simulation result

The presence of DC offset doesn't make any problem for us. We can measure the common mode gain by dividing the peak-peak of output to peak-peak of input voltage. According to Fig. 2.269, the peak-peak of output voltage is 723.33 nV. The peak-peak of input signal is 200 mV. So, according to Fig. 2.270, the common mode gain is 3.6182×10^{-6}.

Fig. 2.269 Common mode gain measurement with cursors

OpampCMRR			✕
Cursor 1	V(out)		
Horz: 8.619469ms		Vert: -809.97568nV	
Cursor 2	V(out)		
Horz: 9.1238938ms		Vert: -86.648004nV	
Diff (Cursor2 - Cursor1)			
Horz: 504.42478µs		Vert: 723.32767nV	
Freq: 1.9824561KHz		Slope: 0.00143397	

```
Command Window                                    ⊙
>> Acm=723.63767e-9/(2*100e-3)

Acm =

   3.6182e-06

fx >> |
```

Fig. 2.270 MATLAB calculation

Now, we need to measure the differential mode gain. Change the schematic to what is shown in Fig. 2.271. Settings of V1 and V2 are shown in Figs. 2.272 and 2.273, respectively. Note that 180° of phase difference exists between V1 and V2. So, the differential voltage that enters the amplifier has the peak value of 100 mV.

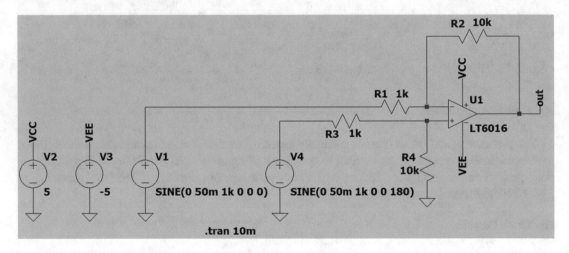

Fig. 2.271 Differential mode gain

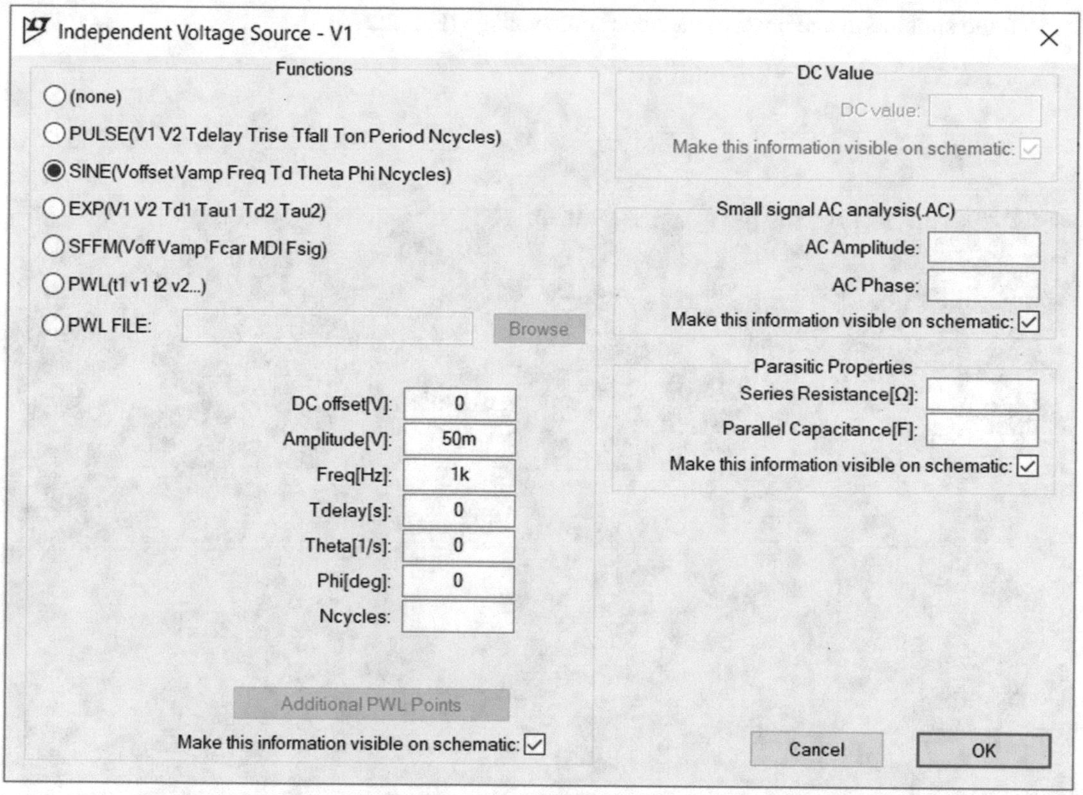

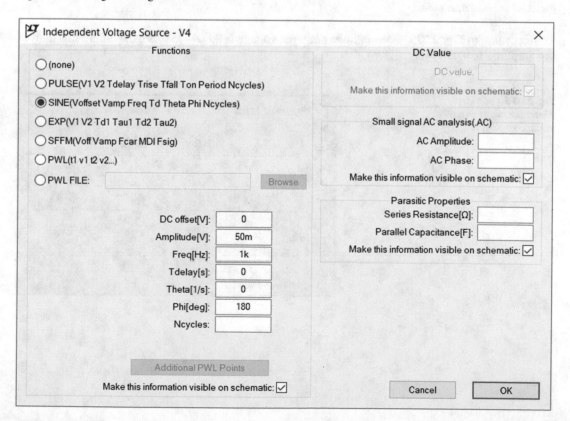

Fig. 2.272 Settings of voltage source V1

Fig. 2.273 Settings of voltage source V4

Run the simulation and observe the node "out" voltage (Fig. 2.274).

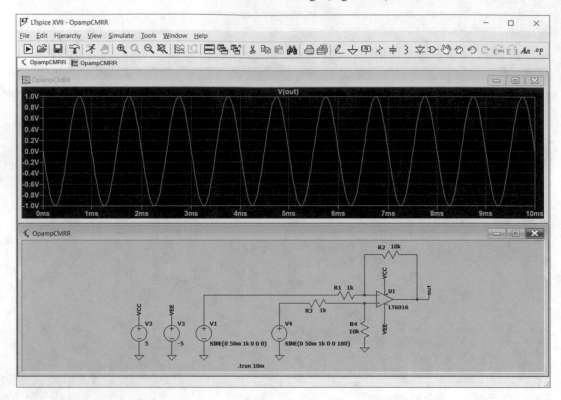

Fig. 2.274 Simulation result

According to Fig. 2.275, the amplitude of output voltage is 997.44 mV. So, the differential gain of the amplifier is 9.9744 (Fig. 2.276).

Fig. 2.275
Measurement of output
voltage amplitude

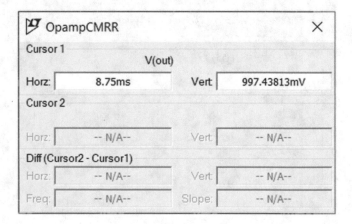

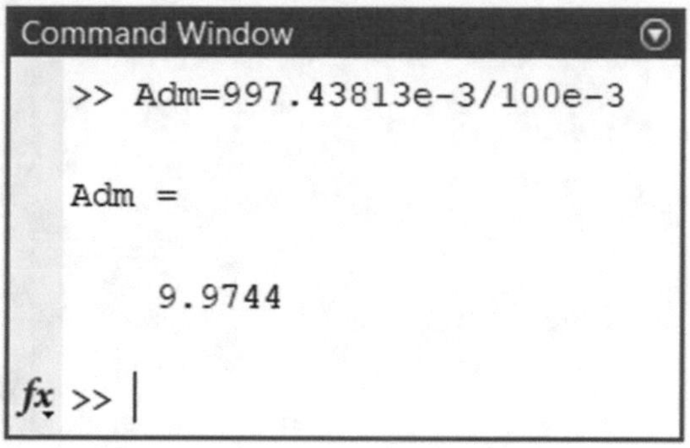

Fig. 2.276 MATLAB calculations

According to Fig. 2.277, the CMRR of the amplifier is 128.81 dB.

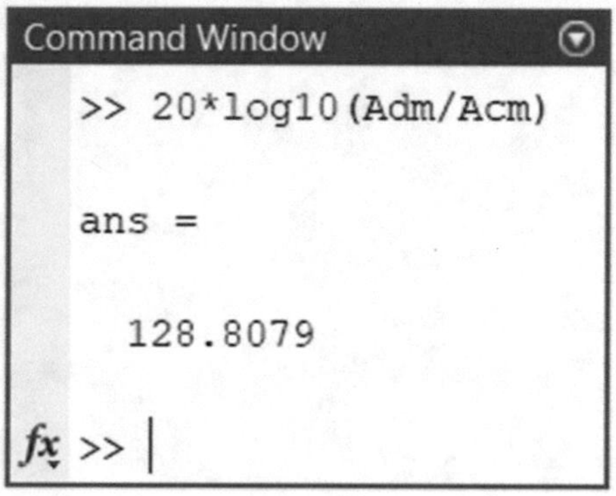

Fig. 2.277 MATLAB calculations

2.25   Example 24: Measurement of CMRR for a Differential Pair Amplifier

In this example, we want to measure the CMRR of differential pair amplifier shown in Fig. 2.278. The two collector resistances are accurate to within ±1%. So, their values changes between $10 \times 0.99 = 9.9$ kΩ and $10 \times 1.01 = 10.1$ kΩ. Other components (transistor Q1 and Q2, resistor RB1 and Rb2, resistor RE1 and RE2) are assumed to be identical.

Fig. 2.278 Circuit for
Example 24. Q1 and Q2
are 2N2222

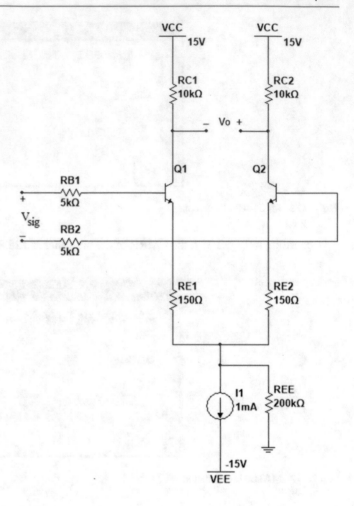

Draw the schematic shown in Fig. 2.279.

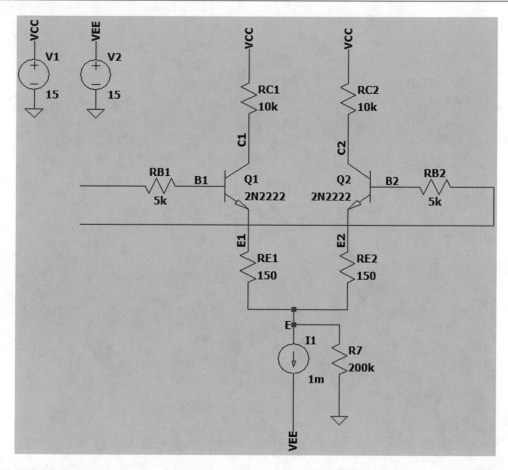

Fig. 2.279 LTspice equivalent of Fig. 2.278

Let's start with a DC analysis of the amplifier. Connect both of the inputs to ground and run the operating point analysis (Fig. 2.280). The result of operating point analysis is shown in Fig. 2.281. The emitter current of transistors and other voltages and currents are equal since the circuit is symmetric.

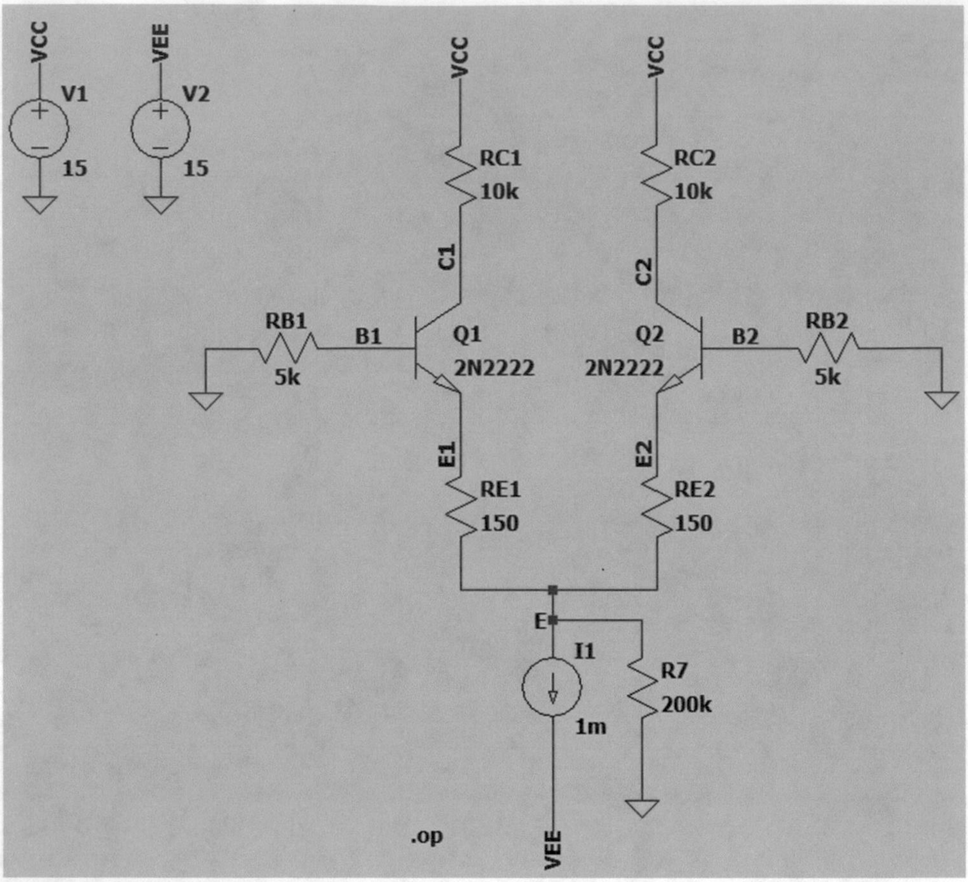

Fig. 2.280 Inputs are connected to ground

Let's measure the common mode gain of the amplifier. In order to do this, add the voltage source V3 to the schematic (Fig. 2.282). Settings of voltage source V3 is shown in Fig. 2.283. Note that in Fig. 2.283, RC1 = 9.9 kΩ and RC2 = 10.1 kΩ. If you run the schematic of Fig. 2.282, with RC1 = RC2 = 10 kΩ, the output (voltage difference between the collectors) becomes zero since the circuit is symmetric and the input is the same for both transistors. When the output for common mode signal is zero, the common mode gain is zero and the CMRR becomes infinity which is desired. However, obtaining completely symmetric circuit is very difficult (if not impossible) in real world.

We set the RC1 to 9.9 kΩ and RC2 to 10.1 kΩ in order to study the worst case (the biggest common mode gain is obtainable for these values).

```
🏴 * C:\Users\farzinasadi\Documents\LTspiceXVII\DifferentailPair.asc                                                    ✕

          --- Operating Point ---

V(c1):            10.0406            voltage
V(b1):            -0.011283          voltage
V(e1):            -0.645946          voltage
V(c2):            10.0406            voltage
V(b2):            -0.011283          voltage
V(e2):            -0.645946          voltage
V(vcc):           15                 voltage
V(e):             -0.720676          voltage
V(vee):           15                 voltage
Ic(Q2):           0.000495942        device_current
Ib(Q2):           2.2566e-006        device_current
Ie(Q2):           -0.000498198       device_current
Ic(Q1):           0.000495942        device_current
Ib(Q1):           2.2566e-006        device_current
Ie(Q1):           -0.000498198       device_current
I(I1):            0.001              device_current
I(R7):            3.60338e-006       device_current
I(Re2):           -0.000498198       device_current
I(Re1):           -0.000498198       device_current
I(Rc2):           -0.000495942       device_current
I(Rc1):           -0.000495942       device_current
I(Rb2):           -2.2566e-006       device_current
I(Rb1):           2.2566e-006        device_current
I(V2):            0.001              device_current
I(V1):            -0.000991883       device_current
```

Fig. 2.281 Simulation result

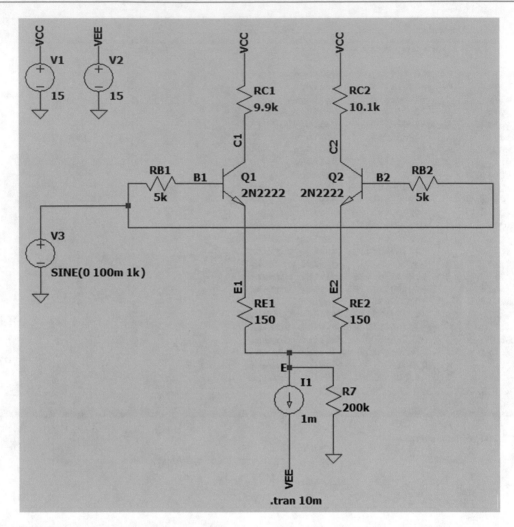

Fig. 2.282 Measurement of common mode gain

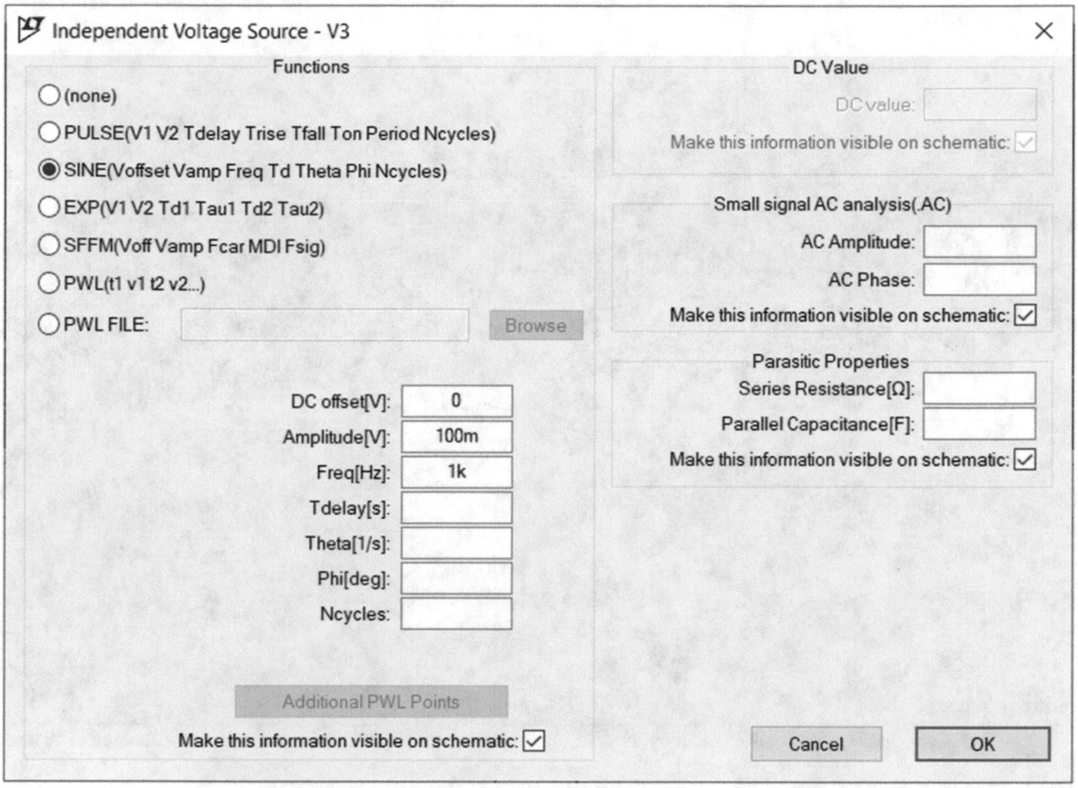

Fig. 2.283 Settings of voltage source V3

Run the simulation and draw the voltage difference between collector of Q2 and Q1. The graph is shown in Fig. 2.284. Note that the graph has an offset about −97.13 mV. Measure the peak-peak of graph. According to Fig. 2.285, the peak-peak voltage is 92.108 µV. So, the (maximum) common mode gain is 9.211×10^{-4} (Fig. 2.286).

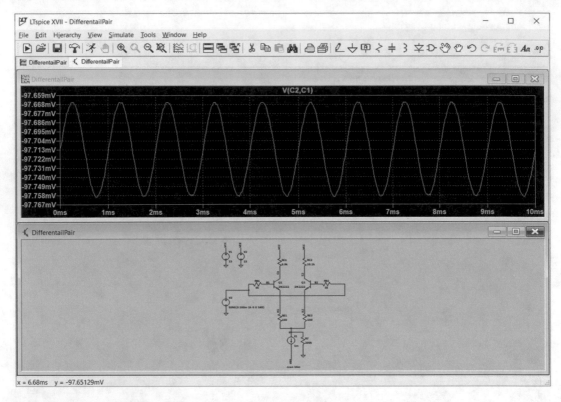

Fig. 2.284 Waveform of voltage difference between collector of Q2 and collector of Q1

Fig. 2.285 Cursors readings

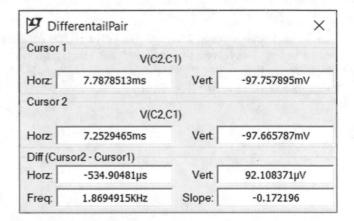

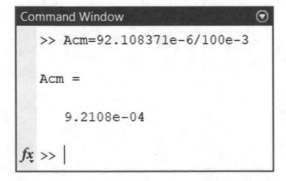

Fig. 2.286 MATLAB calculation

Now it time to measure the differential gain of the amplifier. Add the voltage sources V3 and V4 to the schematic to supply the amplifier with a differential mode input (Fig. 2.287). Settings of V3 and V4 are shown in Figs. 2.288 and 2.289, respectively.

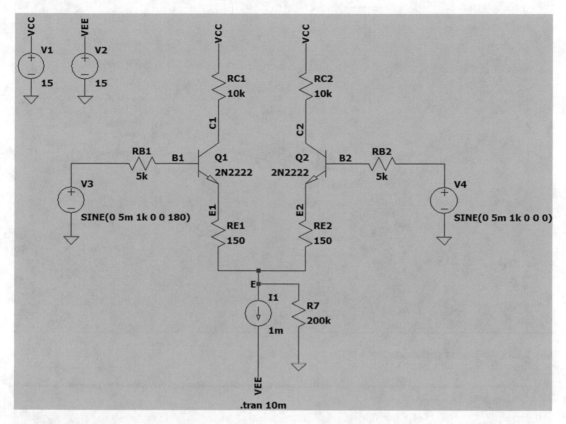

Fig. 2.287   Measurement of differential gain

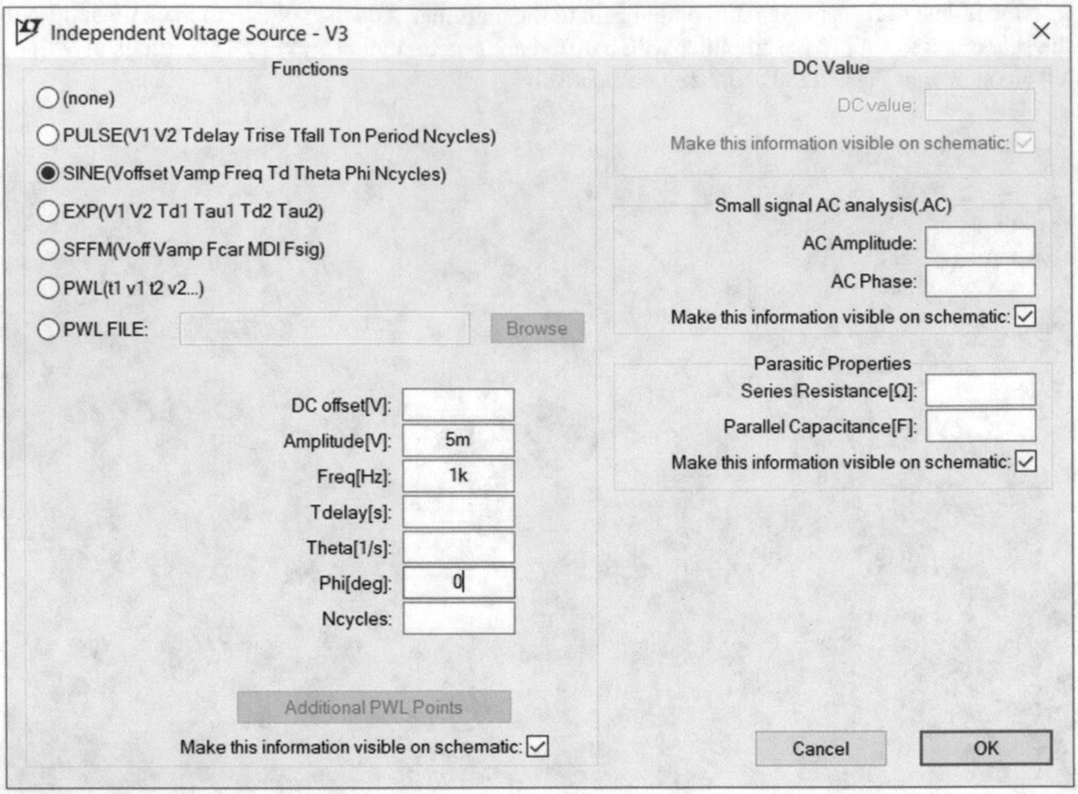

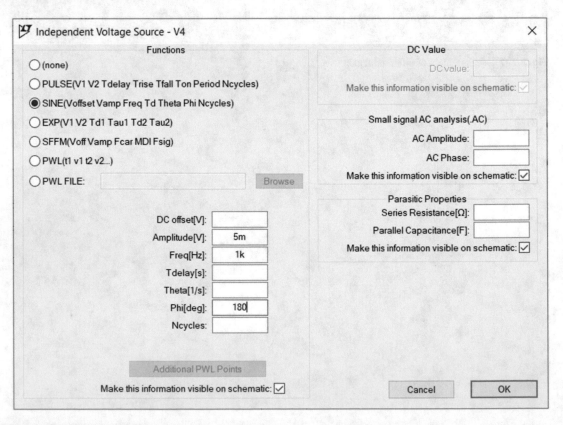

Fig. 2.288 Settings of voltage source V3

Fig. 2.289 Settings of voltage source V4

Run the simulation and draw the voltage difference between the collector of Q2 and collector of Q1 (Fig. 2.290). The amplitude of output is 433.2 mV according to Fig. 2.291. So, the differential mode voltage gain of the amplifier is 43.3 (Fig. 2.292).

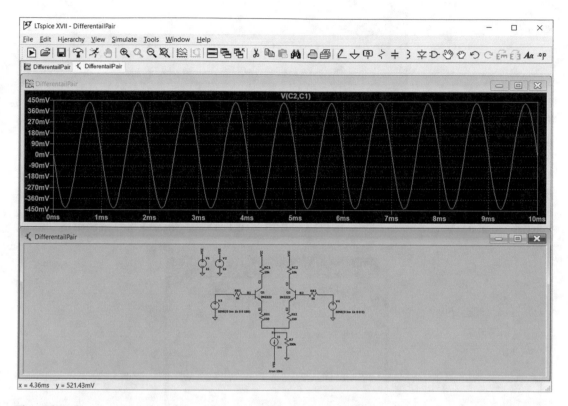

Fig. 2.290 Waveform of voltage difference between collector of Q2 and collector of Q1

Fig. 2.291 Peak of the waveform in Fig. 2.290 is around 433 mV

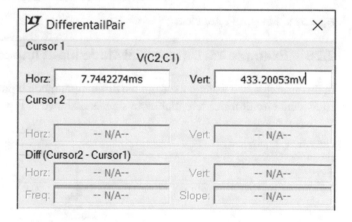

Command Window

```
>> Adm=433.20053e-3/10e-3


Adm =


    43.3201

fx >> |
```

Fig. 2.292 MATLAB calculations

The worst case (minimum) CMRR is 93.45 dB (Fig. 2.293). Note that we used the maximum common mode gain in the calculation of the CMRR. So, minimum CMRR is obtained. When the difference between the RC1 and RC2 decreases, the CMRR becomes bigger than 93.45 dB. For instance, you can simulate the circuit for RC1 = 10 kΩ and RC2 = 10.1 kΩ and observe the increase in CMRR.

Command Window

```
>> 20*log10(Adm/Acm)


ans =


    93.4478

fx >> |
```

Fig. 2.293 MATLAB calculations

2.26 Example 25: Differential Mode Input Impedance of Differential Pair

In this example, we want to measure the differential mode input impedance of previous example. The equivalent circuit for differential input signals is shown in Fig. 2.294.

Fig. 2.294 Equivalent circuit for differential input signals

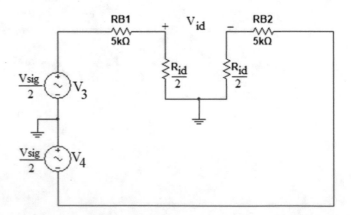

Let's measure the differential mode input impedance for 1 kHz input (Fig. 2.295). Run the simulation and draw the current of through RB1 (Fig. 2.296). Measure the peak-peak of current through resistor RB1. According to Fig. 2.297, the peak-peak of current through RB1 is 206.3 nA.

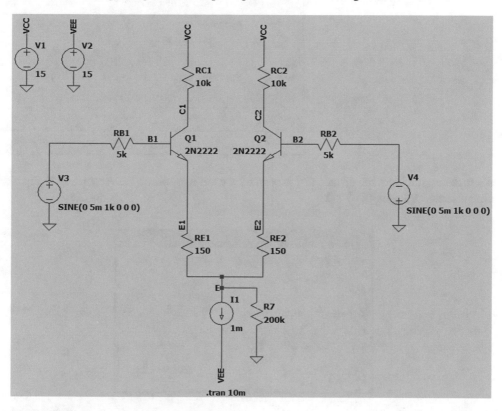

Fig. 2.295   Differential mode input impedance for 1 kHz

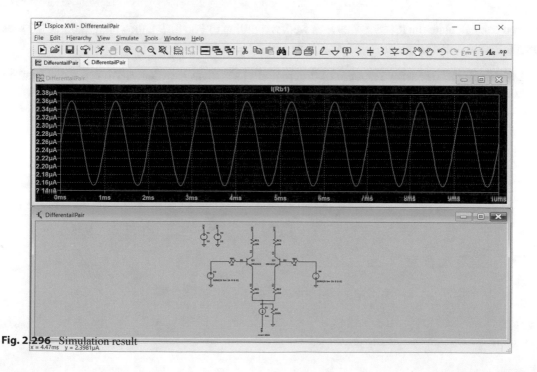

Fig. 2.296   Simulation result

Fig. 2.297 Peak-peak
of current through RB1
is around 206 nA

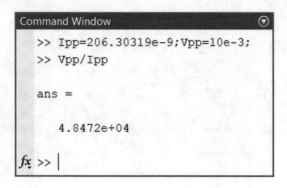

Let's measure the input impedance that is seen by the voltage source V3. According to Fig. 2.298, the voltage source V3 sees 48.47 kΩ.

```
Command Window                              ⊙
  >> Ipp=206.30319e-9;Vpp=10e-3;
  >> Vpp/Ipp

  ans =

     4.8472e+04

fx >> |
```

Fig. 2.298 MATLAB calculations

You can calculate the R_{id} (Fig. 2.294) as well. According to Fig. 2.294, the voltage source V3 sees $RB1 + \dfrac{R_{id}}{2} = 5 + \dfrac{R_{id}}{2} [k\Omega]$. So, $5 + \dfrac{R_{id}}{2} = 48.472$ or $R_{id} = 86.944$ kΩ.

Let's draw the frequency response of differential mode input impedance. We use the schematic shown in Fig. 2.299. The E1 is a voltage-dependent voltage source. Setting of E1 is shown in Fig. 2.300.

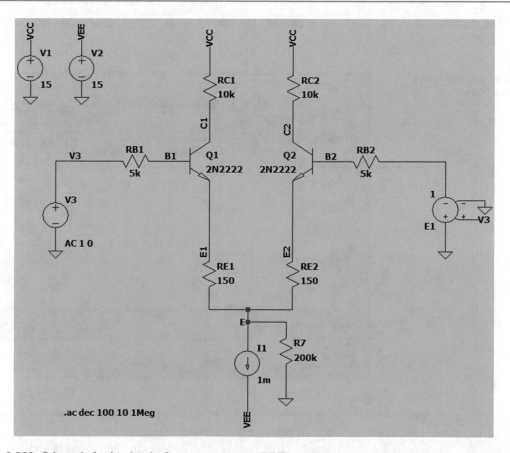

Fig. 2.299 Schematic for drawing the frequency response of differential mode input impedance

Fig. 2.300 Component
Attribute Editor window

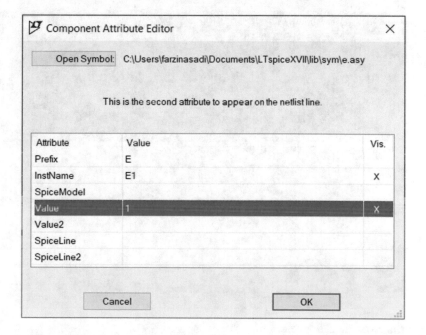

Run the simulation and draw the graph of -V(v3)/I(v3) (Fig. 2.301). The result is shown in Fig. 2.302.

Fig. 2.301 Add Traces to Plot window

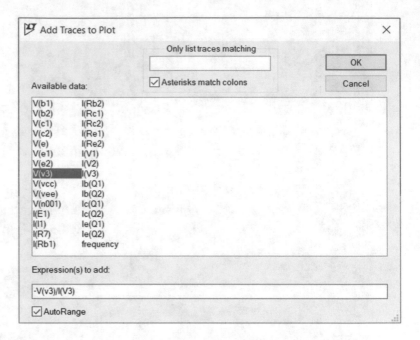

Fig. 2.302 Simulation result

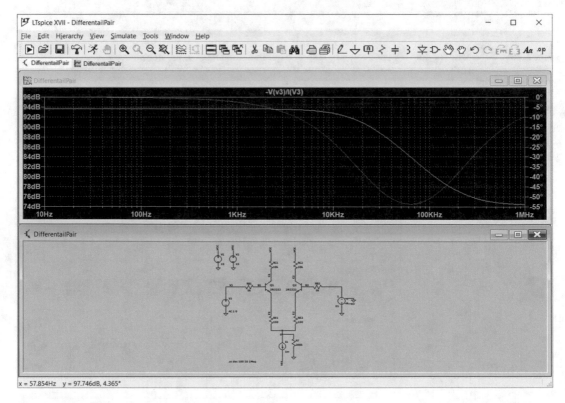

Right click on the vertical axis and select the Linear (Fig. 2.303). The vertical axis unit changes to Ohms (Fig. 2.304). This graph shows the differential mode input impedance of the circuit.

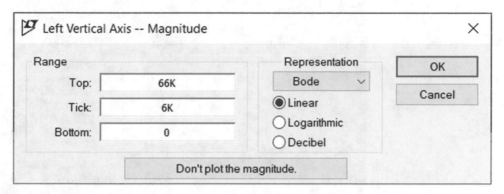

Fig. 2.303 Left Vertical Axis window

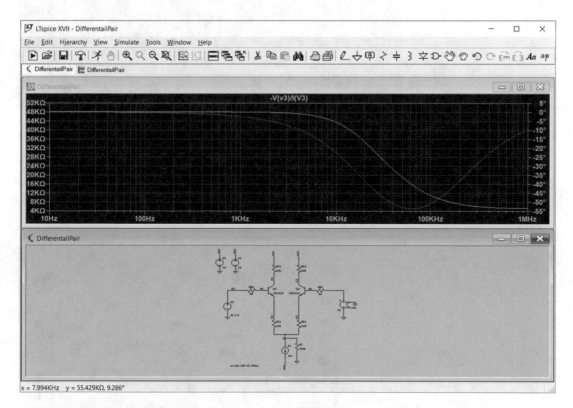

Fig. 2.304 Simulation result

Let's check the obtained graph. Read the impedance at 1 kHz. According to Fig. 2.305, the impedance at 1 kHz is about 48.07 kΩ. This number is quite close to what we found in Fig. 2.298.

Fig. 2.305 Input
impedance at 1 kHz

Fig. 2.305 Input
impedance at 1 kHz

DifferentailPair		✕

Cursor 1

-V(v3)/I(V3)

Freq:	1.006237KHz	Mag:	48.077561KΩ	●
		Phase:	-2.4729888°	○
		Group Delay:	6.8149099µs	○

Cursor 2

Freq:	-- N/A--	Mag:	-- N/A--	○
		Phase:	-- N/A--	○
		Group Delay:	-- N/A--	○

Ratio (Cursor2 / Cursor1)

Freq:	-- N/A--	Mag:	-- N/A--
		Phase:	-- N/A--
		Group Delay:	-- N/A--

2.27 Example 26: Colpitts Oscillator

In this example, we want to simulate a Colpitts oscillator. The schematic of this example is shown in
Fig. 2.306. The calculation shown in Fig. 2.307 shows that the frequency of this oscillator is 31.83 kHz.

Fig. 2.306 Schematic
of Example 26

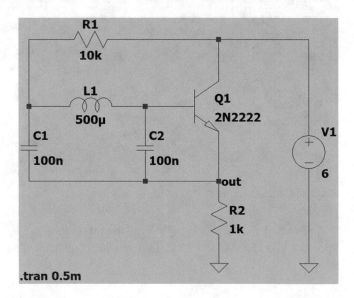

.tran 0.5m

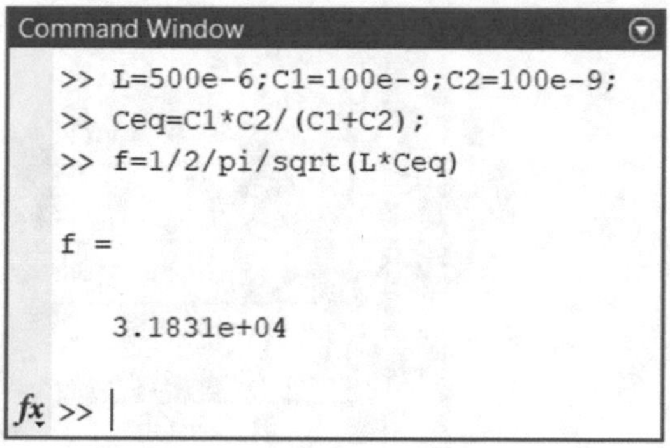

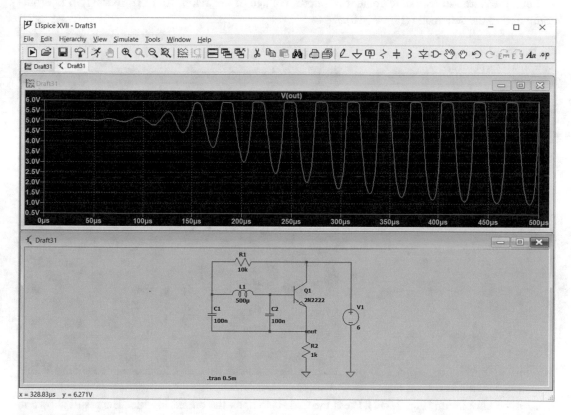

Fig. 2.307 MATLAB calculation

Run the simulation. The result is shown in Fig. 2.308.

Fig. 2.308 Simulation result

Let's measure the frequency of waveform. According to Fig. 2.309, the frequency of this waveform is 31.28 kHz which is quite close to what we obtained.

Fig. 2.309 Coordinates
read by the cursors

Draft31				×
Cursor 1				
	V(out)			
Horz:	386.49573µs		Vert	3.0005481V
Cursor 2				
	V(out)			
Horz:	418.46154µs		Vert	2.9888127V
Diff (Cursor2 - Cursor1)				
Horz:	31.965812µs		Vert	-11.735463mV
Freq:	31.283422KHz		Slope:	-367.125

Let's measure the harmonic content of output voltage of oscillator. Add the .four 31.96 k V(out) command to the schematic (Fig. 2.310).

Fig. 2.310 Measurement of
harmonic content for an
oscillator

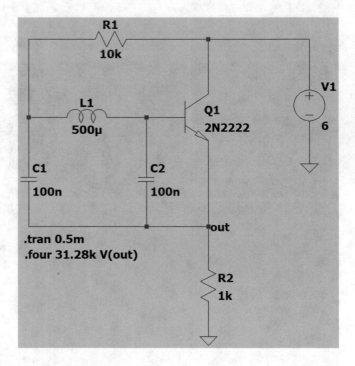

Run the simulation (Fig. 2.311). Press the Ctrl+L to open the output log file. The result shown in Fig. 2.312 appears. According to Fig. 2.312, the THD of the oscillator is 20.526%. Such a high THD is expected since the output (Fig. 2.308) does not resemble a pure sine wave.

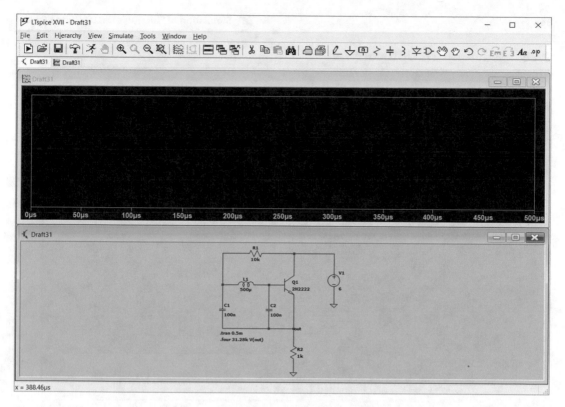

Fig. 2.311 Simulation is run

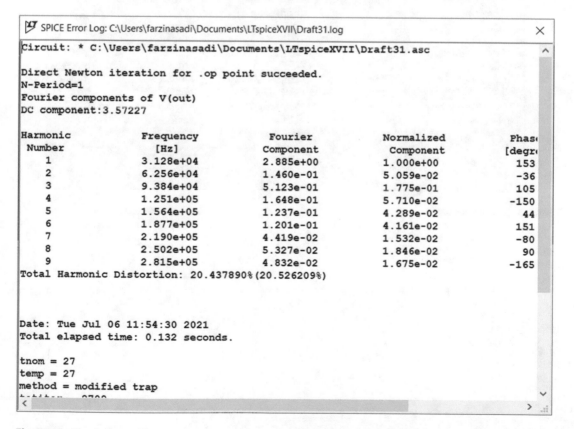

Fig. 2.312 Simulation result

2.28   Example 27: Optocoupler

In this example, we want to simulate a circuit which contains an optocoupler. Optocouplers can be found in the [Optos] section of Select Component Symbol window (Fig. 2.313). Available optocouplers are shown in Fig. 2.314.

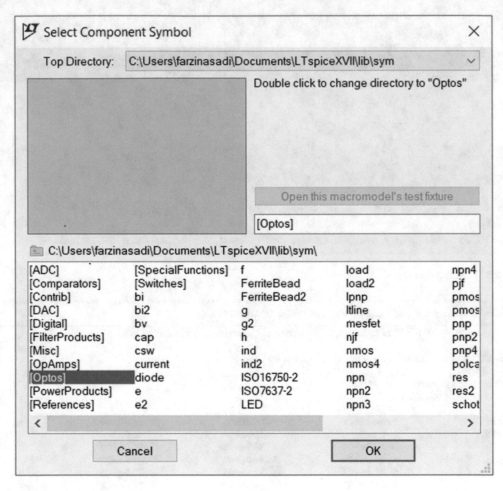

Fig. 2.313 Optos section

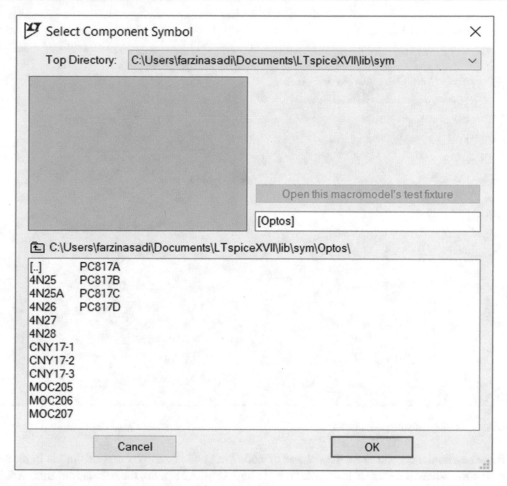

Fig. 2.314 Components inside the Optos section

Let's simulate a circuit which contains an optocoupler. Draw the schematic shown in Fig. 2.315. Settings of voltage source V1 are shown in Fig. 2.316.

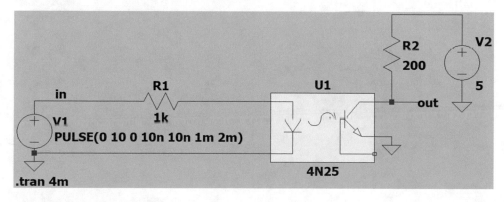

Fig. 2.315 Schematic for Example 27

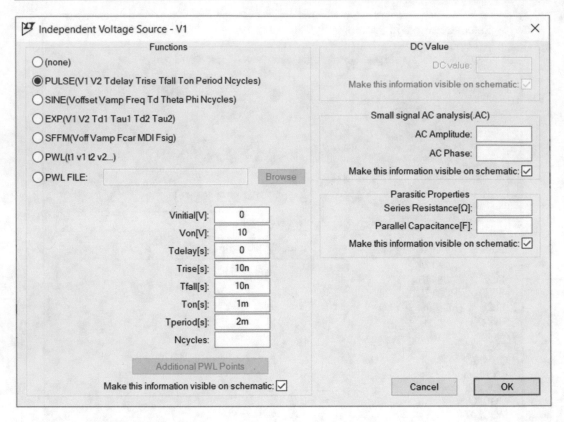

Fig. 2.316 Settings of voltage source V1

Run the simulation and draw the voltage of node "in" and "out." When node "in" is high, node "out" is low. When node "is" is low, node "out" is high (Fig. 2.317). So, the behavior of this circuit is similar to NOT gate.

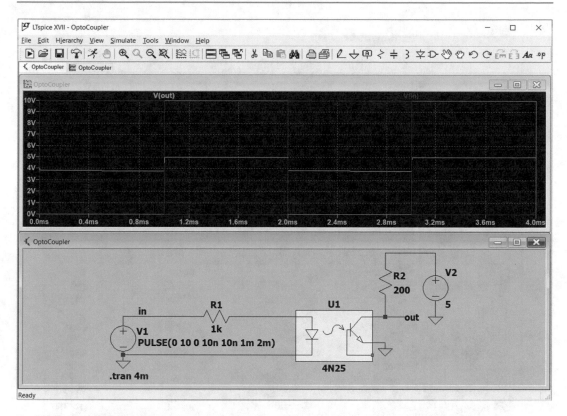

Fig. 2.317 Simulation result

2.29 Example 28: Astable Oscillator with NE 555

In this example, we want to simulate a circuit which contain NE 555 IC. The NE 555 IC can be found in the [Misc] section (Fig. 2.318) of Select Component Symbol window (Fig. 2.319).

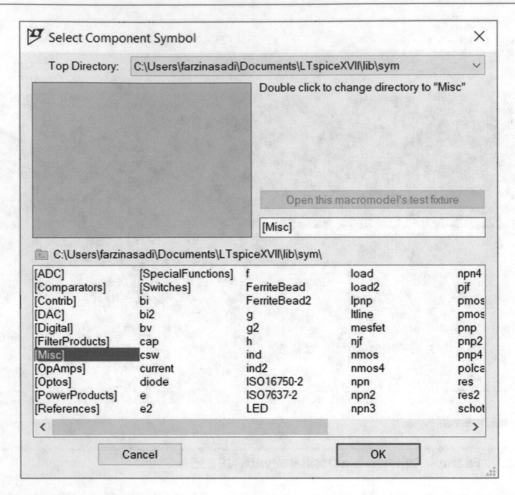

Fig. 2.318 Misc section

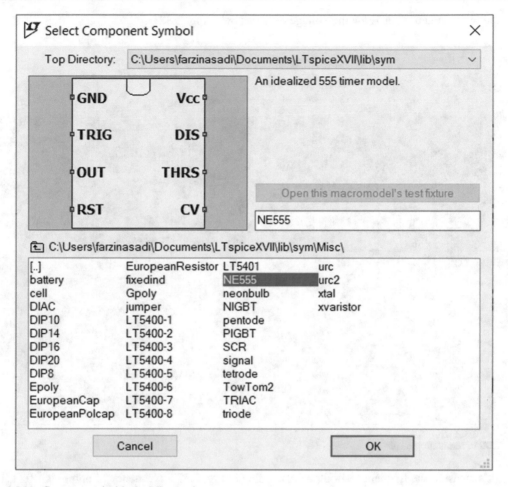

Fig. 2.319 Components inside the Misc section

Draw the schematic shown in Fig. 2.320.

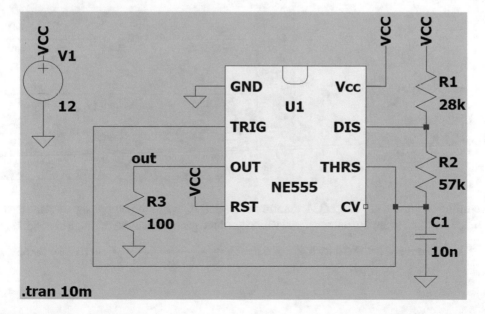

Fig. 2.320 Schematic of Example 28

Run the simulation and draw the voltage of node "out" (Fig. 2.321).

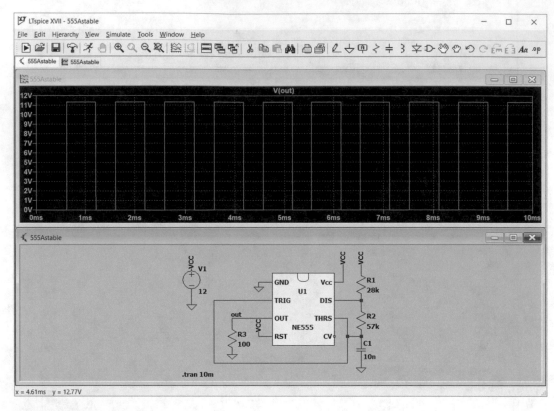

Fig. 2.321 Simulation result

Let's measure the frequency of output. According to Fig. 2.322, frequency of output is 1.02 kHz.

Fig. 2.322 Coordinates
read by the cursors

555Astable			✕
Cursor 1	V(out)		
Horz: 4.1766724ms		Vert: 67.924215µV	
Cursor 2	V(out)		
Horz: 5.1543739ms		Vert: 11.320691V	
Diff (Cursor2 - Cursor1)			
Horz: 977.70154µs		Vert: 11.320623V	
Freq: 1.022807KHz		Slope: 11578.8	

According to Figs. 2.323 and 2.324, duration of High (i.e., when the output signal value is around VCC) and Low (i.e., when the output signal value is around ground) section of graph are 583.19 µs and 394.51 µs, respectively. So, the duty cycle of output is $\dfrac{583.19\,\mu s}{394.51\,\mu s + 583.19\,\mu s} \times 100\% = 59.65\%$.

Fig. 2.323 Duration of high section of waveform shown in Fig. 2.321

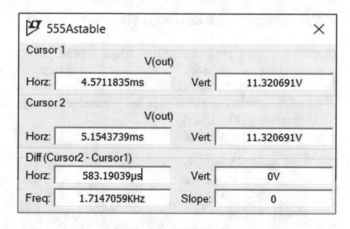

Fig. 2.324 Duration of low section of waveform shown in Fig. 2.321

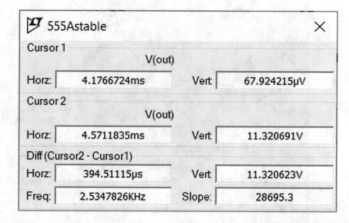

Let's check the obtained results. The calculation shown in Fig. 2.325 shows that LTspice results are correct.

Fig. 2.325 MATLAB calculations

```
Command Window
>> R1=28e3;R2=57e3;C=10e-9;
>> HighDuration_us=0.693*(R1+R2)*C/1e-6

HighDuration_us =

   589.0500

>> LowDuration_us=0.693*R2*C/1e-6

LowDuration_us =

   395.0100

>> frequency=1/((HighDuration_us+LowDuration_us)*1e-6)

frequency =

   1.0162e+03

fx >> |
```

2.30   Example 29: Low Pass Filter

LTspice has ready to use filter block which permits you to simulate the filtering action quite easily without being involved with the details of filter circuit. After obtaining the desired results, you can focus on the design of filter circuit itself.

The filter blocks can be found in the [SpecialFunctions] section (Fig. 2.326) of Select Component Symbol window (Fig. 2.327).

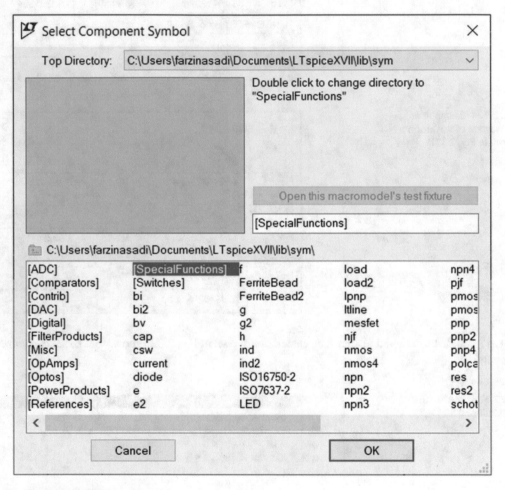

Fig. 2.326   SpecialFunctions section

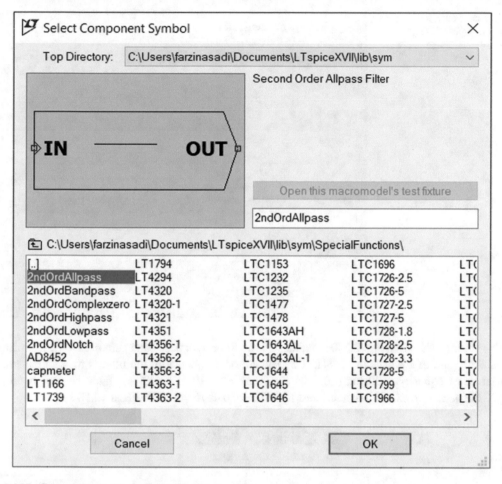

Fig. 2.327 Components inside the SpecialFunctions section

Let's study an example. Draw the schematic shown in Fig. 2.328. Settings of 2ndOrdLowpass filter is shown in Fig. 2.329.

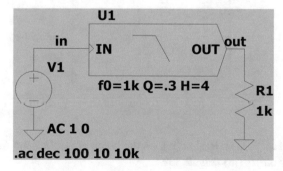

Fig. 2.328 Schematic for Example 29

Fig. 2.329 Settings of
2ndOrdLowpass filter

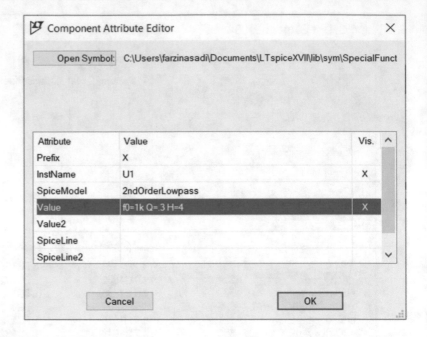

The settings shown in Fig. 2.330 simulates a lows filter with pass band gain of H = 4. According to Fig. 2.330, gain of 4 equals to 12.0412 dB. The cut off frequency of the filter is f0 = 1 kHz and the gain at this frequency is 20 log (Q) = 20 log (0.3) = − 10.4576 dB less than the pass band gain (Fig. 2.330). So, we expect a gain around 12.0412 − 10.4576 = 1.5836 dB at 1 kHz.

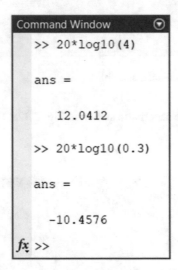

Fig. 2.330 MATLAB calculations

Simulate the circuit (Fig. 2.331). Right click on the black area, click the Add Traces, and draw the frequency response of V(out)/V(in) (Fig. 2.332).

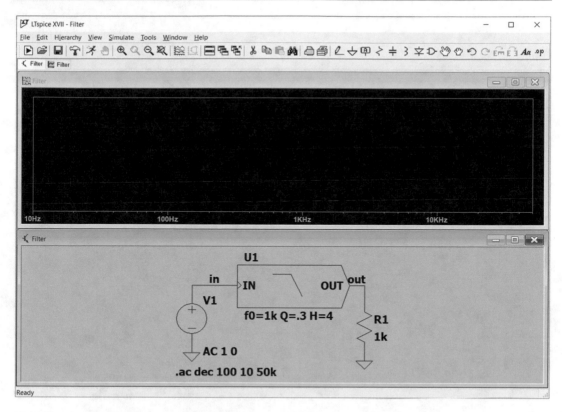

Fig. 2.331 Simulation is run

Fig. 2.332 Add Traces
to Plot window

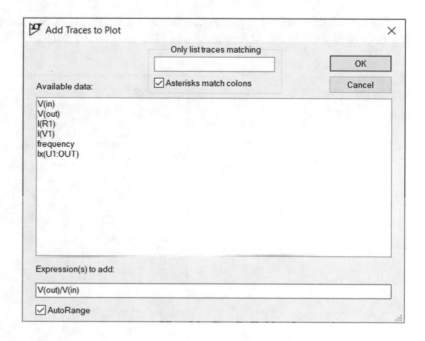

Frequency response of filter (V(out)/V(in)) is shown in Fig. 2.333.

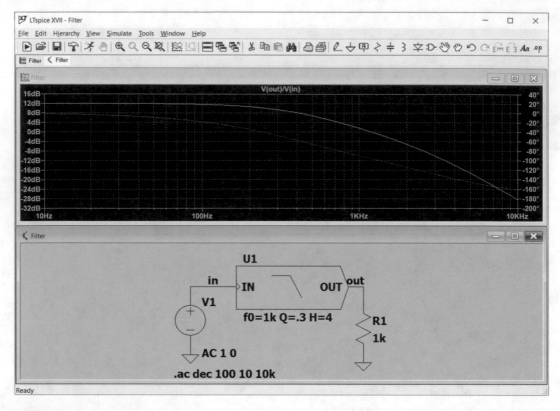

Fig. 2.333 Simulation result

Let's check the obtained results. According to Fig. 2.334, at low frequencies, the pass band gain of the filter is 12.036376 dB. At frequency of 1 kHz, the gain is 1.5469263 dB.

Fig. 2.334 Coordinates
read by the cursors

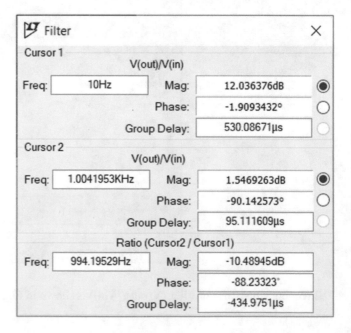

2.31   Example 30: High Pass Filter

In this example, we want to simulate a high pass filter. The schematic of this example is shown in Fig. 2.335. This schematic simulates a high pass filter with pass band gain of H = 4. Gain of 4 equals to 12.0412 dB. The cut off frequency of the filter is f0 = 1 kHz, and the gain at this frequency is 20 log (Q) = 20 log (0.3) = − 10.4576 dB less than the pass band gain. So, we expect a gain around 12.0412 − 10.4576 = 1.5836 dB at 1 kHz.

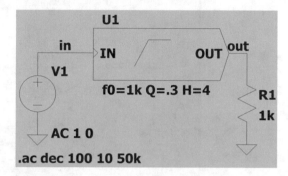

Fig. 2.335 Schematic for Example 30

Run the simulation and draw the frequency response graph of V(out)/V(in) (Fig. 2.336).

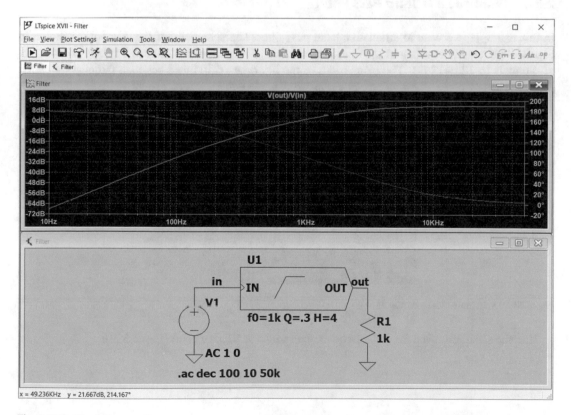

Fig. 2.336 Simulation result

According to Fig. 2.337, the pass band gain is 12.02 dB, and the gain at 1 kHz is around 1.65 dB.

Fig. 2.337 Coordinates read by the cursors

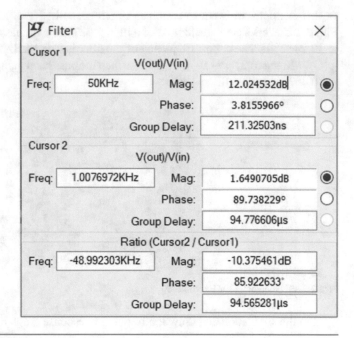

2.32   Example 31: Band Pass Filter

In this example, we want to simulate a band pass filter. The schematic of this example is shown in Fig. 2.338. Center frequency of the filter is f0 = 1 kHz. The gain of the filter at 1 kHz is H = 4. Gain of 4 equals to 12.0412 dB. The band width of the filter, i.e., difference between frequencies which thecenter frequency gain is decreased by 3 dB, is $\dfrac{f0}{Q} = \dfrac{1\,\text{kHz}}{0.3} = 3.33\,\text{kHz}$.

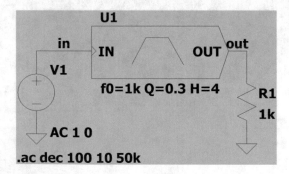

Fig. 2.338 Schematic for Example 31

Run the simulation and draw the frequency response of V(out)/V(in) (Fig. 2.339).

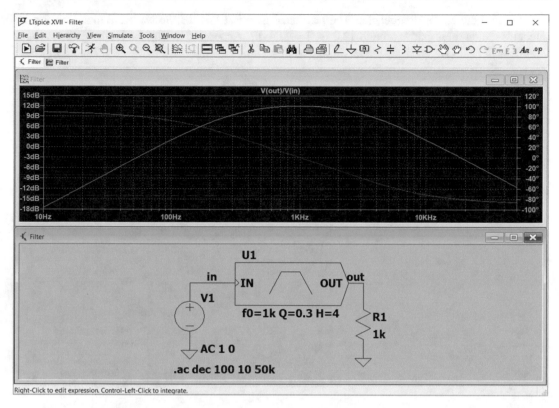

Fig. 2.339 Simulation result

According to Fig. 2.340, the gain at 1 kHz is 12.04 dB.

Fig. 2.340 Measurement of gain for 1 kHz

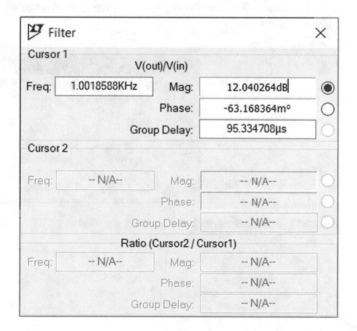

Let's find the cut off frequencies. The pass band gain is 12.04 dB. So, we need to search for frequencies which their gain is around 12.04 − 3 = 9.04 dB. According to Fig. 2.341, the gain is around 9.04 dB at 277.08 Hz and 3.622 kHz. So, the band width of the filter is 3.622 − 0.277 = 3.345 kHz.

Fig. 2.341 Measurement of cutoff frequencies

📐 Filter		✕
Cursor 1		
	V(out)/V(in)	
Freq: 277.07561Hz	Mag: 9.0313718dB	⦿
	Phase: 44.990095°	○
	Group Delay: 334.97933µs	○
Cursor 2		
	V(out)/V(in)	
Freq: 3.6225526KHz	Mag: 9.0129576dB	⦿
	Phase: −45.11194°	○
	Group Delay: 25.596157µs	○
Ratio (Cursor2 / Cursor1)		
Freq: 3.345477KHz	Mag: −18.414208mdB	
	Phase: −90.102035°	
	Group Delay: −309.38318µs	

The quality factor (Q) of the band pass filter is defined as $Q = \dfrac{f_0}{BW}$, where f_0 shows the center frequency and BW shows the bandwidth of the filter. According to Fig. 2.342, the quality factor of this filter is around 0.3.

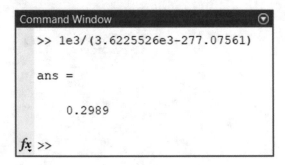

```
Command Window                          ⊙
  >> 1e3/(3.6225526e3-277.07561)

  ans =

      0.2989

fx >>
```

Fig. 2.342 MATLAB calculations

2.33 Exercises

1. Simulate the half wave rectifier circuit (Fig. 2.343) with RL load (R = 10 Ω and L = 10 mH). Compare the result with purely resistive load.

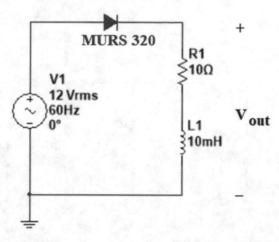

Fig. 2.343 Schematic for Exercise 1

2. Figure 2.344 shows an op-amp clamp circuit with a nonzero reference clamping voltage. The clamping level is at precisely the reference voltage. Use LTspice to simulate the circuit and see the effect of ReferenceVoltage source on output.

Fig. 2.344 Schematic for Exercise 2

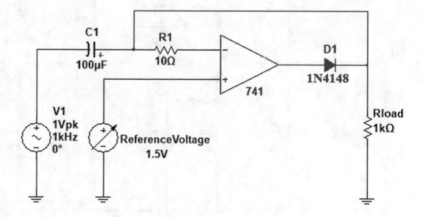

3. Measure the maximum output voltage swing of the circuit shown in Fig. 2.345.

Fig. 2.345 Schematic for Exercise 3

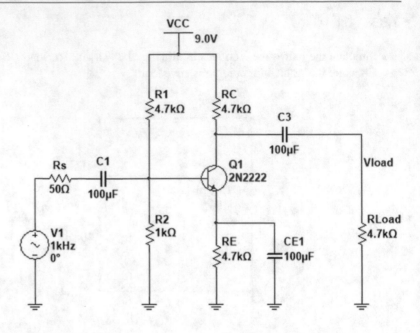

4. Assume the amplifier shown in Fig. 2.346.

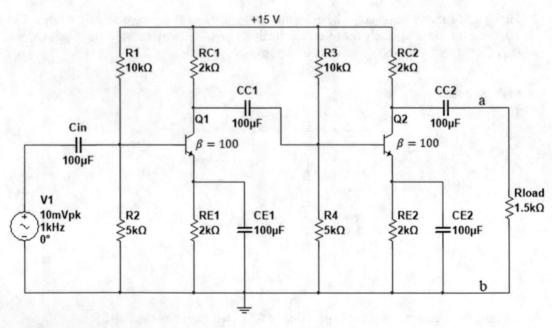

Fig. 2.346 Schematic for Exercise 4

(a) Use hand analysis to calculate the DC voltages (operating point) of the circuit.
(b) Use LTspice to verify results of part (a).
(c) Use hand analysis to calculate the input impedance (impedance seen from source V1) and output impedance (impedance seen from points a and b) of the circuit.
(d) Use LTspice to verify part (c).
(e) Use hand analysis to calculate the overall gain ($\frac{V_{ab}}{V_1}$) of the circuit.
(f) Use LTspice to verify part (e).

5. Assume that both transistors in Fig. 2.346 are 2N2222. Use LTspice to draw the:
   (a) Input impedance as a function of frequency.
   (b) Output impedance as a function of frequency.
   (c) Overall gain of the system a function of frequency.
   (d) Measure the THD of output voltage.

6. Use LTspice to simulate the circuit shown in Fig. 2.347.

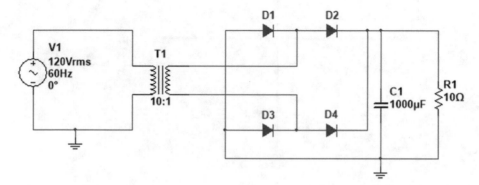

Fig. 2.347 Schematic for Exercise 6

7. (a) Use LTspice to draw the frequency response of the circuit shown in Fig. 2.348.
   (b) Check the result of part (a) with MATLAB.

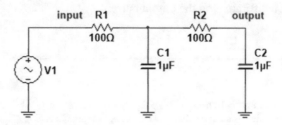

Fig. 2.348 Schematic for Exercise 7

8. Use LTspice to simulate the Wien bridge oscillator circuit shown in Fig. 2.349. Note that oscillation starts when Ra is less than 20 kΩ (R1 = Ra + Rb = 50 kΩ).

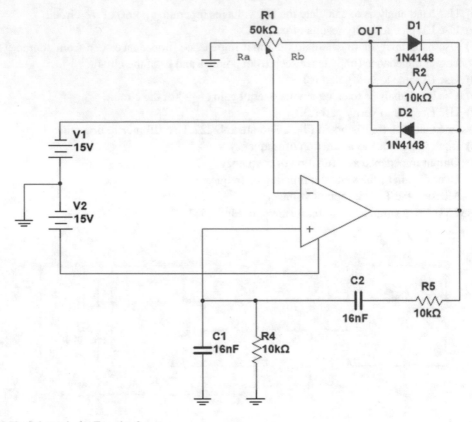

Fig. 2.349 Schematic for Exercise 8

9. Select a common source amplifier circuit from your electronics course textbook and simulate it in LTspice. Draw the frequency response of the selected circuit.
10. Select a current mirror circuit from your electronics course textbook and simulate it in LTspice. Study the effect of temperature on the output current.

References

1. Razavi, B.: Fundamentals of microelectronics, 3rd edition, Wiley (2021)
2. Rashid M.H.: Microelectronic Circuits: Analysis and Design, Cengage Learning (2016)
3. Sedra, A., Smith, K., Carusone, T.C., Gaudet, V.: Microelectronic Circuits, 8th edition, Oxford University Press (2019)

Simulation of Digital Circuits with LTspice®

3

3.1 Introduction

In this chapter, you will learn how to analyze digital circuits in LTspice. The theory behind the studied circuits can be found in any standard digital text book [1–4]. Similar to previous chapters, doing some hand calculations for the given circuits and comparing the hand analysis results with LTspice results are recommended.

3.2 Example 1: Simulation of Logic Circuits

In this example, we want to simulate the logic circuit shown in Fig. 3.1. This is the logic diagram of a two input XOR gate. Truth table of the logic circuit shown in Fig. 3.1 is shown in Table 3.1.

Fig. 3.1 Logic diagram of Example 1

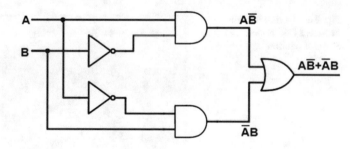

Table 3.1 Truth table of logic circuit shown in Fig. 3.1

A	B	$A\bar{B} + \bar{A}B$
0	0	0
0	1	1
1	0	1
1	1	0

© The Author(s), under exclusive license to Springer Nature Switzerland AG 2023
F. Asadi, *Essential Circuit Analysis using LTspice®*, https://doi.org/10.1007/978-3-031-09853-6_3

The logical gates can be found in the [Digital] section of Select Component Symbol window (Fig. 3.2). After double clicking the [Digital], logic elements appear (Fig. 3.3).

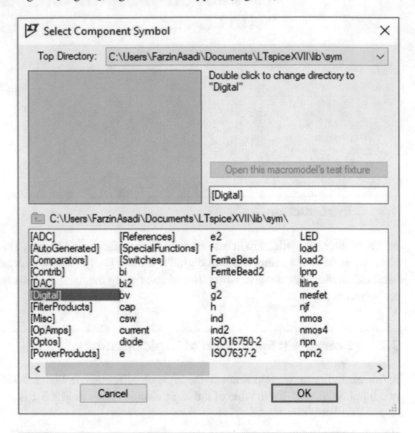

Fig. 3.2 Select Component Symbol window

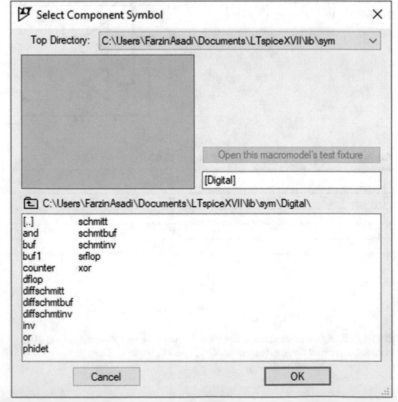

Fig. 3.3 Digital section of Select Component Symbol window

The Special Functions section of LTspice Help is a very good reference for the components shown in Fig. 3.4.

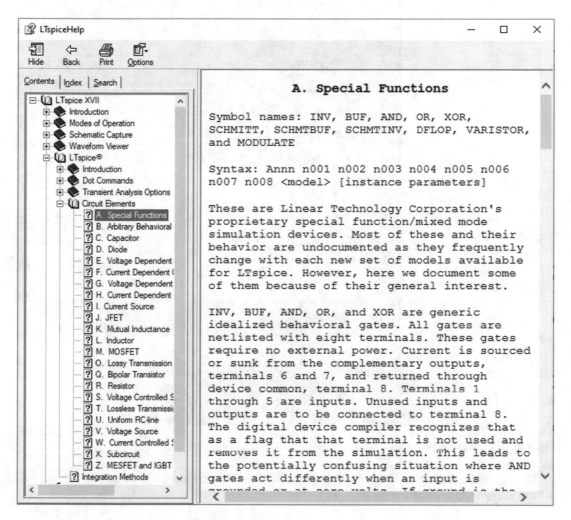

Fig. 3.4 Special Functions sections of LTspice Help

Let's take a closer look to the logic gates. Consider the And gate (Fig. 3.5) shown in Fig. 3.6. It can accept up to five inputs (terminals 1 through 5 are inputs). Note that LTspice ignores the unused inputs. So, when you don't connect anything to an input terminal of a gate, that input is ignored automatically. Terminal 6 is the output of And gate and terminal 7 is the complement of terminal 6. So, with the aid of terminal 7, you can simulate a NAND gate easily. Terminal 8 can be connected to the circuit ground or it can be leaved floating. By default, low input level is [0,0.5 V] and high input level is [0.5 V, 1 V]. Default values of low output and high output are 0 V and 1 V, respectively. However, the user can change these default values. Note that digital components have no supply terminal.

Fig. 3.5 And gate

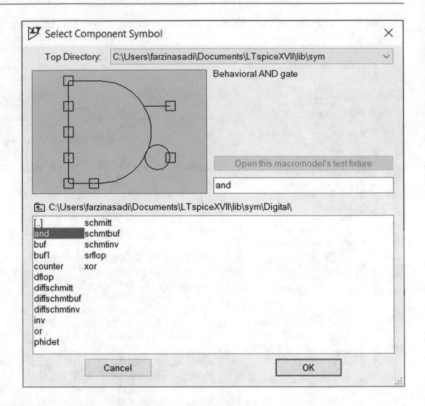

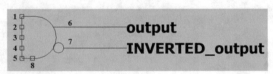

Fig. 3.6 AND gate with five inputs

Let's simulate the logic circuit shown in Fig. 3.1. Draw the schematic shown in Fig. 3.7. A4 and A5 are Behavioral Inverter blocks (Fig. 3.8).

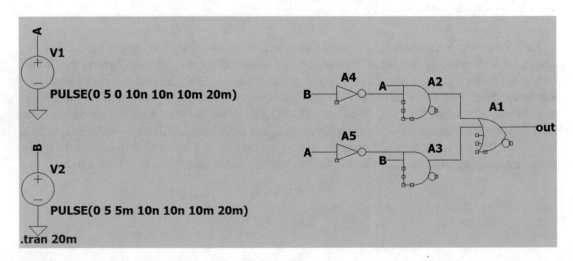

Fig. 3.7 LTspice equivalent of Fig. 3.1

Fig. 3.8 Inverter gate

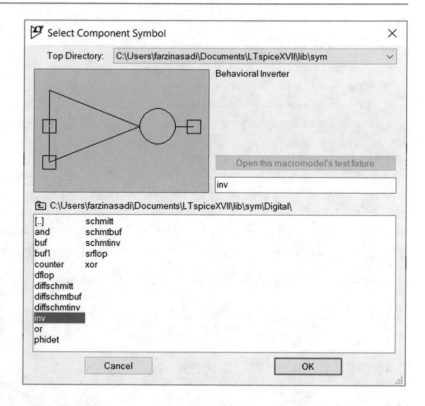

Settings of V1 and V2 are shown in Figs. 3.9 and 3.10, respectively.

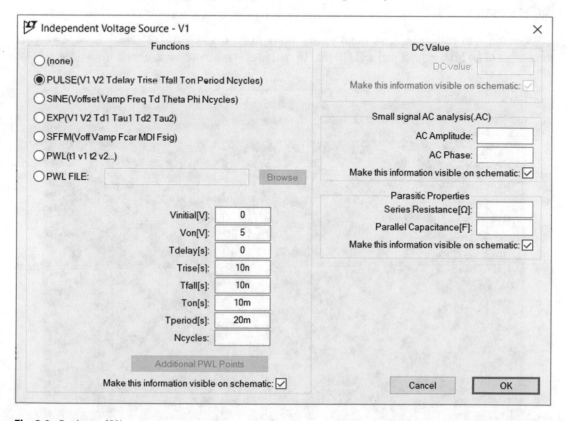

Fig. 3.9 Settings of V1

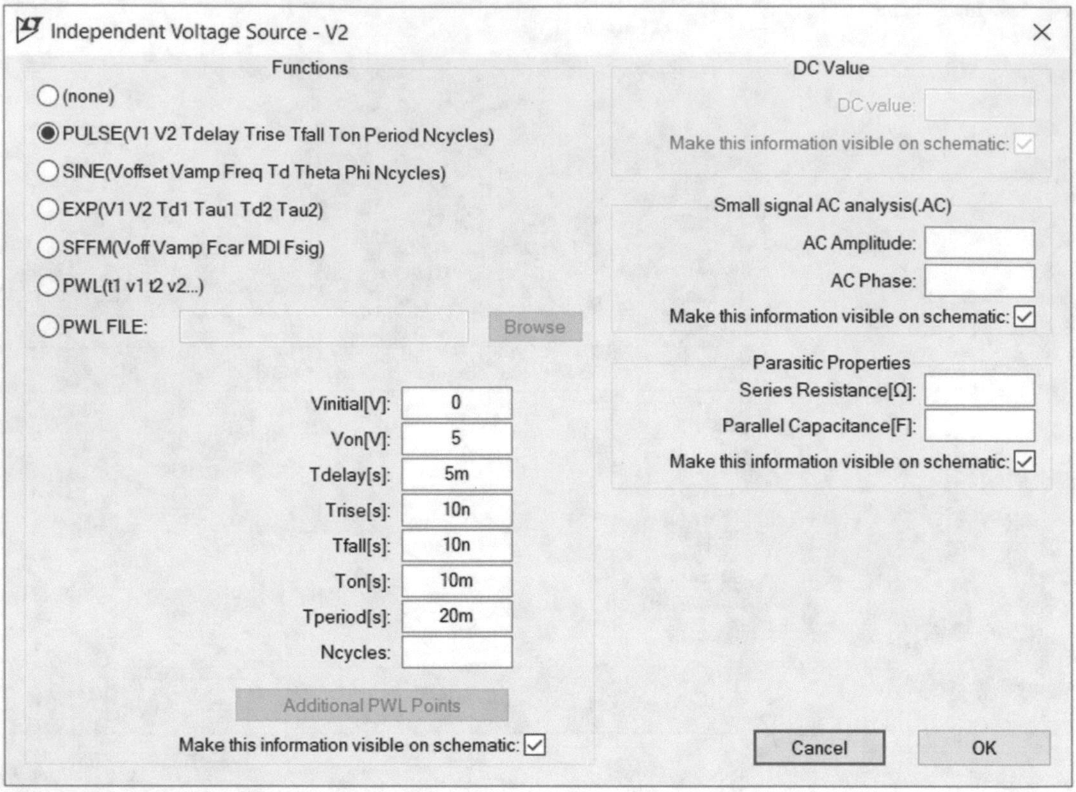

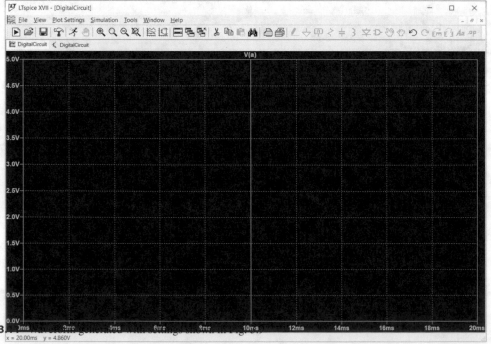

Fig. 3.10 Settings of V2

Waveforms of voltage source V1 and V2 are shown in Figs. 3.11 and 3.12, respectively. We expect the output to be high at [0, 5 ms] and [10 ms, 15 ms] intervals according to circuit truth table.

Fig. 3.11 Waveform generated with settings shown in Fig. 3.9

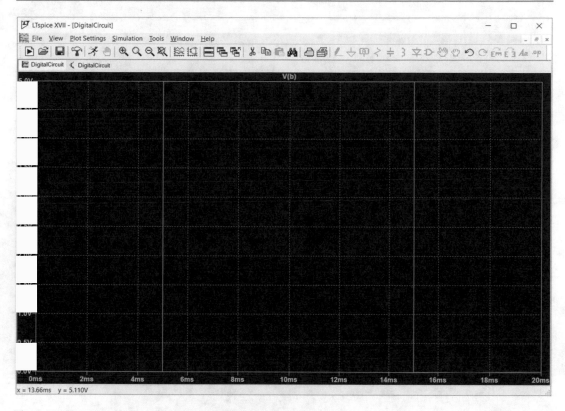

Fig. 3.12 Waveform generated with settings shown in Fig. 3.10

Run the simulation and draw the graph of inputs and outputs (Fig. 3.13). The input is high at [0, 5 ms] and [10 ms, 15 ms] intervals as expected. Note that the low level and high level of output are 0 V and 1 V, respectively.

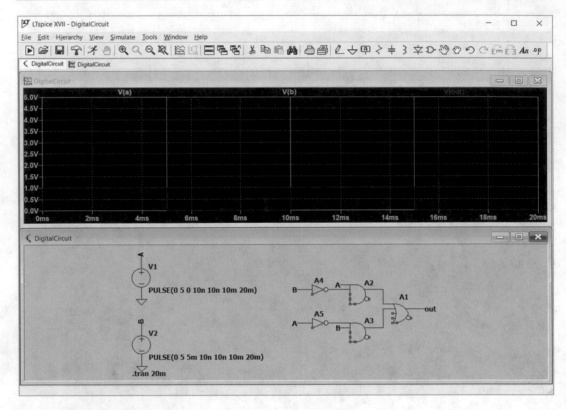

Fig. 3.13 Simulation result

You can change the low and high levels of output easily. Just right click on the gate and enter the desired low and high values to the Value box (Fig. 3.14). After determining new values for low level and high level of output, the $[0, \frac{V_{low}+V_{high}}{2}]$ interval shows the low input, and $[\frac{V_{low}+V_{high}}{2}, V_{high}]$ interval shows the high input. For instance, if you enter "Vlow=0 Vhigh=5" to the Value box (Fig. 3.14), then [0, 2.5 V] interval shows the low input, and [2.5 V, 5 V] shows the high input.

Fig. 3.14 Determining the intervals for low and high signals

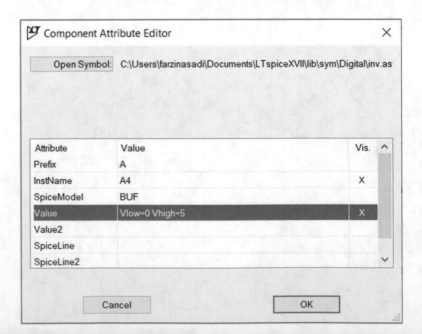

Right click on the logic gates of Fig. 3.13 and enter "Vlow=0 Vhigh=5" to their Value box (Fig. 3.15).

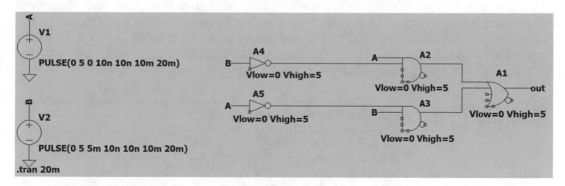

Fig. 3.15 "Vlow=0 Vhigh=5" is entered to the logic gates

Run the schematic shown in Fig. 3.15. Result is shown in Fig. 3.16. According to Fig. 3.16, the low and high levels of output are 0 V and 5 V, respectively.

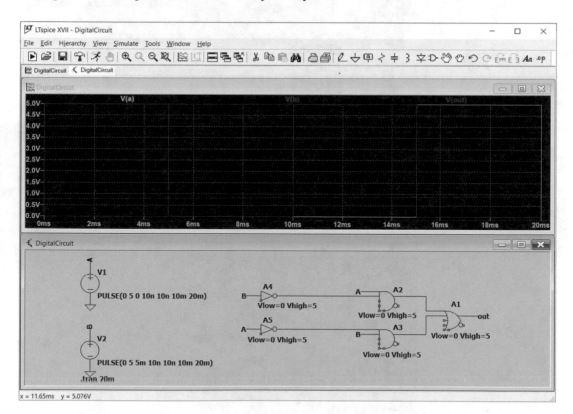

Fig. 3.16   Simulation result

3.3   Example 2: Schmitt-Triggered Buffer Block

In this example, we want to study the Schmitt-Triggered buffer block (Fig. 3.17). The schematic of this example is shown in Fig. 3.18. Waveform of "OutB" is the complement of waveform of "OutA." Settings of the Schmitt-Triggered buffer block and voltage source V1 is shown in Figs. 3.19 and 20, respectively. The settings in Fig. 3.20 simulate a Schmitt trigger with low level of $Vt - Vh = 0.5 - 0.3 = 0.2$ V and high level of $Vt + Vh = 0.5 + 0.3 = 0.8$ V.

Fig. 3.17 Schmitt-
Triggered buffer

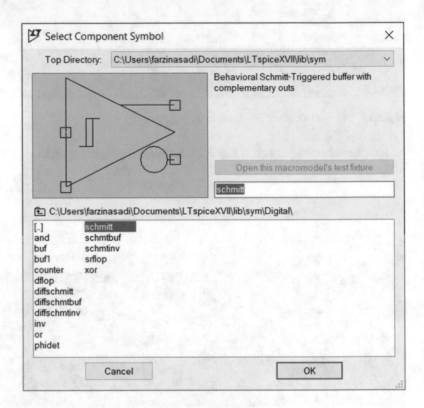

Fig. 3.18 LTspice Schematic for Example 2

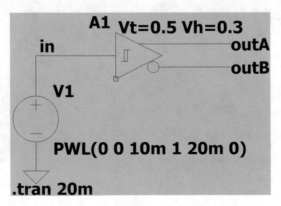

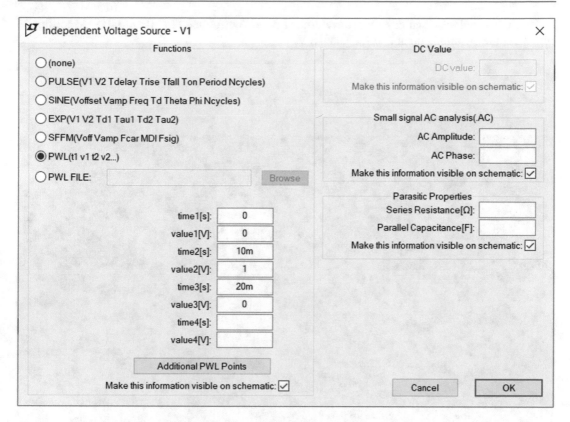

Fig. 3.19 Settings of V1

Fig. 3.20 Settings of
Schmitt trigger

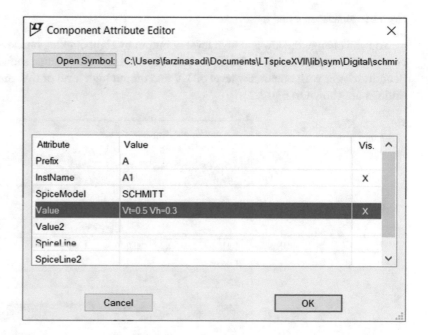

Run the simulation and draw the waveform of node "outA" and "in" (Fig. 3.21). According to Fig. 3.21, the switching point is 0.2 V and 0.8 V.

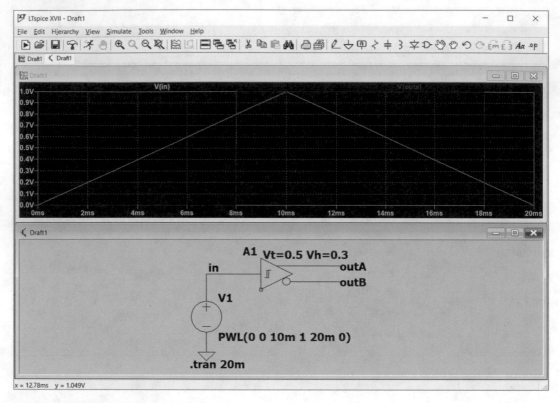

Fig. 3.21   Simulation result

You can change the low and high level of output by right clicking on the Schmitt trigger block and entering the desired levels to the Value box. For instance, the settings shown in Fig. 3.22 simulate a Schmitt trigger with output low level of 1 V and output high level of 6 V. Simulation results for these settings are shown in Fig. 3.23.

Fig. 3.22   New settings of Schmitt trigger

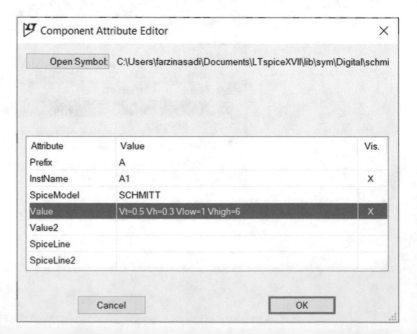

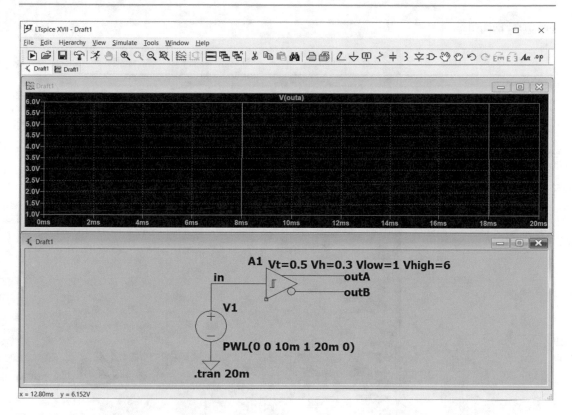

Fig. 3.23 Simulation result

3.4   **Example 3: Flip Flop Blocks**

In this example, we want to use D flip flops to divide the frequency of an input signal by four. D flip flops and RS flip flops are available in LTspice (Figs. 3.24 and 3.25).

Fig. 3.24 D flip flop

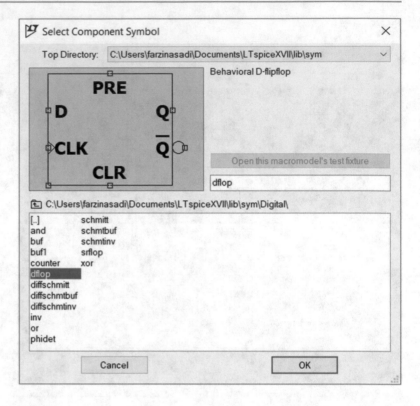

Fig. 3.25 SR flip flop

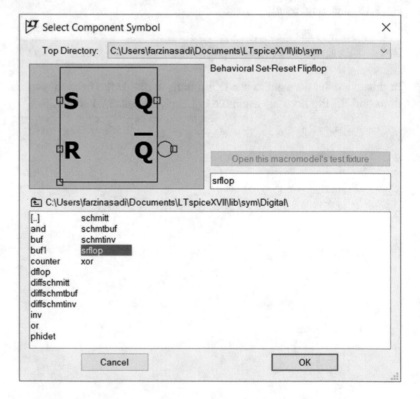

A D flip flop can be used to divide the frequency of an input square wave by factor of two (Fig. 3.26). Two D flip flops can be used to divide the frequency of input signal by four.

Fig. 3.26 Dividing the input frequency by two

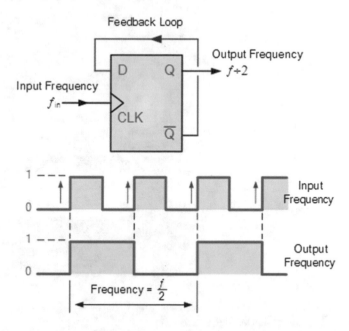

Draw the schematic shown in Fig. 3.27. Settings of flip flops and voltage source V1 are shown in Figs. 3.28 and 3.29, respectively. Voltage source V1 generates a square wave with frequency of 1 kHz and duty cycle of 50%.

Fig. 3.27 Schematic of Example 3

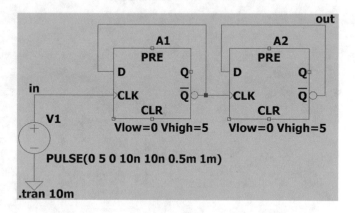

Fig. 3.28 Settings of flip flop

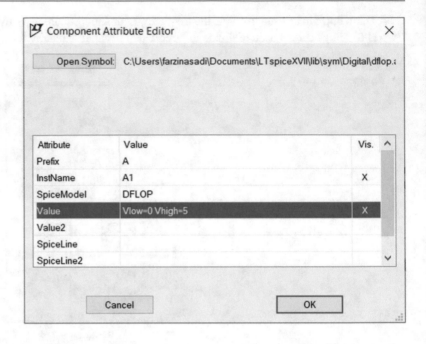

Fig. 3.29 Settings of voltage source V1

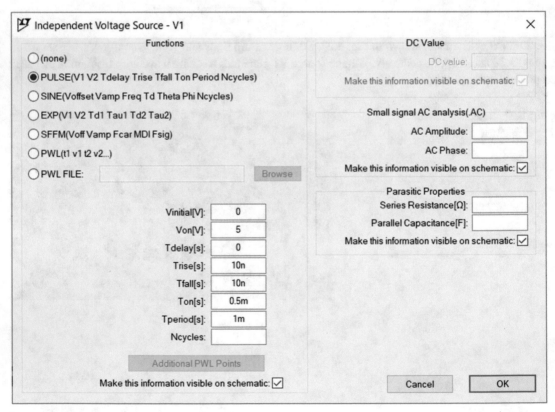

Run the simulation. The waveforms of node "in" and "out" are shown in Figs. 3.30 and 3.31, respectively. According to Fig. 3.32, the frequency of node "out" is 250 Hz. If you cascade n D flip flops, the frequency of output is $\frac{1}{2^n}$ of frequency of input.

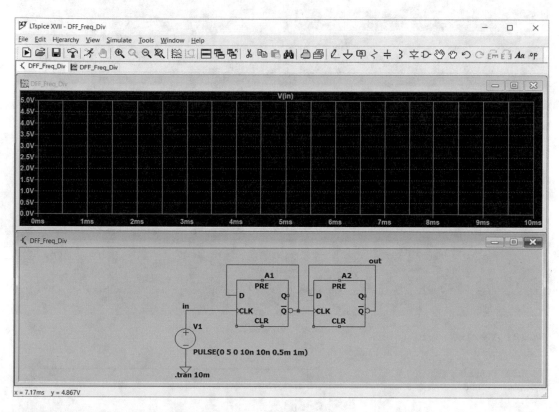

Fig. 3.30 V(in) waveform

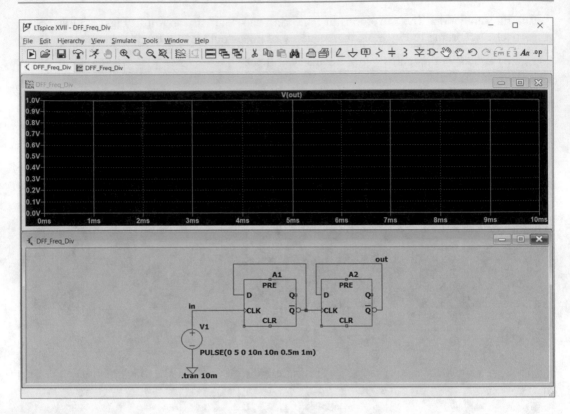

Fig. 3.31 V(out) waveform

Fig. 3.32 Frequency of
node out is around
250 Hz

3.5   Example 4: Counter Block

In the previous example, we saw how D flip flop can be used to divide the frequency of the input signal by four. You can do the frequency division with the aid of Counter block (Fig. 3.33) as well.

Fig. 3.33 Counter
block

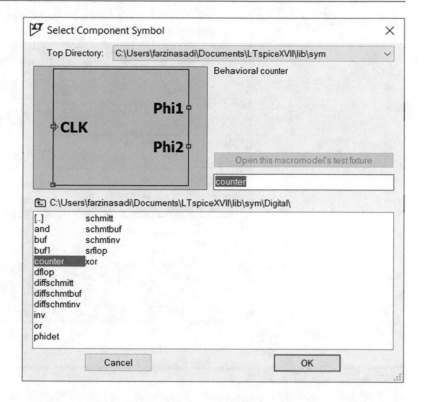

Draw the schematic shown in Fig. 3.34. Settings of Counter block A1 and voltage source V1 are shown in Figs. 3.35 and 3.36, respectively. Cycles = 4 (Fig. 3.35) cause the frequency of the input signal to be divided by four. Voltage source V1 generates a signal with frequency of 1 kHz and duty cycle of 50%.

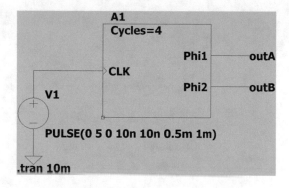

Fig. 3.34 Schematic of Example 4

Fig. 3.35 Settings of
the counter block

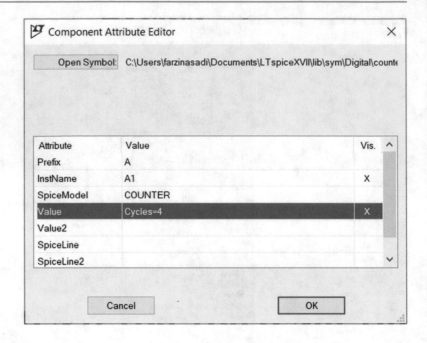

Fig. 3.35 Settings of
the counter block

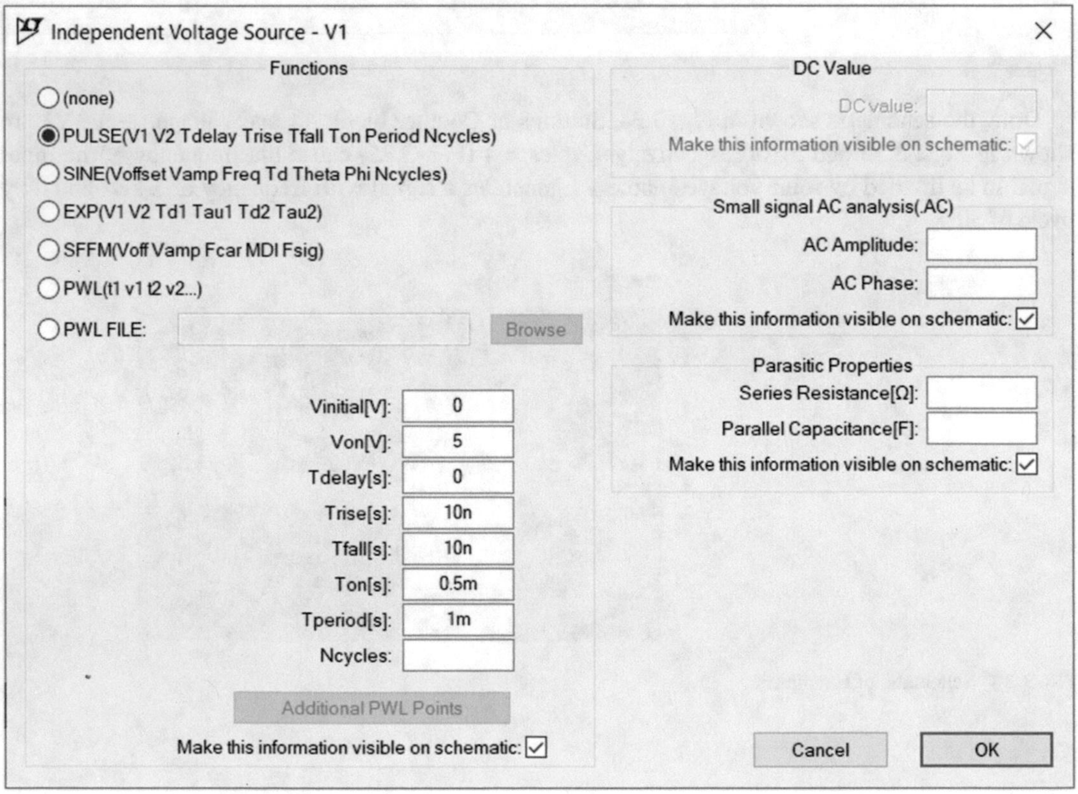

Fig. 3.36 Settings of V1 block

Run the simulation. Voltage of node "outA" is shown in Fig. 3.37. According to Fig. 3.38, the frequency of the waveform shown in Fig. 3.37 is 250 Hz.

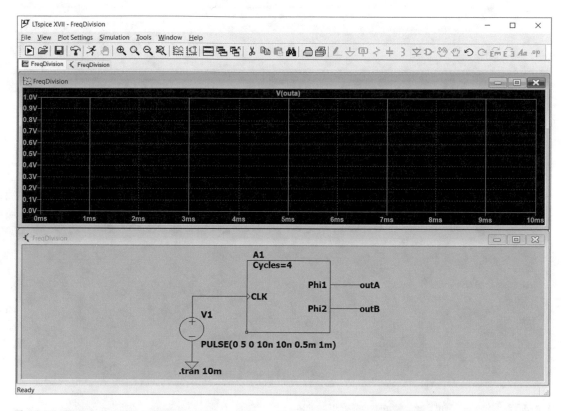

Fig. 3.37 Waveform of V(outA)

Fig. 3.38 Frequency of node outA is around 250 Hz

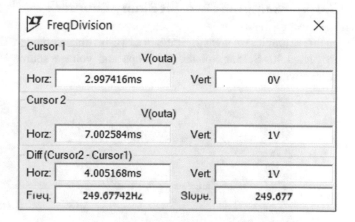

When value of Cycles parameter of the Counter block is an even number, the "outA" and "outB" outputs are complement of each other (Fig. 3.39).

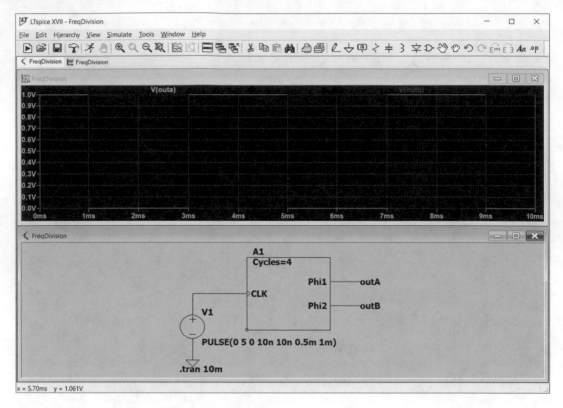

Fig. 3.39 OutA and OutB are complement of each other

3.6   Example 5: Two-Bit Binary Counter

In this example, we want to simulate a two-bit binary counter. The schematic of this example is shown in Fig. 3.40. Settings of the flip flops and voltage source V are shown in Figs. 3.41 and 3.42, respectively.

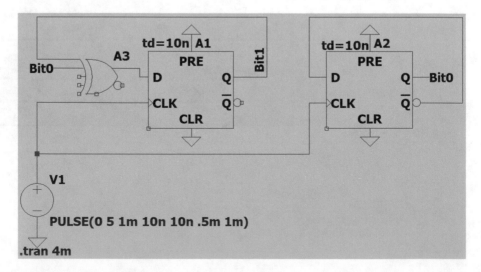

Fig. 3.40 Schematic of Example 5

Fig. 3.41 D flip flop
settings

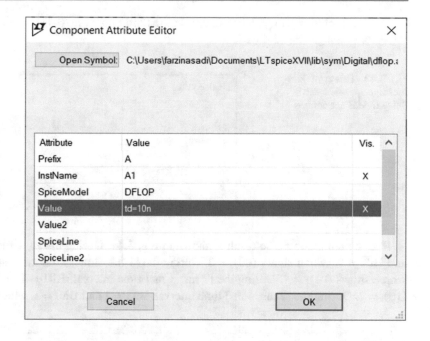

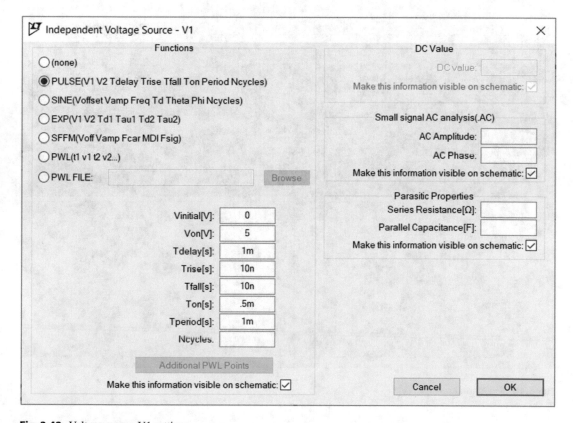

Fig. 3.42 Voltage source V1 settings

Don't forget to enter the delay time (td) of the flip flops (Fig. 3.41). Otherwise you get the following error message (Fig. 3.43).

Fig. 3.43 This error message appears when delay time is not defined

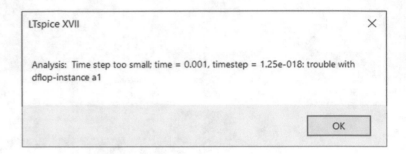

Run the simulation. The result is shown in Fig. 3.44. During the [0, 1 ms] time interval, Bit1 = 0 and Bit0 = 0, which shows $(00)_2 = 0$. During the [1 ms, 2 ms] time interval, Bit1 = 0 and Bit0 = 1, which shows $(01)_2 = 1$. During the [2 ms, 3 ms] time interval, Bit1 = 1 and Bit0 = 0, which shows $(10)_2 = 2$. During the [3 ms, 4 ms] time interval, Bit1 = 1 and Bit0 = 1, which shows $(11)_2 = 3$.

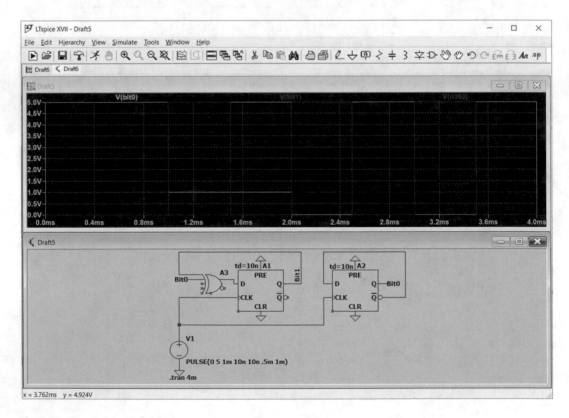

Fig. 3.44 Simulation result

3.7   Exercises

1. Implement the following Boolean function with Multisim.

$$F = (A + B)(\bar{A}B + A\bar{C} + \bar{B}C)$$

2. Simulate a full adder circuit.
3. (a) Simulate an 8×1 multiplexer in LTspice and test it.
    (b) Simulate a 4-bit shift register in LTspice and test it. Use D flip flop.
4. Simulate a 4-bit Johnson counter in LTspice and test it.
5. An state diagram is given in Fig. 3.45.

Fig. 3.45 State diagram
of Exercise 5

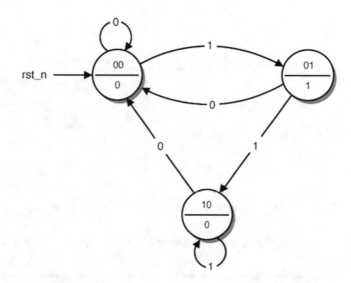

The state table of this state diagram is shown in Table 3.2.

Table 3.2 State table of Fig. 3.45

Current state		Input I	Next state		Outputs
A	B		A_{next}	B_{next}	Y
0	0	0	0	0	0
0	0	1	0	1	0
0	1	0	0	0	1
0	1	1	1	0	1
1	0	0	0	0	0
1	0	1	1	0	0
1	1	0	X	X	X
1	1	1	X	X	X

(a) Use hand analysis to ensure that the circuit in Fig. 3.46 implements the given state diagram.

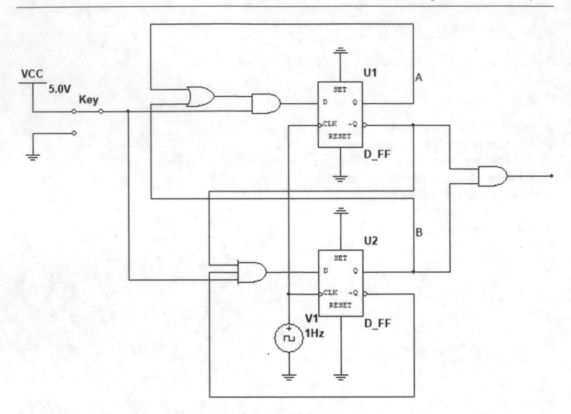

Fig. 3.46 D flip flop realization of state diagram shown in Fig. 3.45

(b) Simulate the circuit shown in Fig. 3.46 in LTspice and ensure that it works similar to Table 3.2.

References

1. Mano, M., Ciletti, M.D: Digitla Design, 6th edition, Pearson (2018)
2. Nelson, V., Nagle, H.T., Irvin, H.T., Carrol, B.D.: Digital Logic Circuit Analysis and Design, Pearson (1995)
3. Floyd, T.: Digital Fundaments, 11th edition, Pearson (2014)
4. Marcovitz, A.B.: Introduction to Logic Design, 3rd edition, Mc-Graw Hill (2009)

Simulation of Power Electronics Circuits with LTspice®

4

4.1 Introduction

In this chapter, you will learn how to analyze power electronics circuits in LTspice. The theory behind the studied circuits can be found in any standard power electronics text book [1–4]. Similar to previous chapters, doing some hand calculations for the given circuits and comparing the hand analysis results with LTspice results are recommended.

4.2 Example 1: Buck Converter (I)

In this example, we want to simulate a buck converter. Draw the schematic shown in Fig. 4.1. An NMOS block (Fig. 4.2) is used to simulate the MOSFET switch. MOSFET switch is controlled with voltage source V2. Settings of the voltage source V2 is shown in Fig. 4.3. The frequency and duty cycle of the pulse applied to the MOSFET is determined by the Tperiod[s] and Ton[s] boxes shown in Fig. 4.3. According to the settings of Fig. 4.3, the switching frequency is $\dfrac{1}{40\mu} = 25\,\mathrm{kHz}$ and duty cycle of the pulse applied to the MOSFET is $\dfrac{24\mu}{40\mu} = 0.6$.

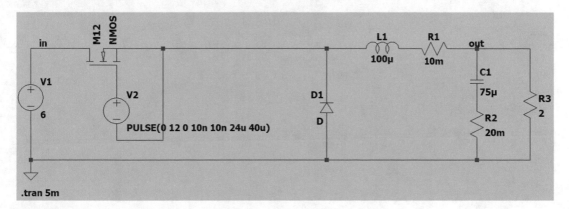

Fig. 4.1 Schematic for Example 1

© The Author(s), under exclusive license to Springer Nature Switzerland AG 2023
F. Asadi, *Essential Circuit Analysis using LTspice®*, https://doi.org/10.1007/978-3-031-09853-6_4

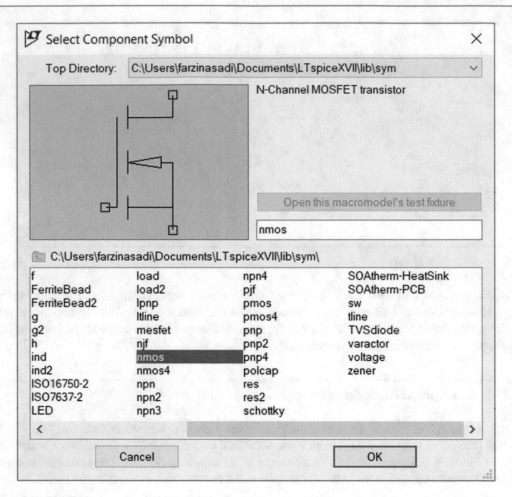

Fig. 4.2 NMOS block

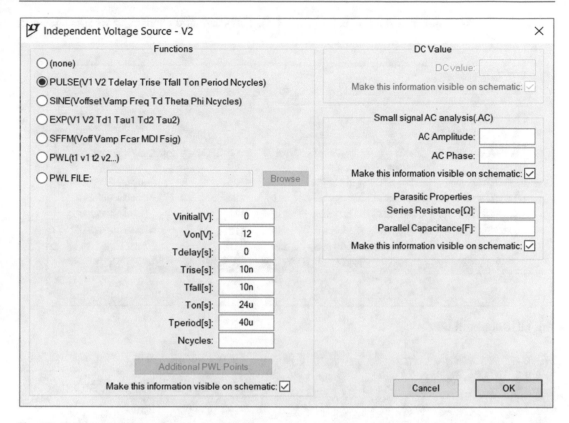

Fig. 4.3 Voltage source V2 settings

Right click on the MOSFET and click the Pick New MOSFET button (Fig. 4.4).

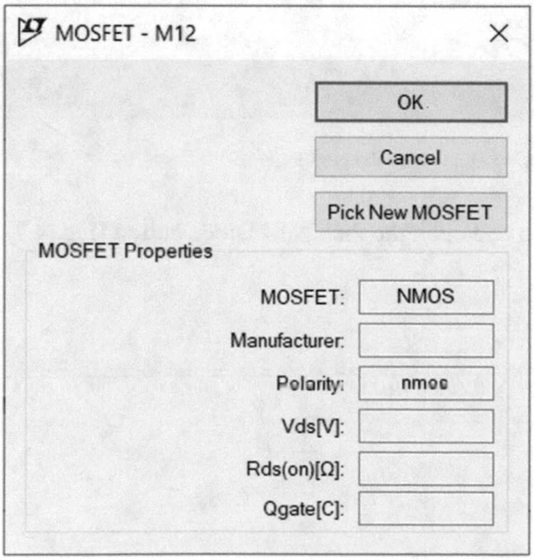

Fig. 4.4 MOSFET window

Select IRFZ44V and click the OK button (Fig. 4.5). The schematic changes to what is shown in Fig. 4.6.

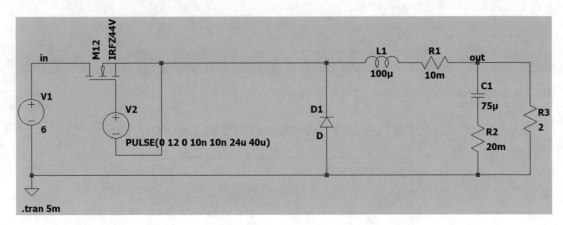

Fig. 4.5 Selection of IRFZ44V

Part No.	Manufacturer	Polarity	Vds[V]	Ron[mΩ]	Gate Chg[nC]	SPICE Model
IRFP250N	International Rectifier	N-chan	200.0	75.0	123	.model IRFP250N \
IRFR120Z	International Rectifier	N-chan	200.0	150.0	7	.model IRFR120Z \
IRFR2307Z	International Rectifier	N-chan	75.0	12.8	50	.model IRFR2307Z
IRFR2607Z	International Rectifier	N-chan	75.0	17.6	34	.model IRFR2607Z
IRFR2905Z	International Rectifier	N-chan	55.0	11.1	29	.model IRFR2905Z
IRFZ44N	International Rectifier	N-chan	55.0	13.9	63	.model IRFZ44N VI
IRFZ44V	International Rectifier	N-chan	55.0	13.9	67	.model IRFZ44V VI
IRFZ46N	International Rectifier	N-chan	55.0	12.8	72	.model IRFZ46N VI

Fig. 4.6 Schematic after applying changes to the components

Right click on the diode and click the Pick New Diode button (Fig. 4.7).

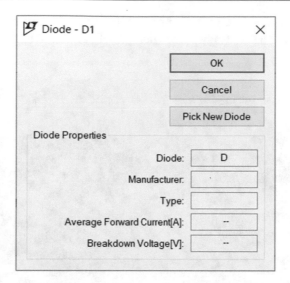

Fig. 4.7 Diode window

Select the 1N5818 and click the OK button (Fig. 4.8). The schematic changes to what is shown in Fig. 4.9. The schematic is ready to do the simulation.

Part No.	Mfg.	type	Vbrkdn[V]	Iave[A]	SPICE Model
1N914	OnSemi	silicon	75.0	0.20	.model 1N914 D(Is=2.52n Rs
1N4148	OnSemi	silicon	75.0	0.20	.model 1N4148 D(Is=2.52n F
MMSD4148	Onsemi	silicon	100.0	0.20	.model MMSD4148 D(Is=2.5
1N5817	OnSemi	Schottky	20.0	1.00	.model 1N5817 D(Is=31.7u F
1N5818	OnSemi	Schottky	30.0	1.00	.model 1N5818 D(Is=31.7u F
1N5819	OnSemi	Schottky	40.0	1.00	.model 1N5819 D(Is=31.7u F
BAT54	Vishay	Schottky	30.0	0.30	.model BAT54 D(Is=.1u Rs=.

Fig. 4.8 Select Diode window

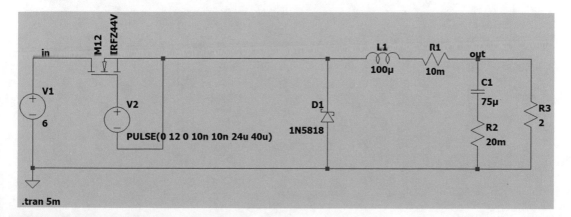

Fig. 4.9 Schematic after applying changes to the components

Run the simulation and draw the voltage of node "out." The voltage of node "out" is composed of two regions: Transient region and steady-state region (Fig. 4.10). The transient is finished after 1.5 ms.

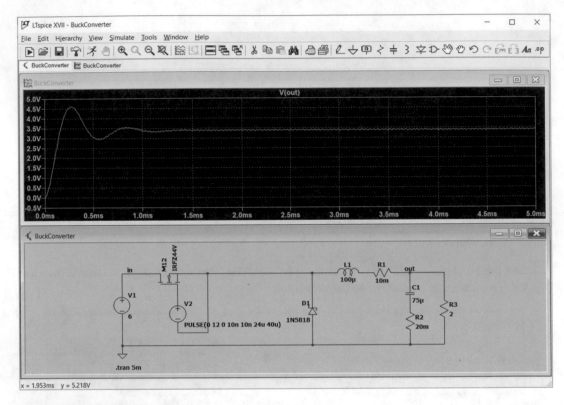

Fig. 4.10 Graph of V(out) for [0, 5 ms]

Let's measure the output voltage ripple. Zoom into the steady-state region of the graph (Fig. 4.11).

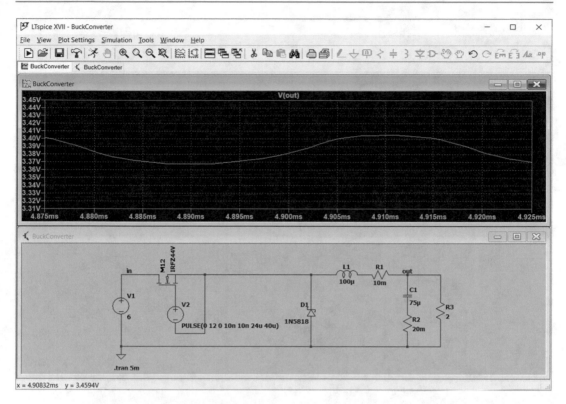

Fig. 4.11 Graph of V(out) for [4.875 ms, 4.925 ms]

Use two cursors to measure the output voltage ripple (Fig. 4.12). According to Fig. 4.13, the output voltage ripple is around 37 mV. You can calculate the average of output voltage by averaging the maximum and minimum of output voltage. The average of output voltage is $\dfrac{3.367 + 3.405}{2} = 3.386\,\mathrm{V}$. Remember that output voltage of an ideal buck converter (i.e., with 100% efficiency) operating in Continuous Conduction Mode (CCM) is given by $V_o = D \times V_{in}$, where D and V_{in} show the duty cycle of the pulse applied to the MOSFET and input voltage, respectively. For the given converter, $V_{in} = 6$ V and $D = 0.6$. So, we expect the average of output voltage to be around 3.6 V. The average output voltage of the simulation is a little bit less than 3.6 V since the efficiency of the converter is not 100%.

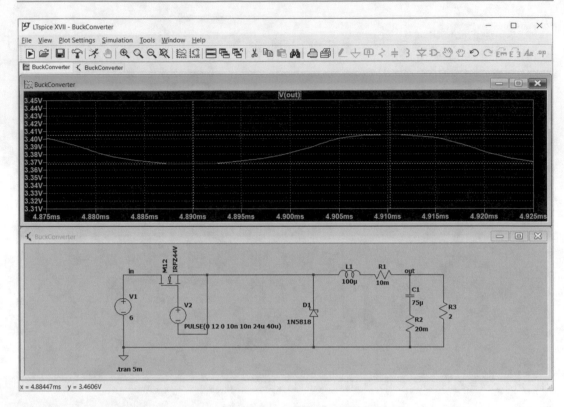

Fig. 4.12   Two cursors are used to measure the minimum and maximum of the graph

Fig. 4.13   Values
measured by cursors

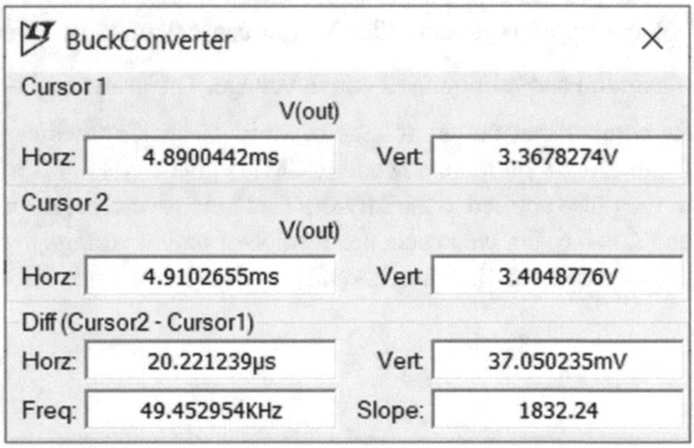

Let's measure the frequency of the ripple. According to Fig. 4.14, the frequency of the output voltage ripple is around the switching frequency of MOSFET (25 kHz).

Fig. 4.14 Frequency of
ripple is around 25 kHz

BuckConverter			✕
Cursor 1	V(out)		
Horz: 4.9097619ms		Vert:	3.404799V
Cursor 2	V(out)		
Horz: 4.949709ms		Vert:	3.4051083V
Diff (Cursor2 - Cursor1)			
Horz: 39.94709µs		Vert:	309.3713µV
Freq: 25.033113KHz		Slope:	7.74453

4.3   Example 2: Buck Convert (II)

In the previous example we learned how to simulate a buck converter with the aid of a pulse voltage source. In this example, we want to use a pulse width modulator to produce the control pulses of the MOSFET.

The modulator produces the gate pulses by comparing a reference signal with a high frequency saw tooth (ramp) carrier (Fig. 4.15). When the reference signal is bigger than the carrier signal, output of the comparator is high. When the reference signal is less than the carrier signal, output of the comparator is low. We want to simulate the modulator of Fig. 4.15 in LTspice.

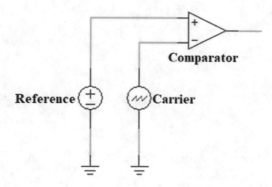

Fig. 4.15   Generation of PWM signal

Draw the schematic shown in Fig. 4.16. Settings of voltage source V2 is shown in Fig. 4.17. The voltage source V2 produces the saw tooth carrier signal. Amplitude of the produced ramp is 1 V. Frequency of voltage source V2 determines the frequency of output pulses. Voltage source V3 determines the duty cycle of output. For instance, when V3 equals to 0.6 V, the duty cycle of the output pulses is 0.6.

Voltage source B1 is an arbitrary behavioral voltage source (Fig. 4.18). It's settings are shown in Fig. 4.19. According to the settings in Fig. 4.19, when the reference signal (voltage source V3 in Fig. 4.16) is bigger than the carrier signal (voltage source V2 in Fig. 4.16), the output pulse has the value which is determined by variable Vpmax. When the reference signal is less than the carrier signal, the output pulse has 0 value. Note that value of Vpmax must be bigger than the gate-source threshold voltage of the MOSFET. Otherwise it cannot turn on the MOSFET.

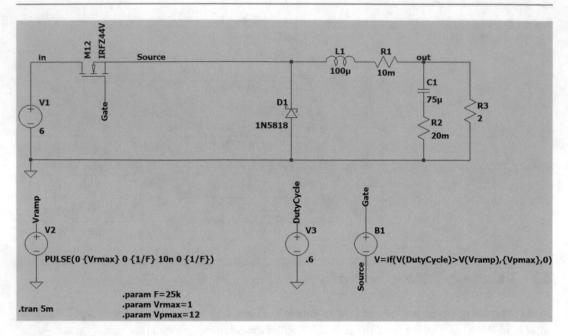

Fig. 4.16 Buck converter

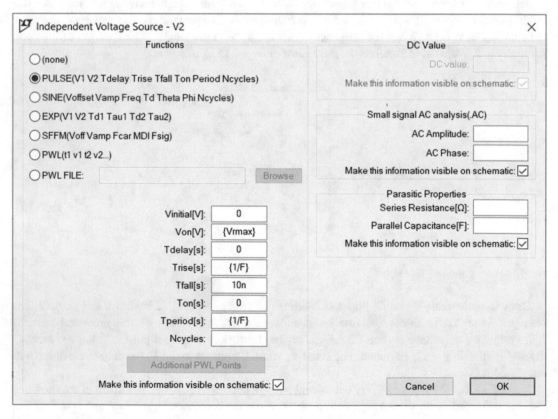

Fig. 4.17 Settings of voltage source V2

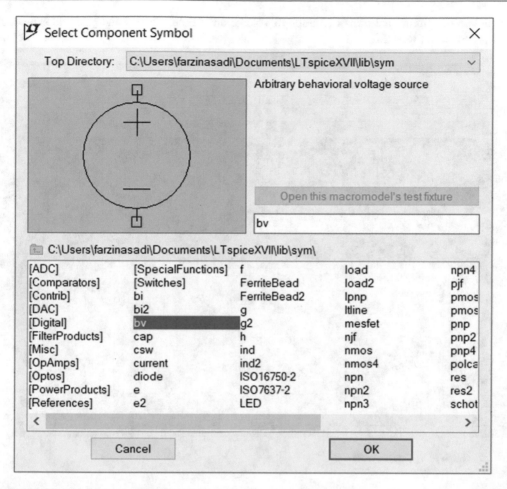

Fig. 4.18 Arbitrary behavioral voltage source block

Fig. 4.19 Settings of
arbitrary behavioral
voltage source block

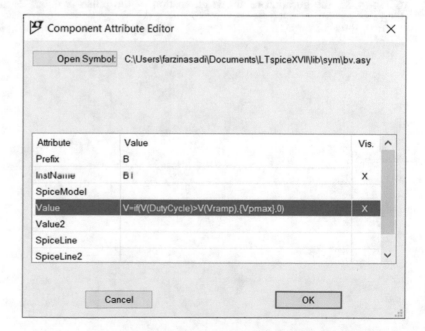

Simulate the circuit and draw the graph of voltage difference between the gate and source. The result is shown in Fig. 4.20.

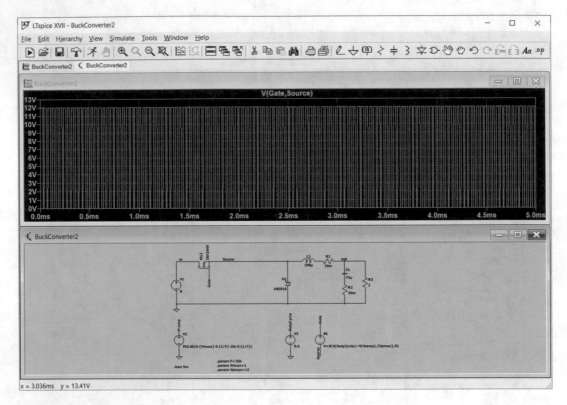

Fig. 4.20 Graph of V(Gate, Source)

Zoom in the graph (Fig. 4.21). Let's measure the duty cycle of the gate-source pulses. According to Fig. 4.22, the duration of the High section of the pulse is about 24 μs. So, the duty cycle is $\frac{24\,\mu s}{40\,\mu s} = 0.6$.

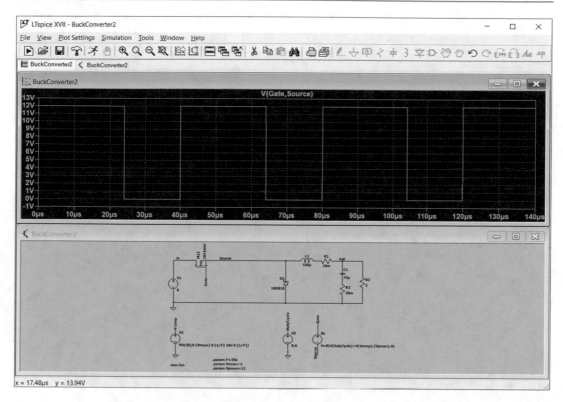

Fig. 4.21 Zoomed graph of V(Gate, Source)

Fig. 4.22 Measurement
of high and low section
of graph

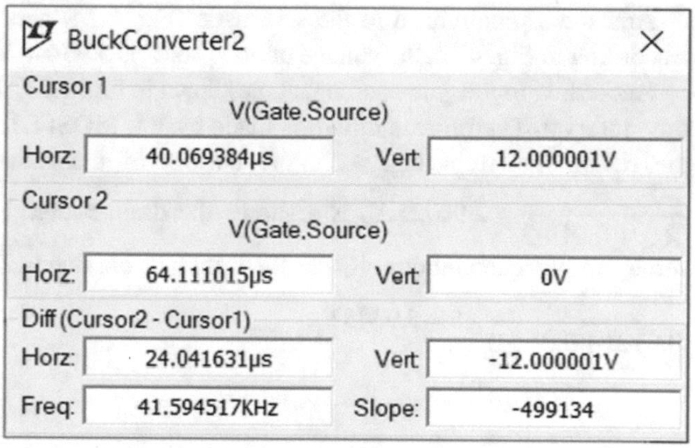

Waveform of node "out" voltage is shown in Fig. 4.23. Note that output voltage is around 6 V at t = 0. Let's see why output voltage starts from around 6 V.

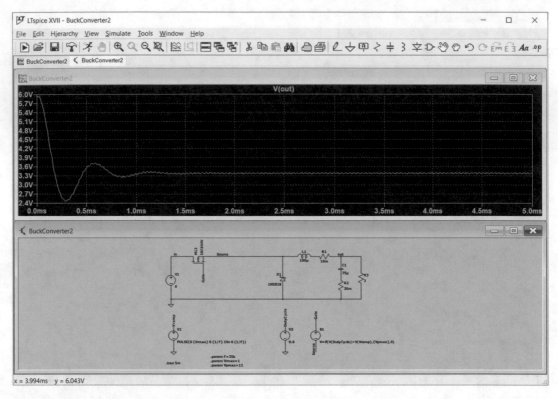

Fig. 4.23 Graph of V(out)

Add the .op command to the schematic (Fig. 4.24) and run it. The result is shown in Fig. 4.25. According to Fig. 4.25, the voltage of node "out" is 5.93446 V and current of inductor L1 is 2.96723 A.

According to the gate-source voltage shown in Fig. 4.21, the MOSFET is closed in the [0, 24.04 µs] time interval. Therefore, a circuit is made by V1, MOSFET, inductor and load resistor. According to the result shown in Fig. 4.25, the current of the inductor is 2.96723 A. So, we can write $\dfrac{6}{R_{DS}+0.01+2}=2.96723\,\text{A}$. R_{DS} shows the drain-source resistance of the MOSFET. After doing some simple calculations, $R_{DS}=12.1\,\text{m}\Omega$ is obtained. Therefore, the voltage of node "out" is $\dfrac{2}{0.0121+0.01+2}\times 6=5.9345\,\text{V}$.

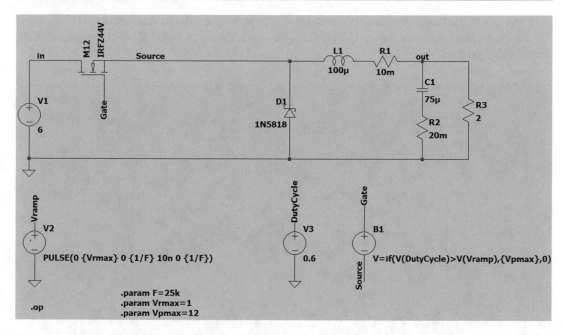

Fig. 4.24   Addition of .op command to the schematic

Fig. 4.25 Simulation
result

```
* C:\Users\farzinasadi\Documents\LTspiceXVII\BuckConverter2.asc        ×

         --- Operating Point ---

V(in):            6                 voltage
V(source):        5.9671            voltage
V(n001):          5.96413           voltage
V(out):           5.93446           voltage
V(n002):          8.90169e-018      voltage
V(gate):          17.9671           voltage
V(vramp):         0                 voltage
V(dutycycle):     0.6               voltage
I(B1):            1.77636e-015      device_current
Id(M12):          2.96726           device_current
Ig(M12):          -3.2e-009         device_current
Is(M12):          -2.96726          device_current
I(C1):            4.45085e-016      device_current
I(D1):            -3.17e-005        device_current
I(L1):            2.96723           device_current
I(R3):            2.96723           device_current
I(R2):            4.45085e-016      device_current
I(R1):            2.96723           device_current
I(V3):            0                 device_current
I(V2):            0                 device_current
I(V1):            -2.96726          device_current
```

You can add the initial condition command to force the output voltage starts from 0 V (Fig. 4.26). After running the schematic shown in Fig. 4.26, the result shown in Fig. 4.27 is obtained. Now, the output voltage starts from zero.

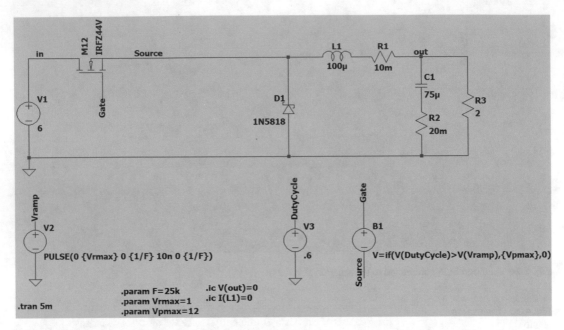

Fig. 4.26 Applying the zero initial conditions to the simulation

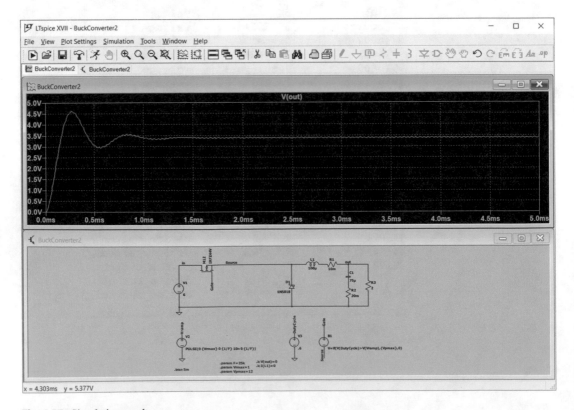

Fig. 4.27 Simulation result

You can force the output to start from zero with the aid of another method: Right click on the Arbitrary behavioral voltage source B1 and change the Value box to what is shown in Fig. 4.28 and click the OK button. Now, the schematic changes to what is shown in Fig. 4.29.

Fig. 4.28 New settings for the arbitrary behavioral voltage source B1

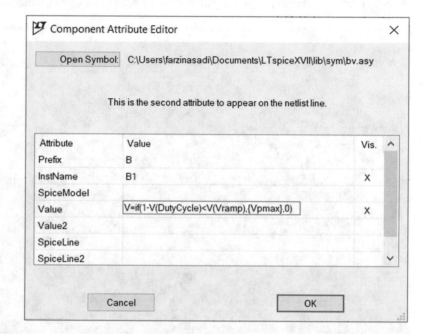

Fig. 4.29 Schematic after applying changes

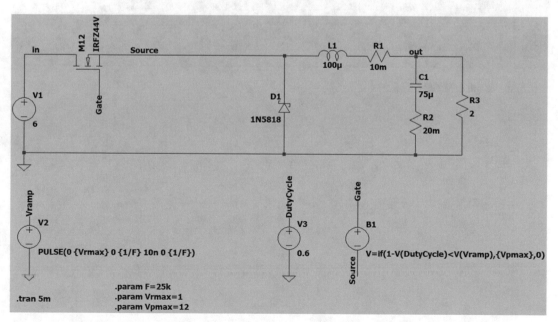

Run the simulation and draw the graph of gate-source voltage (Fig. 4.30). Now, the gate-source voltage starts from 0 V. In other words, the MOSFET is open at t = 0 and voltage source V1 cannot affect the node "out" voltage. If you draw the voltage of node "out," you can see that the voltage starts from 0 V automatically (Fig. 4.31) and there is no need to any .ic command.

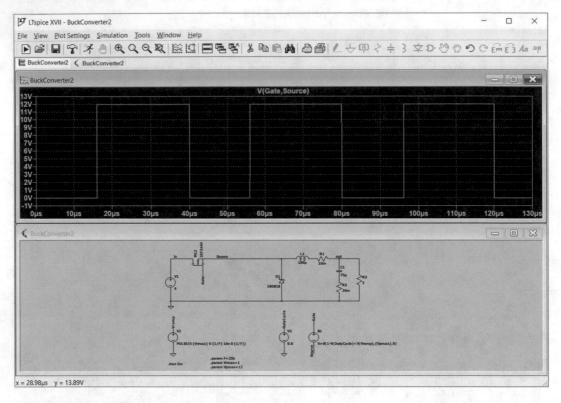

Fig. 4.30 Graph of V(Gate, Source)

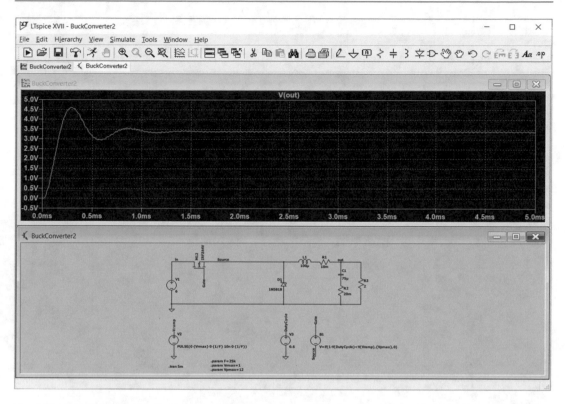

Fig. 4.31   Graph of V(out)

4.4   Example 3: Making New Blocks

In the previous example, we saw how to generate the PWM signal required to control the MOSFET. If you need the pulse width modulator in another schematic, you need to redraw it or copy paste the blocks from previous example to the new schematic.

There is another way to reuse a group of blocks which is better than the previous two methods: Making a new block from them. The generated new block is added to LTspice components and you have access to it in all of the schematics.

Let's convert the pulse width modulator of previous example into a new block called PWM. The PWM block has one input: The desired duty cycle and one output which applies the generated pulses to the circuit.

Let's start. Draw the schematic shown in Fig. 4.32 and save as it with the name PWM.asc.

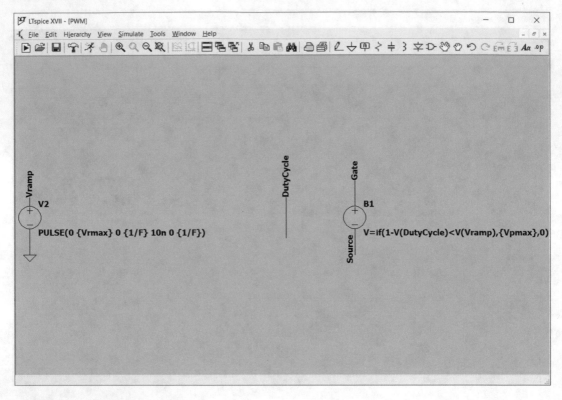

Fig. 4.32 First sketch for Example 3

Click the File> New Symbol (Fig. 4.33). After clicking the File> New Symbol, the window changes to what is shown in Fig. 4.34.

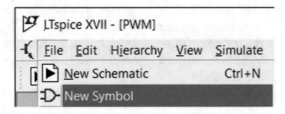

Fig. 4.33 File> New Symbol

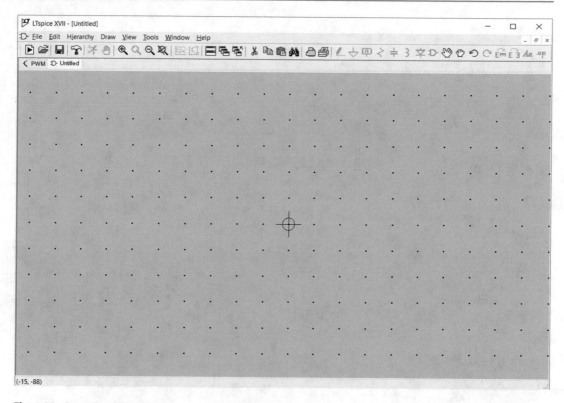

Fig. 4.34 Opened window

Click the Draw> Rect (Fig. 4.35) and draw a rectangle (Fig. 4.36).

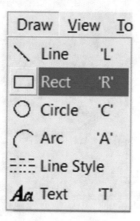

Fig. 4.35 Draw> Rect

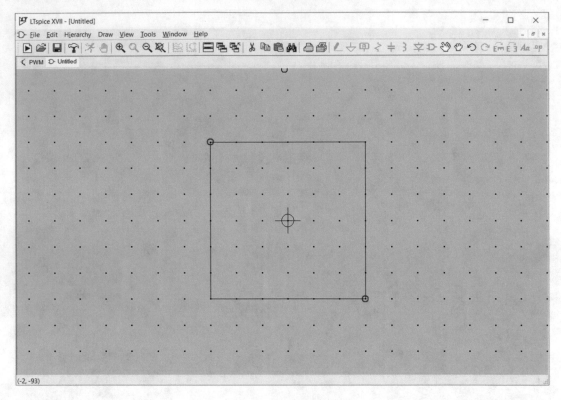

Fig. 4.36 Drawing a rectangle

Click the Draw> Line (Fig. 4.37) and connect three lines to the drawn rectangle (Fig. 4.38).

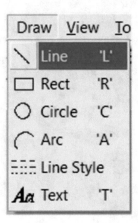

Fig. 4.37 Draw> Line

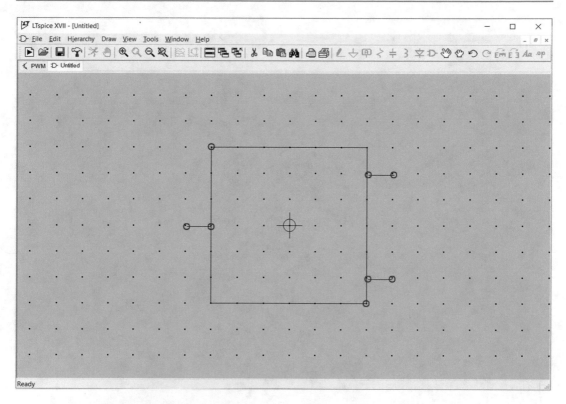

Fig. 4.38 Three line are connected to the drawn rectangle

Use the Draw> Text (Fig. 4.39) to add the "PWM" label to the drawn rectangle (Fig. 4.40).

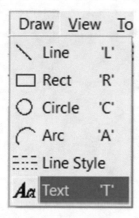

Fig. 4.39 Draw> Text

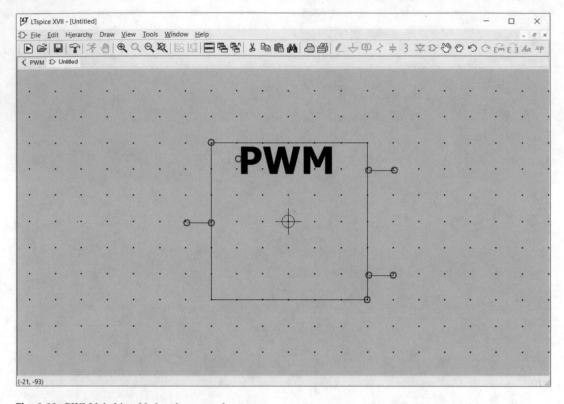

Fig. 4.40 PWM label is added to the rectangle

Click the Window> Tile Vertically (Fig. 4.41) to see the drawn symbol and drawn schematic simultaneously (Fig. 4.42).

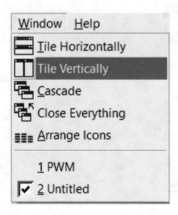

Fig. 4.41 Window> Tile Vertically

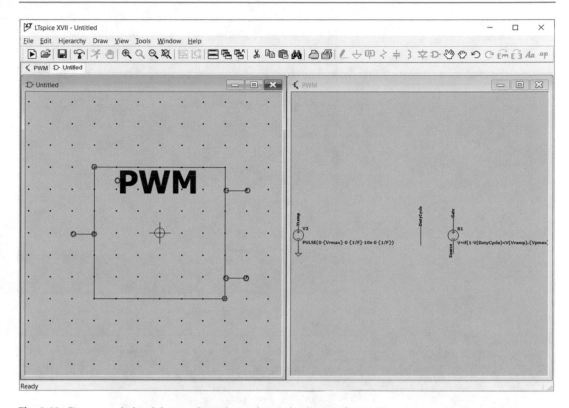

Fig. 4.42   Drawn symbol and drawn schematic are shown simultaneously

Click the Edit> Add Pin/Port (Fig. 4.43). After clicking the Edit> Add Pin/Port, the Pin/Port properties window is opened. Do the settings similar to Fig. 4.44 and click the OK button. After clicking the OK button, a pin is attached to the mouse pointer. Click on the left end of the left line to attach the pin to it (Fig. 4.45). Note that only the labels defined in the schematic can be entered to the Label box of Pin/Port Properties window (Fig. 4.44).

Fig. 4.43   Edit> Add Pin/Port

Fig. 4.44 Pin/Port
Properties window

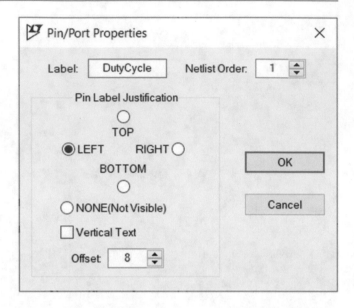

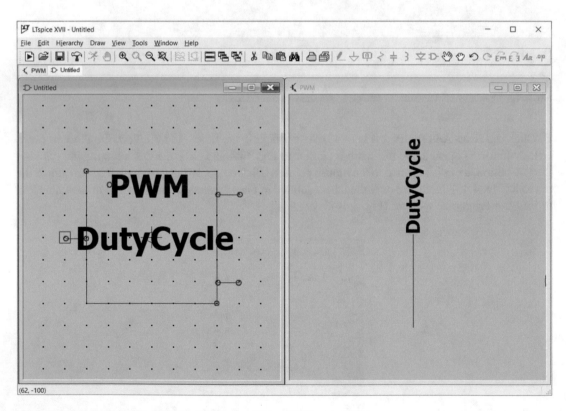

Fig. 4.45 DutyCycle label is added to the left pin of the rectangle

Add two more pins to the right ends of right lines (Fig. 4.46).

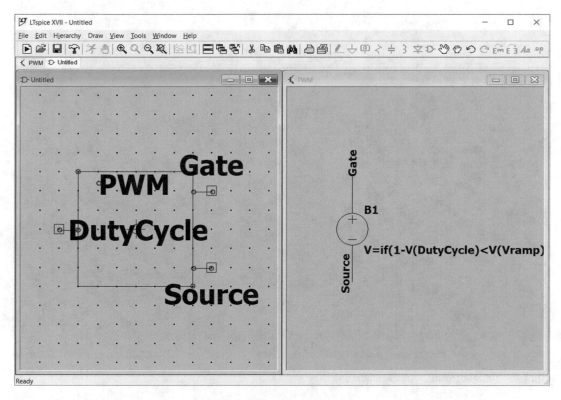

Fig. 4.46   Gate and Source labels are added to the right pins of the rectangle

Click the Edit> Attributes> Attributes Window (Fig. 4.47). This opens the Attributes Window to Add window (Fig. 4.48). Click on the InstName to highlight it. Then click the OK button. Now add the <InstName> to the drawn symbol (Fig. 4.49).

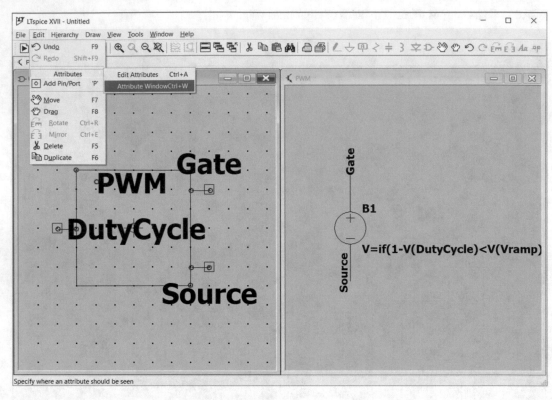

Fig. 4.47 Edit> Attributes> Attributes Window

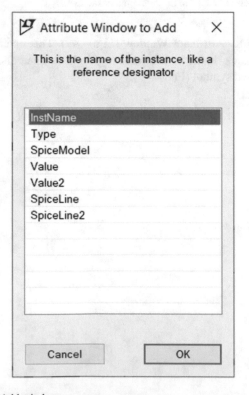

Fig. 4.48 Attribute Window to Add window

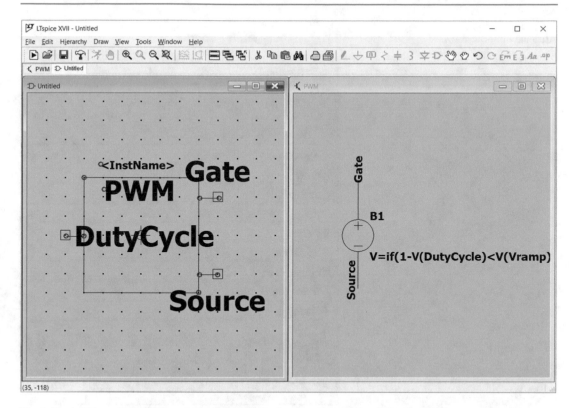

Fig. 4.49 Addition of <InstName> to the drawn rectangle

Click the Edit> Attributes> Edit Attributes (Fig. 4.50) and type F=50k Vrmax=1 Vpmax=12 into the SpiceLine box (Fig. 4.51). This line specifies the default value of carrier frequency (remember that frequency of output signal is the same as the frequency of the carrier), amplitude of carrier signal (amplitude of saw tooth signal) and high level value of output pulses.

Fig. 4.50   Edit>
Attributes> Edit
Attributes

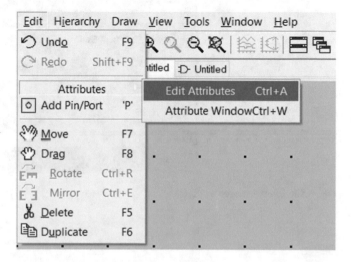

Fig. 4.51 Symbol
Attribute Editor window

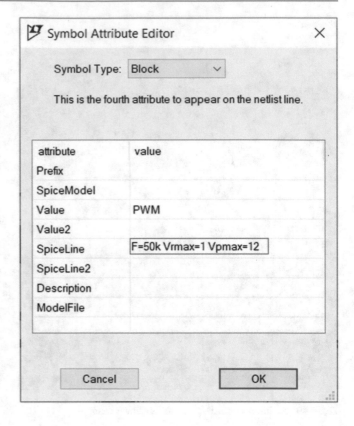

Click the File> Save As (Fig. 4.52) and save the file with the name PWM.asy (Fig. 4.53). Note that the entered name must be the same as the name of schematic file.

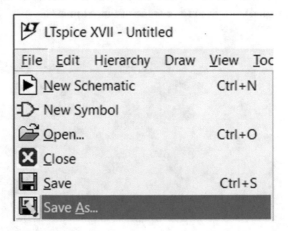

Fig. 4.52 File> Save As

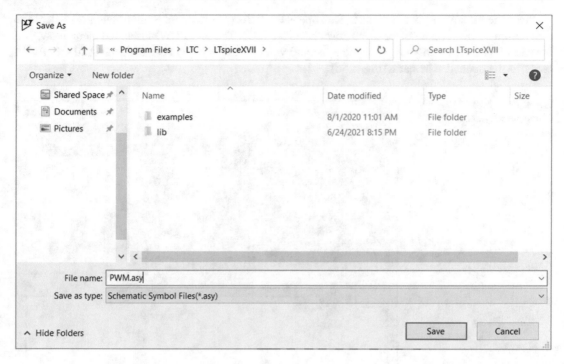

Fig. 4.53 Save As window

Now, open a new schematic (Fig. 4.54).

Fig. 4.54 A new schematic is opened

Press the F2 key of your keyboard. This opens the Select Component Symbol window. Select the first option of the "Top Director" drop down list (Fig. 4.55). The block PWM is added there. Now, you have the PWM block in your component list and you can use it in your simulations. Add one PWM block to the opened schematic (Fig. 4.56).

Fig. 4.55 Select Component Symbol window

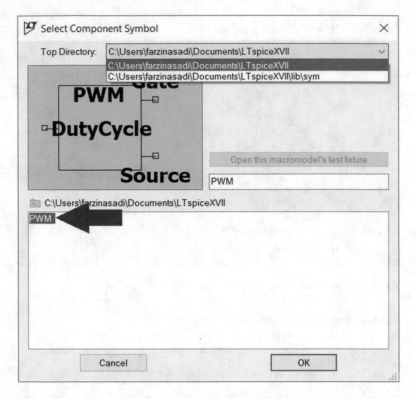

Fig. 4.56 PWM block is added to the schematic

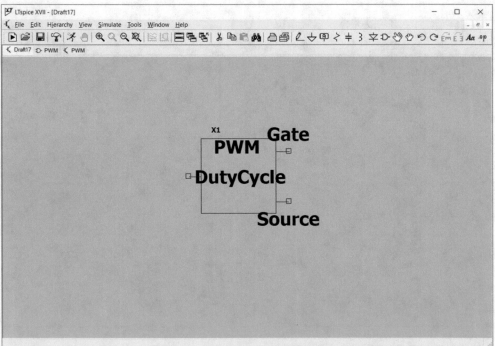

Right click on the PWM block. This opens the window shown in Fig. 4.57 and permits you to enter the values that you want.

Fig. 4.57 Navigate/Edit Schematic Block window

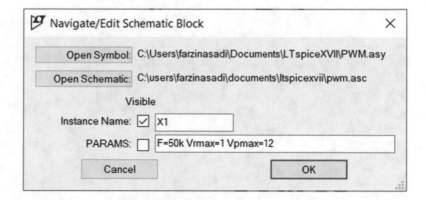

Change the frequency to 25 kHz and click the OK button (Fig. 4.58).

Fig. 4.58 Frequency is changed to 25 kHz

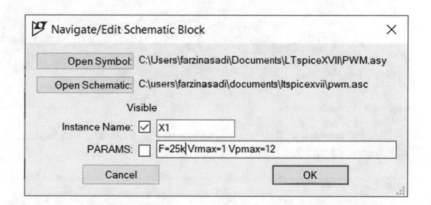

Add the required parts to the schematic and draw the schematic shown in Fig. 4.59.

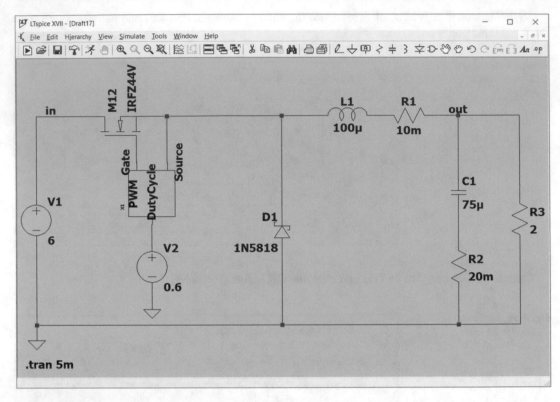

Fig. 4.59 Drawn schematic

Run the simulation. The result is shown in Fig. 4.60. The obtained result is the same as the result of Examples 1 and 2.

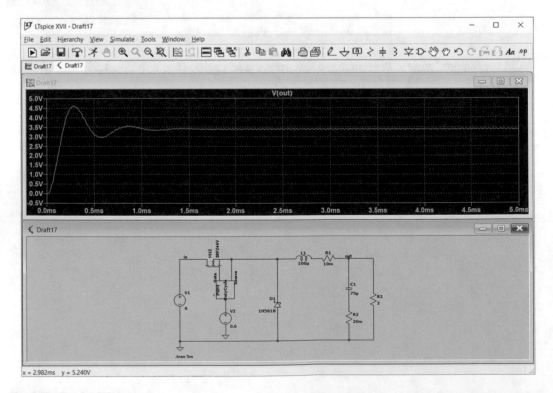

Fig. 4.60 Graph of V(out)

4.5   Example 4: Operating Mode of DC–DC Converter

In this example, we want to determine the operating mode (Continuous Conduction Mode (CCM) or Discontinuous Conduction Mode (DCM)) of the Buck converter studied in Example 1. In order to do that, run the schematic of Example 1 and draw the inductor current (Fig. 4.61).

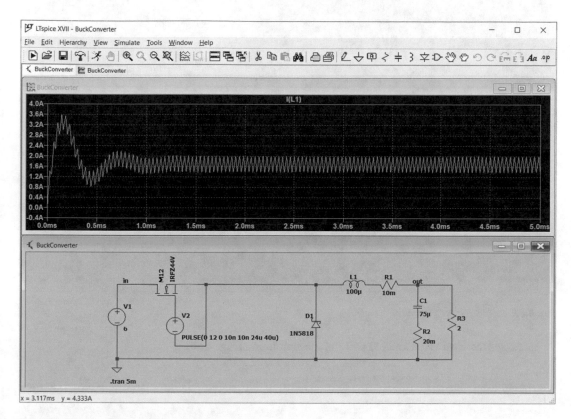

Fig. 4.61   Graph of I(L1)

Zoom into the steady-state region (Fig. 4.62). According to Fig. 4.62, minimum of the inductor current is bigger than zero. So, the converter is operated in CCM.

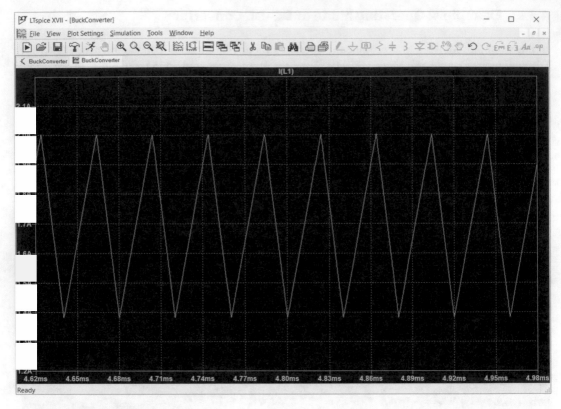

Fig. 4.62 Zoomed inductor current

Increases the load value to 25 Ω and run the simulation (Fig. 4.63).

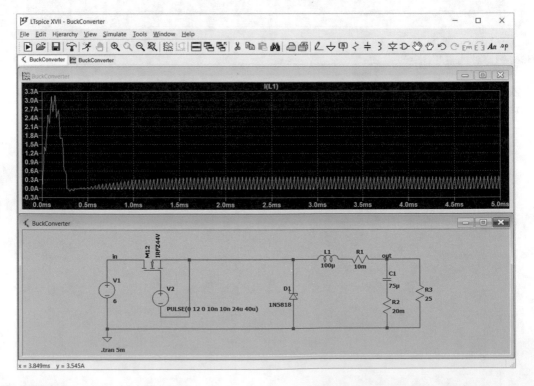

Fig. 4.63 Inductor current for 25 Ω load

Zoom into the steady-state region of the graph (Fig. 4.64). According to Fig. 4.64, the converter is operated in DCM.

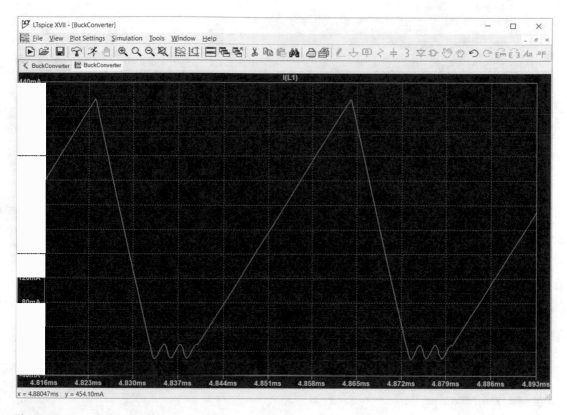

Fig. 4.64 Zoomed inductor current

4.6   **Example 5: Efficiency of the Converter**

In this example, we want to measure the efficiency of the buck converter of Example 1 for different output loads. Change the schematic of Example 1 to what is shown in Fig. 4.65. The .step param Iload 1 3 0.1 command changes the value of Iload from 1 A to 3 A with 0.1 A steps.

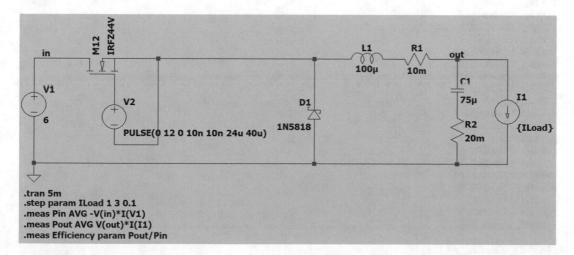

Fig. 4.65 Schematic of Example 5

Run the simulation (Fig. 4.66).

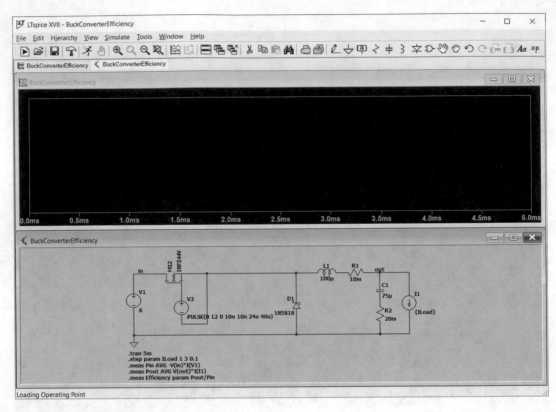

Fig. 4.66 Simulation is run

Press the Ctrl+L. This opens the log file (Fig. 4.67).

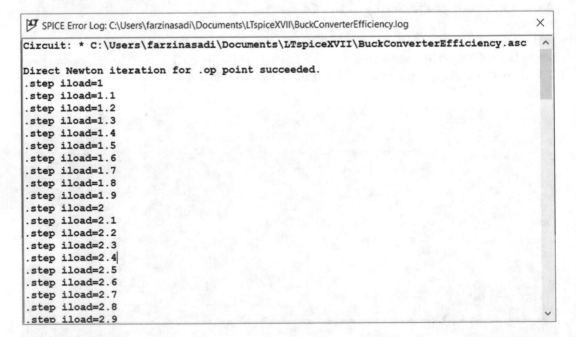

Fig. 4.67 Log file

Scroll down until you see the values of efficiency (Fig. 4.68). For instance, according to Fig. 4.68, efficiency is 92.14% for step 1. According to Fig. 4.67, Iload=1 A for step 1, Iload=1.1 A for step 2, etc.

SPICE Error Log: C:\Users\farzinasadi\Documents\LTspiceXVII\BuckConverterEfficiency.log	✕

Measurement: efficiency

step	pout/pin
1	0.92144
2	0.922677
3	0.923327
4	0.923464
5	0.92334
6	0.922864
7	0.922197
8	0.921472
9	0.92052
10	0.919571
11	0.918395
12	0.917193
13	0.916142
14	0.91503
15	0.91389
16	0.912696
17	0.911424
18	0.91019
19	0.908881
20	0.907556

Fig. 4.68 Log file

Right click on the output log file shown in Fig. 4.68. Then click the Plot .step'ed .meas data (Fig. 4.69). After clicking the Plot .step'ed .meas data, the window shown in Fig. 4.70 appears.

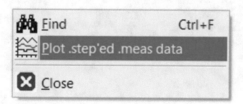

Fig. 4.69 Plot .step'ed .meas data

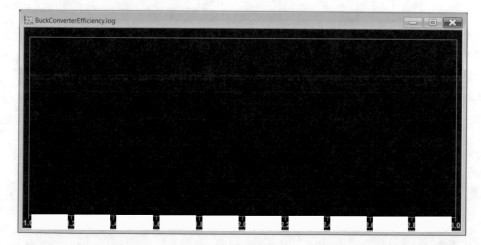

Fig. 4.70 Appeared window

Right click on the window shown in Fig. 4.70 and click the Add Traces (Fig. 4.71).

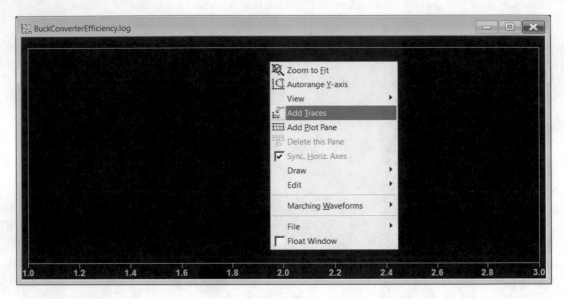

Fig. 4.71 Add Traces

Enter efficiency*100 to the Expression to add box (Fig. 4.72) and click the OK button. After clicking the OK button, the plot shown in Fig. 4.73 appears. This is the graph of efficiency vs. output current.

Fig. 4.72 Add Traces to
Plot window

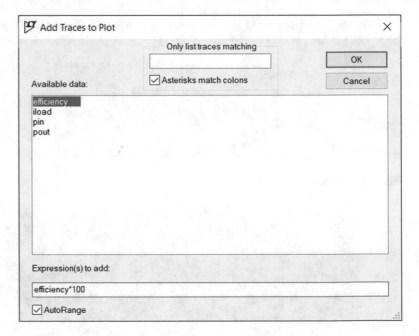

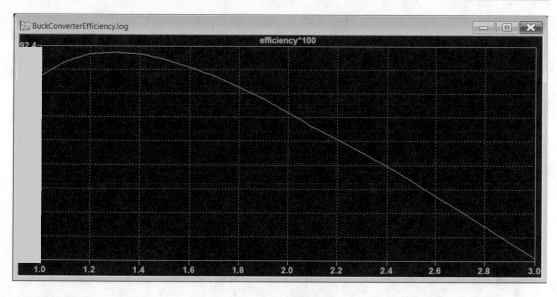

Fig. 4.73 Graph of efficiency

4.7 Example 6: Simulation of Circuits Containing LT IC's

LTspice has the model of a great deal of IC's made by Linear Technology. Let's study an example which contains an IC. Double click the [PowerProducts] section (Fig. 4.74).

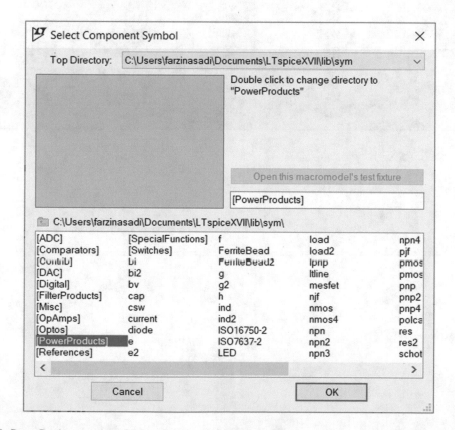

Fig. 4.74 PowerProducts section of Select Component Symbol window

Select the LT3467A (Fig. 4.75). Then click the Open this macromodel's test fixture button. After clicking the Open this macromodel's test fixture, the schematic shown in Fig. 4.76 appears. This is a sample circuit to show the basic operation of the IC.

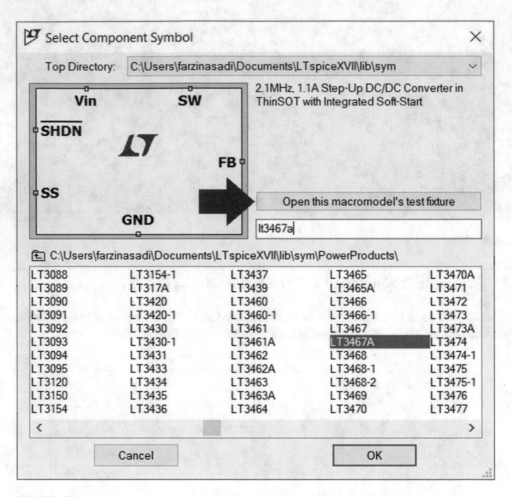

Fig. 4.75 LT3467A

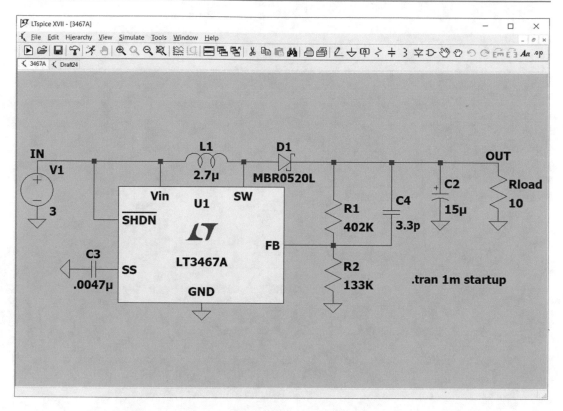

Fig. 4.76 Opened test fixture

There is another way to open the test fixture of an IC: Place the IC on the schematic (Fig. 4.77). Then right click on it. After right clicking on the IC, the menu shown in Fig. 4.78 appears. Click the Open this macromodel's test fixture button to see the IC's test fixture.

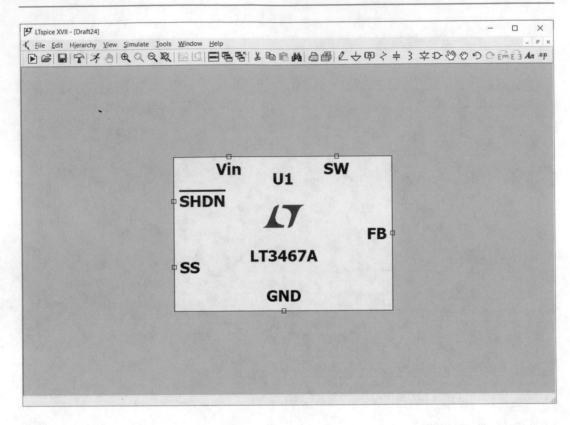

Fig. 4.77 LT3467A is added to the schematic

Fig. 4.78 Click the
Open this macromodel's
test fixture to open the
given sample circuit

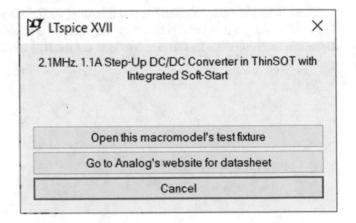

You are not limited to the provided sample circuits (test fixtures). For instance, in Fig. 4.79, a SEPIC converter with coupled inductors is simulated with LT3467 A.

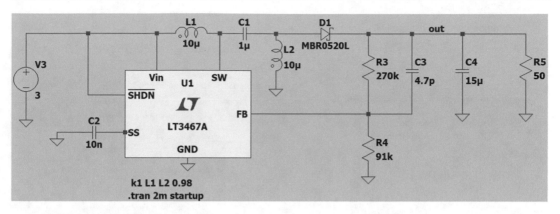

Fig. 4.79   SEPIC converter with LT3467A

Run the simulation and draw the voltage of node "out" (Fig. 4.80). According to Fig. 4.80, the output voltage is about 4.90 V.

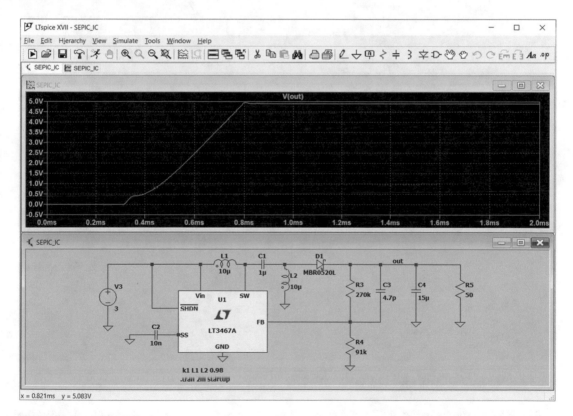

Fig. 4.80   Graph of V(out)

Let's study the effect of input voltage changes on the output voltage of the converter. In order to that, right click on the voltage source V3 and do the settings similar to Fig. 4.81.

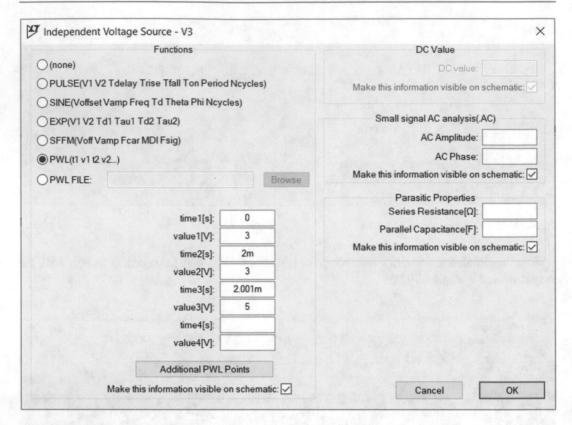

Fig. 4.81 Settings of V3 voltage source

After clicking the OK button in Fig. 4.81, the schematic changes to what is shown in Fig. 4.82.

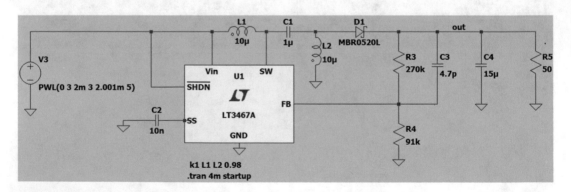

Fig. 4.82 Voltage of V3 jumps from 3 V to 5 V at t = 2 ms

The settings shown in Fig. 4.81 cause a jump from 3 V to 5 V at t = 2 ms in the input voltage of the converter. The input voltage of the converter is shown in Fig. 4.83.

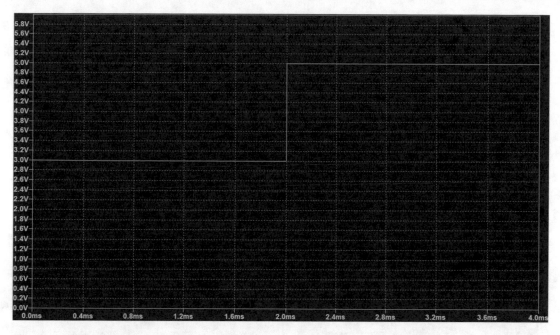

Fig. 4.83 Graph of V3 voltage

Run the simulation and draw the graph of output voltage (Fig. 4.84).

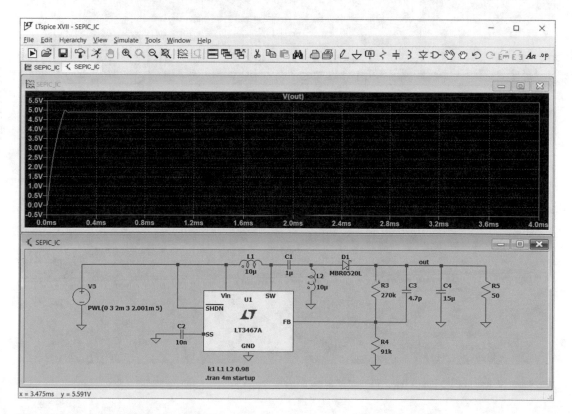

Fig. 4.84 Graph of V(out) for closed loop converter

Zoom in around the time instant that input voltage changed (Fig. 4.85). According to Fig. 4.85, the effect of input voltage change on the output voltage is very negligible. The schematic shown in Fig. 4.79 used feedback control to keep the output constant. That is why we see such a small change in the output.

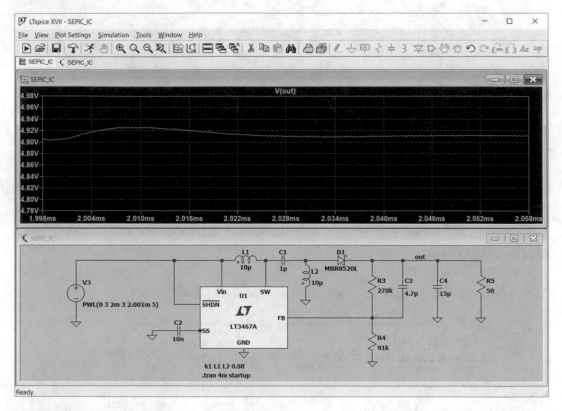

Fig. 4.85   Graph of V(out) around t = 2 ms

Let's apply the same change to an open loop converter and see the results. The schematic shown in Fig. 4.86 is used for this purpose. Settings of voltage source V1 is shown in Fig. 4.87.

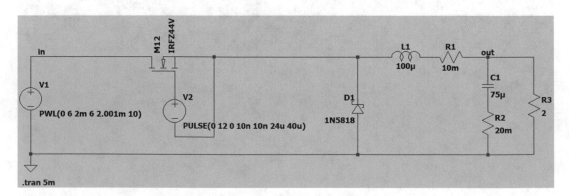

Fig. 4.86   Voltage of V3 jumps from 3 V to 5 V at t = 2 ms

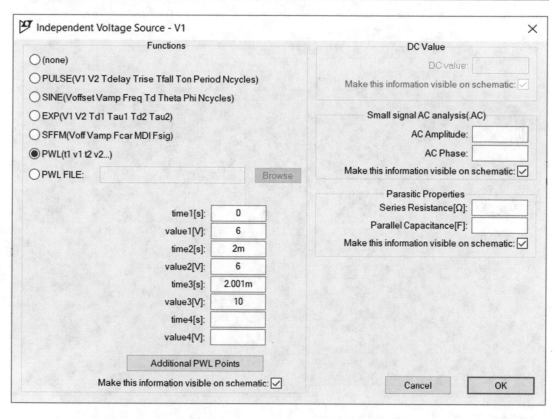

Fig. 4.87 Settings of voltage source V1

Run the simulation and draw the graph of output voltage. The result is shown in Fig. 4.88. Output voltage change is about 2.1 V.

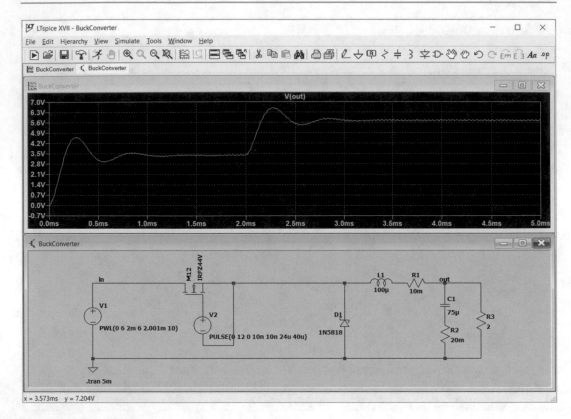

Fig. 4.88   Graph of V(out) for open loop converter

4.8   Example 7: Voltage Regulator Circuit

In this example, we want to simulate a voltage regulator circuit. We use LT1086 IC (Fig. 4.89) which is an adjustable voltage regulator.

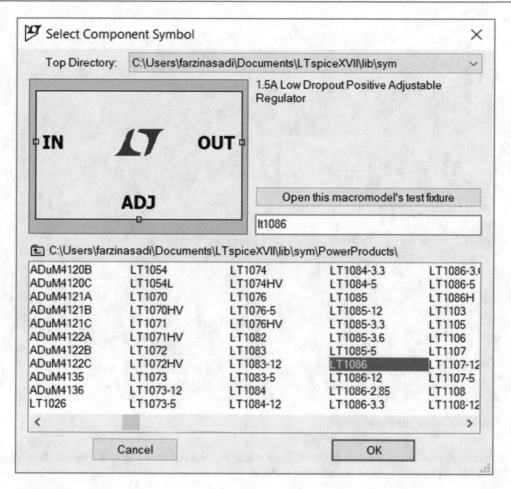

Fig. 4.89 LT1086

Draw the schematic shown in Fig. 4.90. Voltage of node "IN" increased linearly from 0 to 10 V in 1 s.

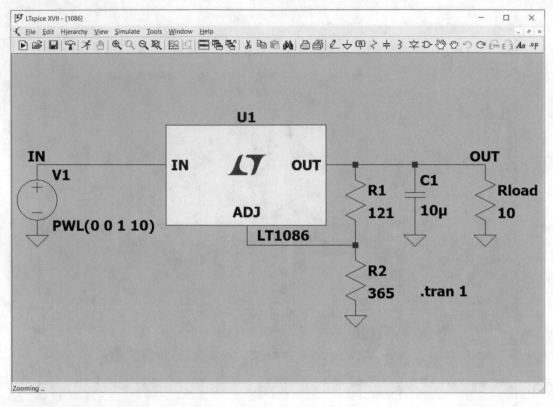

Fig. 4.90 Schematic of Example 7

The formula for output voltage of the IC is shown in Fig. 4.91. According to the calculations shown in Fig. 4.92, the output voltage must be around 5 V.

Output Voltage

The LT1086 develops a 1.25V reference voltage between the output and the adjust terminal (see Figure 1). By placing resistor R1 between these two terminals, a constant current is caused to flow through R1 and down through R2 to set the overall output voltage. Normally this current is chosen to be the specified minimum load current of 10mA. Because I_{ADJ} is very small and constant when compared with the current through R1, it represents a small error and can usually be ignored. For fixed voltage devices R1 and R2 are included in the device.

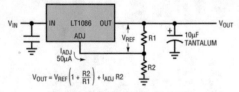

Figure 1. Basic Adjustable Regulator

Fig. 4.91 Formula for output voltage of LT1086

Fig. 4.92 MATLAB
commands

```
Command Window                                          ⊙
   >> Vref=1.25;R1=121;R2=365;Iadj=50e-6;
   >> Vout=Vref*(1+R2/R1)+Iadj*R2

   Vout =

        5.0389

fx >> |
```

Run the simulation. The result shown in Fig. 4.93 is obtained. When input voltage is bigger than
6 V, the output voltage is constant. Value of output voltage is about 5.004 V which is quite close to the
value calculated in Fig. 4.92.

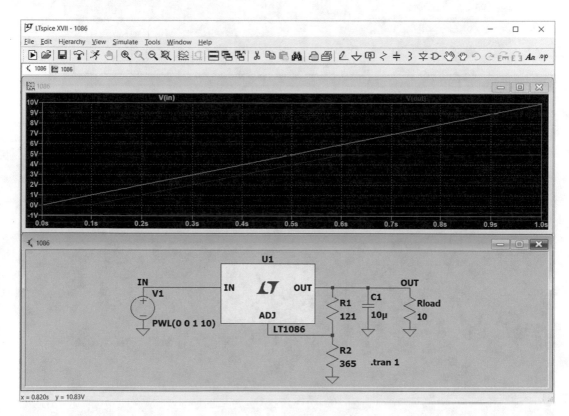

Fig. 4.93 Simulation result

4.9   Example 8: Measurement of Voltage Regulation

Let's measure the voltage regulation of the circuit studied in Example 7. The voltage regulation of a regulator is defined as $\dfrac{V_O}{V_{IN}}$ which ΔV_O and ΔV_{IN} show the output voltage changes and input voltage changes, respectively. Lower value of voltage regulation is better since it shows less output changes vs. the input changes.

Change the input voltage source V1 into a DC voltage source (Fig. 4.94).

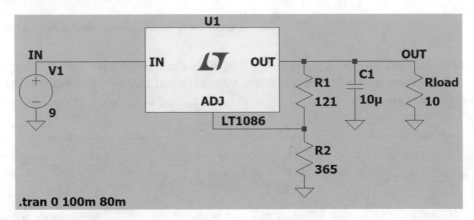

Fig. 4.94 Schematic of Example 8

Run the simulation (Fig. 4.95). According to Fig. 4.96, the output voltage is 5.0037 V for 9 V input.

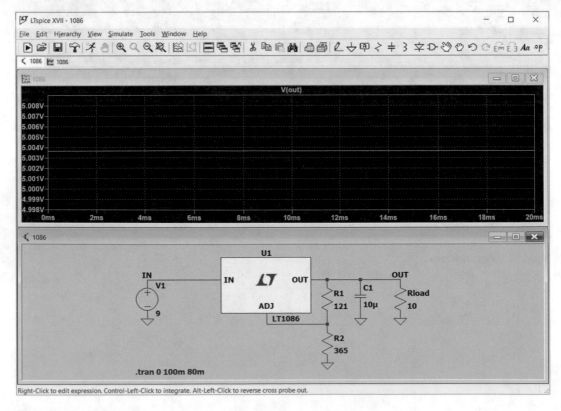

Fig. 4.95 Graph of V(out)

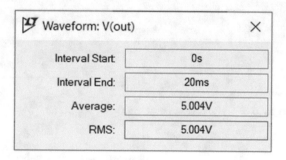

Fig. 4.96 Average and RMS values of V(out) for [0, 20 ms] interval

Change the voltage of V1 into 10 V (Fig. 4.97) and run the simulation. According to Fig. 4.98, the output voltage changes to 5.004 V.

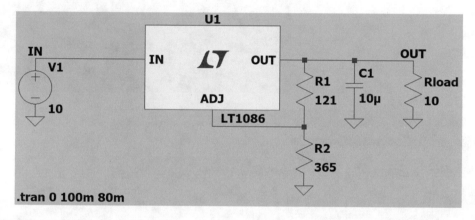

Fig. 4.97 V1 value is changed to 10 V

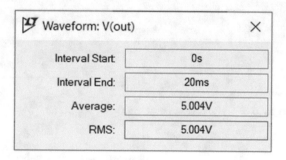

Fig. 4.98 Average and RMS values of V(out) for [0, 20 ms] interval

The voltage regulation of the circuit is $\dfrac{5.004 - 5.0037}{10 - 9} = 0.0003 \ \% \text{ or } 0.03\%.$

4.10 Example 9: Dimmer Circuit

In this example, we want to simulate a dimmer circuit. A dimmer circuit permits you to control the RMS of the voltage which is applied to the load. You can use the dimmer circuit to control the intensity of incandescent light bulbs and small universal motors. The schematic of this example is shown in Fig. 4.99. Diac and triac can be found in the [Misc] section of Select Component Symbol window (Figs. 4.100, 4.101, and 4.102).

Fig. 4.99 Schematic of Example 9

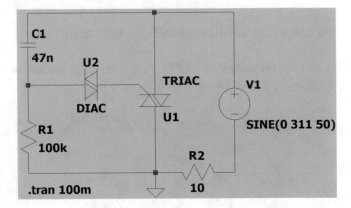

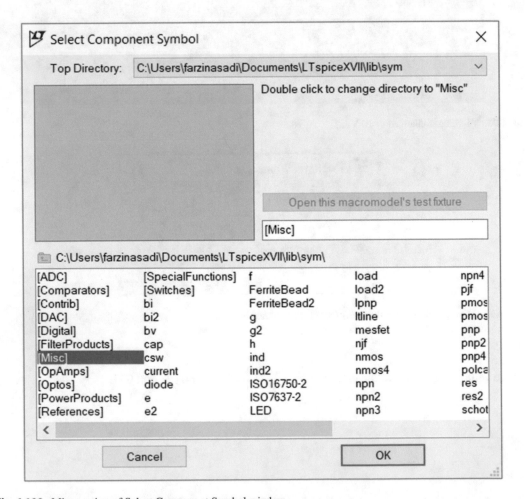

Fig. 4.100 Misc section of Select Component Symbol window

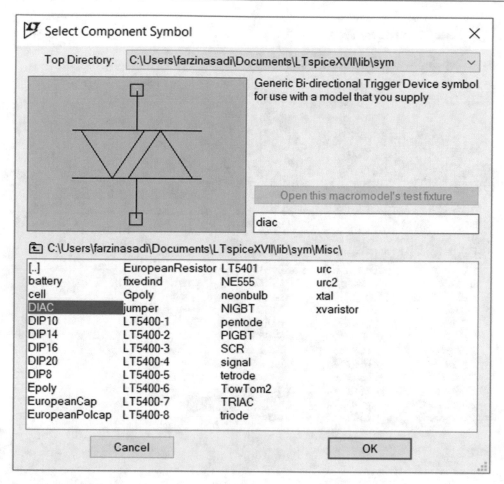

Fig. 4.101 DIAC block

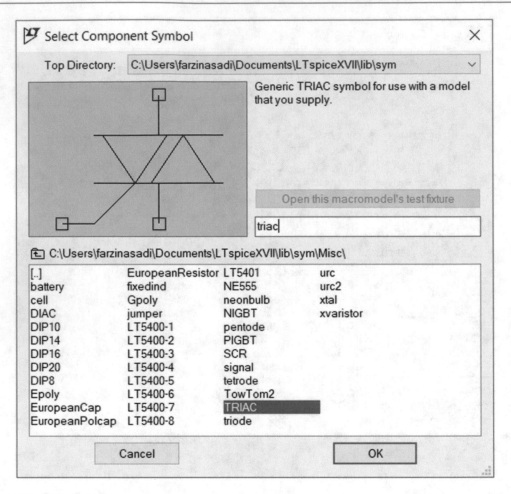

Fig. 4.102 TRIAC block

If you try to run the schematic shown in Fig. 4.99, the error message shown in Fig. 4.103 appears. Note that no simulation model is assigned to the triac and diac in Fig. 4.99. That is why we cannot simulate the circuit. We need to give the simulation models of triac and diac to LTspice.

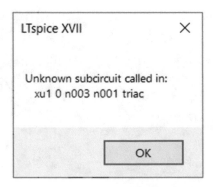

Fig. 4.103 Error message

Go to the ST Microelectronic website and search for "triac pspice model" (Fig. 4.104). Download the "Standard and Snubberless Triac PSpice model (.lib) and symbols (.olb)" file. You can visit other companies' websites in order to download their simulation models as well.

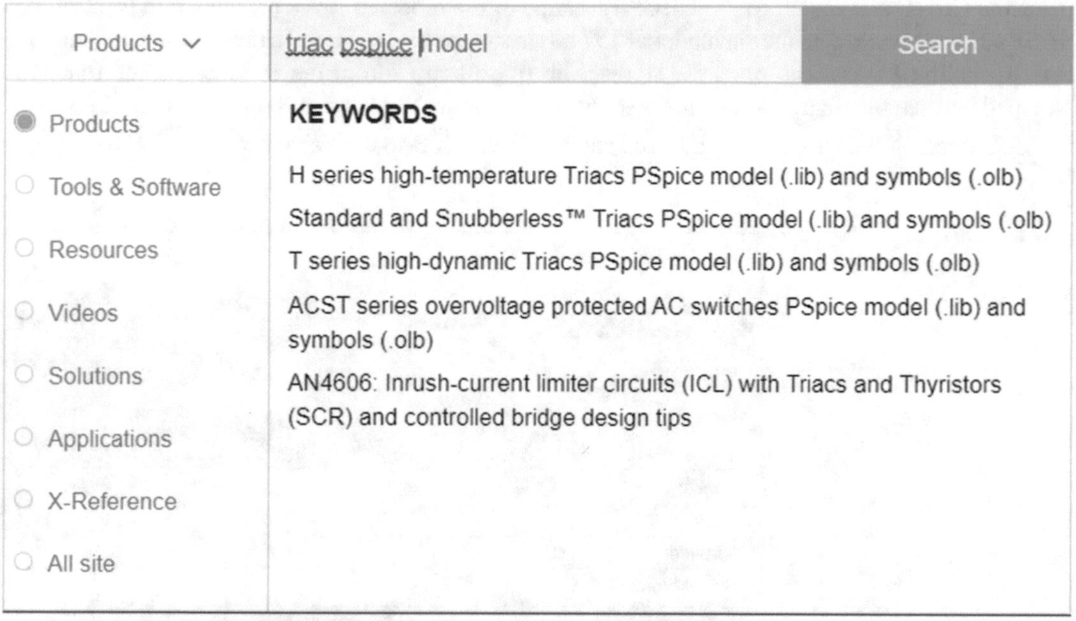

Products ⌄	triac pspice model	Search
● Products	**KEYWORDS**	
○ Tools & Software	H series high-temperature Triacs PSpice model (.lib) and symbols (.olb)	
○ Resources	Standard and Snubberless™ Triacs PSpice model (.lib) and symbols (.olb)	
○ Videos	T series high-dynamic Triacs PSpice model (.lib) and symbols (.olb)	
	ACST series overvoltage protected AC switches PSpice model (.lib) and symbols (.olb)	
○ Solutions		
○ Applications	AN4606: Inrush-current limiter circuits (ICL) with Triacs and Thyristors (SCR) and controlled bridge design tips	
○ X-Reference		
○ All site		

Fig. 4.104   Search for triac Pspice model

Now search for "diac pspice" and download the "Diacs PSpice model (.lib) and symbols (.olb)" file (Fig. 4.105).

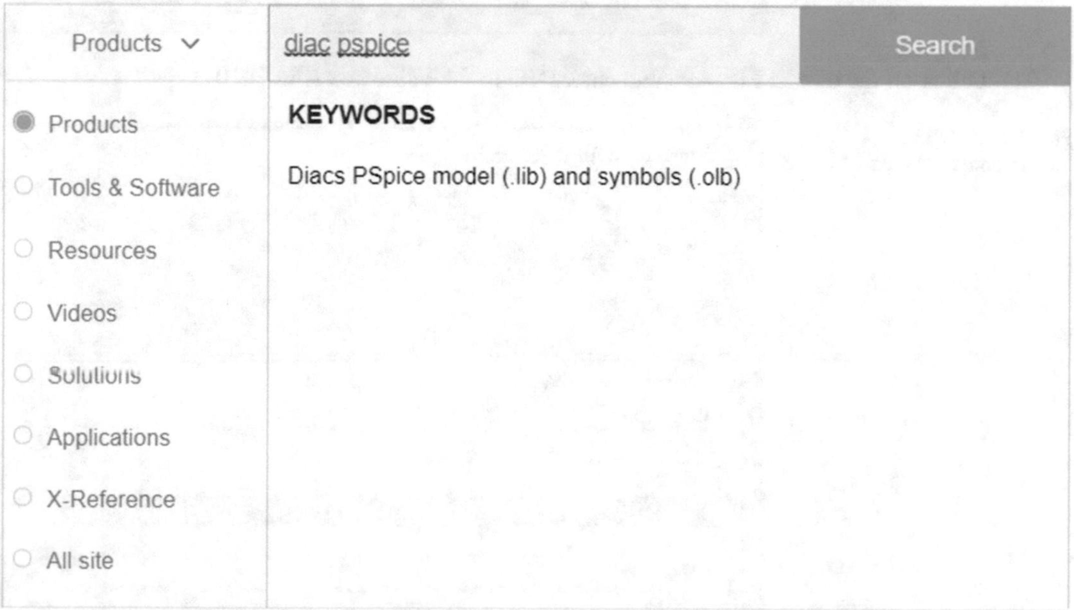

Products ⌄	diac pspice	Search
● Products	**KEYWORDS**	
○ Tools & Software	Diacs PSpice model (.lib) and symbols (.olb)	
○ Resources		
○ Videos		
○ Solutions		
○ Applications		
○ X-Reference		
○ All site		

Fig. 4.105   Search for diac Pspice model

Unzip the downloaded files. This produces two folders for you: en.standard_snubberless_triacs_ pspice and en.diacs_pspice. Copy these two folders to the Lib directory of LTspice.

You can use Notepad to open the st_standard_snubberless_triacs.lib (the library file in the en.standard_snubberless_triacs_pspice folder). After opening the st_standard_snubberless_triacs.lib file, press the Ctrl+F and search for ".SUBCKT." Name of the modeled triacs are shown in front of the .SUBCKT command. Simulation model of 157 different triacs is available in the st_standard_snubberless_triacs.lib file. You can open the st_diacs.lib (the library file in the en.diacs_pspice folder) in Notepad and see the diacs that are modeled there. 7 diacs are modeled in the st_diacs.lib file.

Right click on the triac in Fig. 4.99 and enter BTA12-600B to the Value box (Fig. 4.106).

Fig. 4.106 BTA12-600B is entered into the Value box

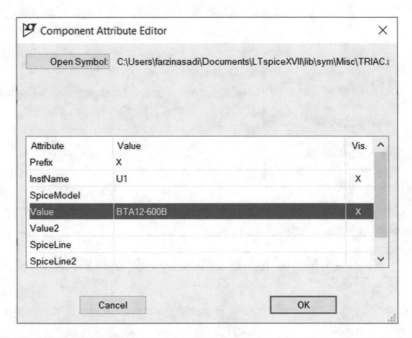

Right click on the diac in Fig. 4.99 and enter DB3 to the Value box (Fig. 4.107).

Fig. 4.107 DB3 is entered into the Value box

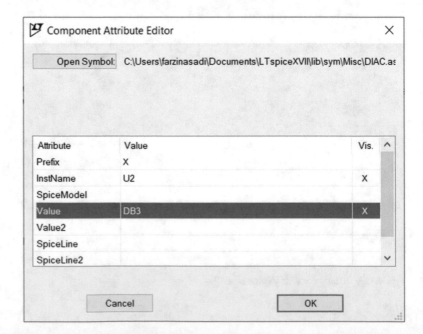

Add the .inc commands shown in Fig. 4.108. These two commands permit your schematic to use the "st_standard_snubberless_triacs.lib" and "st_diacs.lib" files.

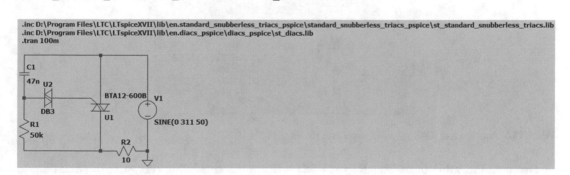

Fig. 4.108 Addition of .inc commands to the schematic

Run the simulation. The result is shown in Fig. 4.109. According to Fig. 4.110, the RMS of load R2 is about 140.77 V.

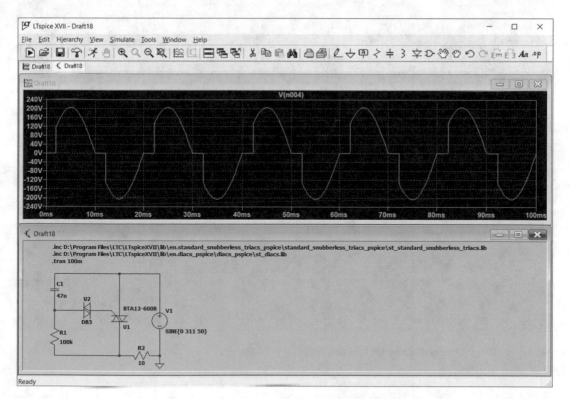

Fig. 4.109 Graph of voltage across the load resistor R2

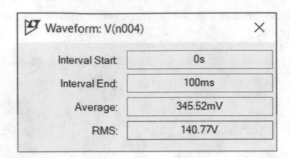

Fig. 4.110 Average and RMS values of waveform shown in Fig. 4.109

Decrease the value of resistor R1 to 50 kΩ and run the simulation (Fig. 4.111). According to Fig. 4.112, the RMS of load R2 is about 143.7 V.

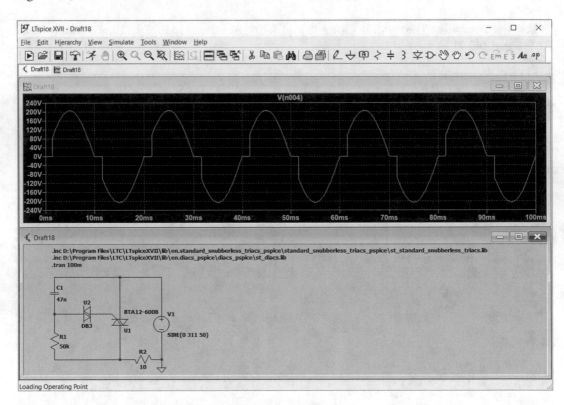

Fig. 4.111 Graph of voltage across the load resistor R2

Fig. 4.112 Average and RMS values of waveform shown in Fig. 4.111

Increase the value of resistor R1 to 300 kΩ and run the simulation (Fig. 4.113). According to Fig. 4.114, RMS of the load R2 is about 121.96 V.

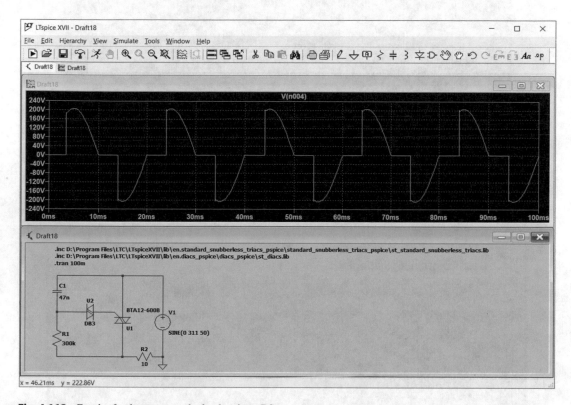

Fig. 4.113   Graph of voltage across the load resistor R2

Fig. 4.114   Average and RMS values of waveform shown in Fig. 4.113

So, we can control the RMS of load by changing the value of resistor R1: When you decrease the value of R1, the RMS of load R2 increases. When you increase the value of R1, the RMS of load R2 decreases.

4.11   Example 10: Single-Phase Half Wave Controlled Rectifier

In this example, we want to simulate a single-phase half wave controlled (thyristor) rectifier. The schematic of this example is shown in Fig. 4.115. The SCR can be found in the [Misc] section (Fig. 4.116) of Select Component Symbol window (Fig. 4.117).

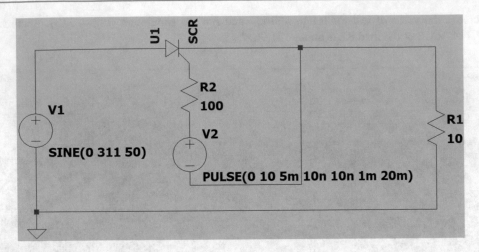

Fig. 4.115 Schematic of Example 10

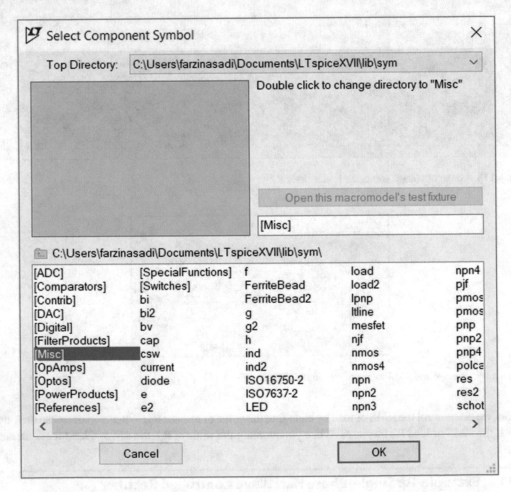

Fig. 4.116 Misc section of Select Component Symbol window

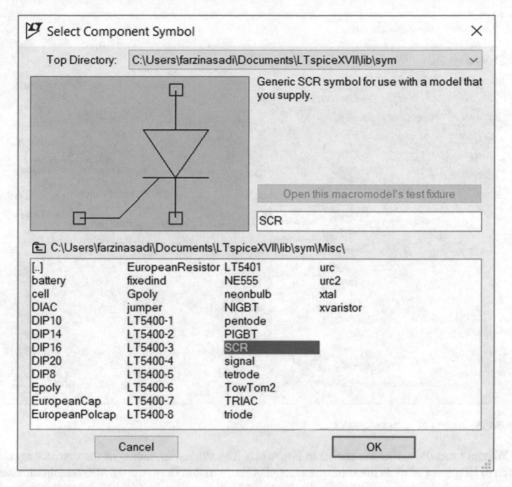

Fig. 4.117 SCR block

Settings of voltage source V2 is shown in Fig. 4.118. The value entered into the Tdelay[s] box determines the firing angle of the thyristor. For instance, for Tdelay = 5 ms, the firing angle is $\frac{5\,\text{m}}{20\,\text{m}} \times 360° = 90°$. Note that frequency of the pulse applied to the gate of the thyristor is equal to the frequency of the AC source V1.

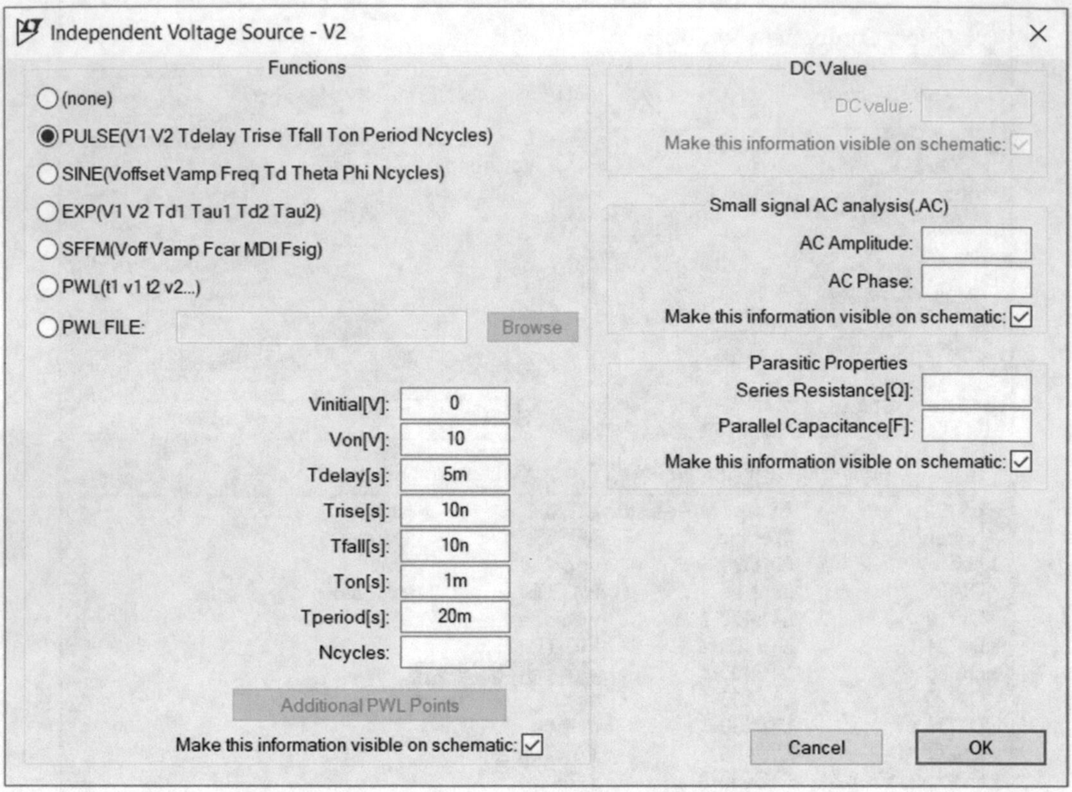

Fig. 4.118 Settings of voltage source V2

We can't run the schematic shown in Fig. 4.115. The simulation model of the thyristor must be given to LTspice. Let's add the simulation model of the thyristor. Go to the ST Microelectronics website and search for "scr pspice" (Fig. 4.119). Download the "Standard and sensitive SCRs PSpice model (.lib) and symbols (.olb)."

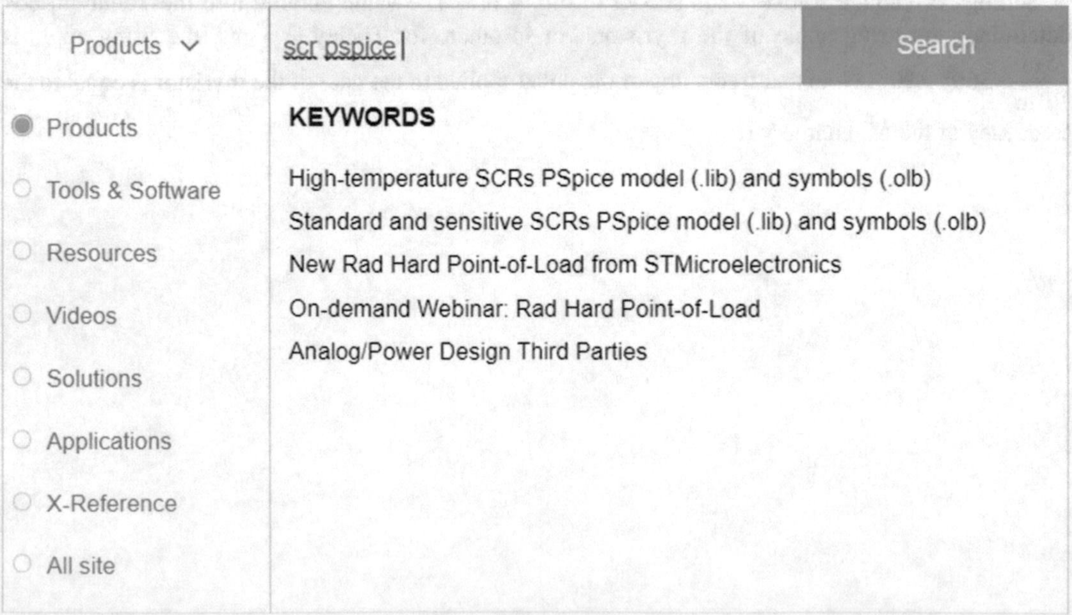

Fig. 4.119 Searching for scr Pspice

After downloading the file, unzip it. This produces the "en.standard_sensitive_scr_pspice" folder for you. Copy this folder in to the Lib folder of LTspice. Use the Notepad to open the "st_standard_sensitive_scr.lib" and search for ".subckt TXN825RG." Figure 4.120 shows the simulation model of TXN825RG.

```
.subckt TXN825RG A K G
X1 A K G SCR_ST params:
+ Vdrm=600v
+ Igt=40ma
+ Ih=50ma
+ Rt=0.014
* 2021 / ST / Rev 0
.ends
*$
```

Fig. 4.120 Simulation model of TXN825RG

According to Fig. 4.120, anode is the terminal 1 of the symbol (because letter A is in the first place after the component name), cathode is the terminal 2 of the symbol (because letter K is in the second place after the component name) and gate is terminal 3 of the symbol (because letter G is in the third place after the component name).

Right click on the thyristor U1 in Fig. 4.115 and enter TXN825RG to the Value box (Fig. 4.121). After clicking the OK button, the schematic changes to what is shown in Fig. 4.122.

Fig. 4.121 TXN825RG is entered to the Value box

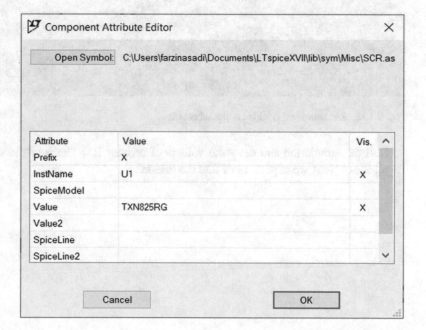

Fig. 4.122 Applied changes are shown on the schematic

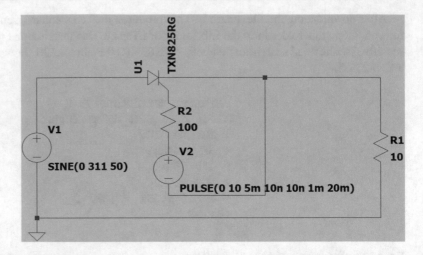

Add the commands shown in Fig. 4.123 to the schematic.

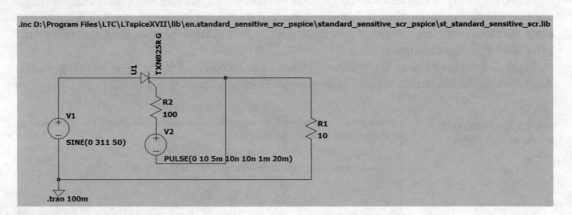

Fig. 4.123 .inc command is added to the schematic

Run the simulation and draw the voltage of resistor R1. The result is shown in Fig. 4.124. The result is not what we expect. Let's find the reason.

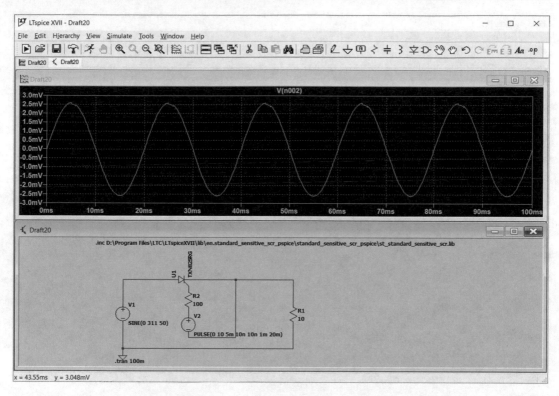

Fig. 4.124 Voltage of load resistor R1

Right click on the thyristor U1 and click the Open Symbol button (Fig. 4.125). This opens the SCR symbol (Fig. 4.126).

Fig. 4.125 TXN825RG is entered to the Value box

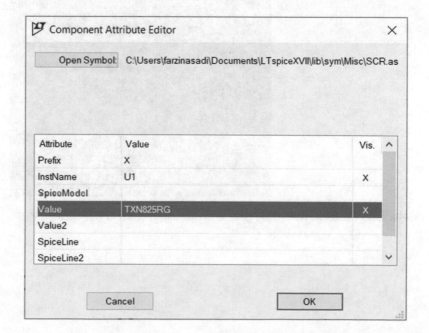

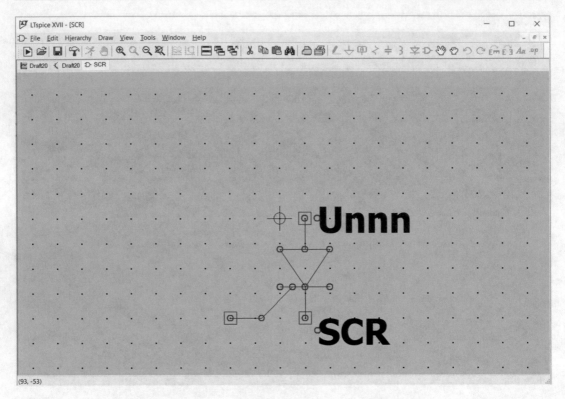

Fig. 4.126 SCR model is opened

Let's give the label A, B and C to the terminals of the symbol (Fig. 4.127).

Fig. 4.127 Label A, B and C are given to the terminals of SCR in Fig. 4.126

If you right click on the terminal A in Fig. 4.127, the window shown in Fig. 4.128 appears. According to this figure, terminal A (gate terminal) is the terminal number two of the symbol.

Fig. 4.128 Pin/Port
Properties window for
terminal A in Fig. 4.127

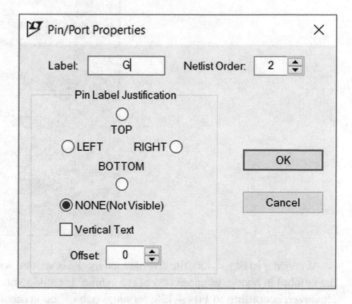

If you right click on the terminal B in Fig. 4.127, the window shown in Fig. 4.129 appears. According to this figure, terminal B (anode terminal) is the terminal number one of the symbol.

Fig. 4.129 Pin/Port
Properties window for
terminal B in Fig. 4.127

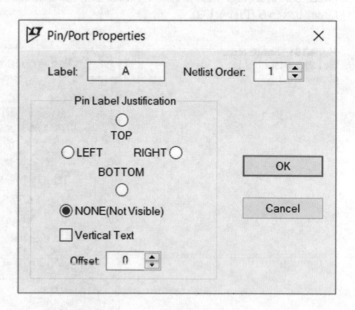

If you right click on the terminal C in Fig. 4.127, the window shown in Fig. 4.130 appears. According to this figure, terminal C (cathode terminal) is the terminal number three of the symbol.

Fig. 4.130 Pin/Port
Properties window for
terminal C in Fig. 4.127

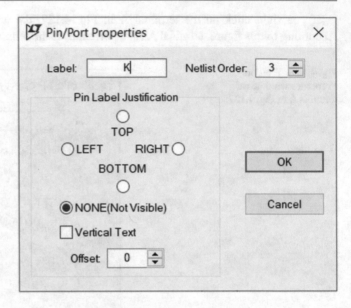

According to Fig. 4.120, the downloaded model is written with the assumption that terminal one of the symbol is anode, terminal two of the symbol is cathode and terminal three of the symbol is gate. However, according to Fig. 4.128, terminal two of the symbol is gate, and according to Fig. 4.130, terminal three of the symbol is cathode.

In order to solve the problem, right click on the terminal C in Fig. 4.127 and change the Netlist Order box to 2 (Fig. 4.131). Then right click on the terminal A in Fig. 4.127 and change the Netlist Order box to 3 (Fig. 4.132).

Fig. 4.131 Netlist order
for cathode terminal is
changed to 2

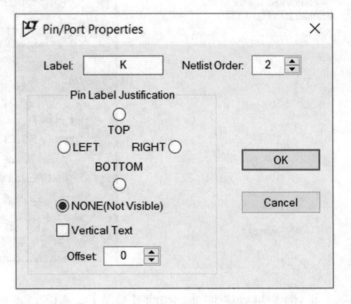

Fig. 4.132 Netlist order
for gate terminal is
changed to 3

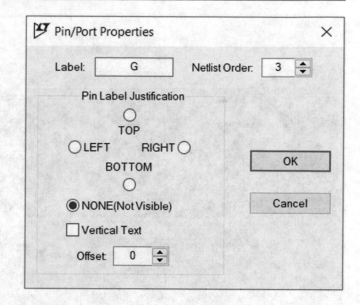

After applying the changes, close the SCR window (Fig. 4.133).

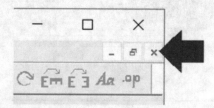

Fig. 4.133 Use the shown button to close the SCR window

Run the simulation. The result is shown in Fig. 4.134. This is the correct result. Let's measure the average and RMS of output voltage. According to Fig. 4.135, the average and RMS of output voltage are around 49.235 V and 109.44 V, respectively.

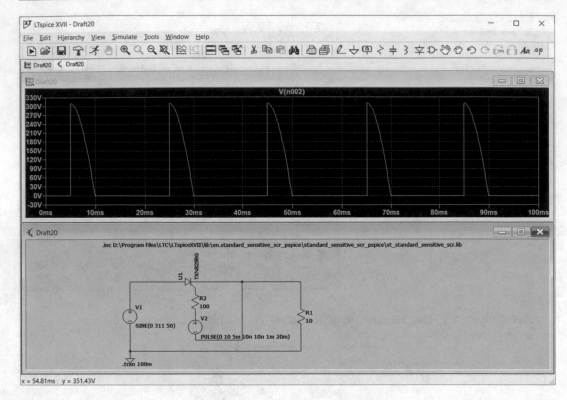

Fig. 4.134 Graph of load R1 voltage

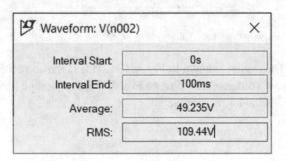

Fig. 4.135 Average and RMS values of waveform shown in Fig. 4.134

Let's check the obtained results. The MATLAB calculations shown in Fig. 4.136 shows that LTspice results are correct. Note that in MATLAB calculations, the voltage drop of the thyristor is ignored. So, LTspice results (which consider the voltage drop of the thyristor and other non-linear effects) are more accurate.

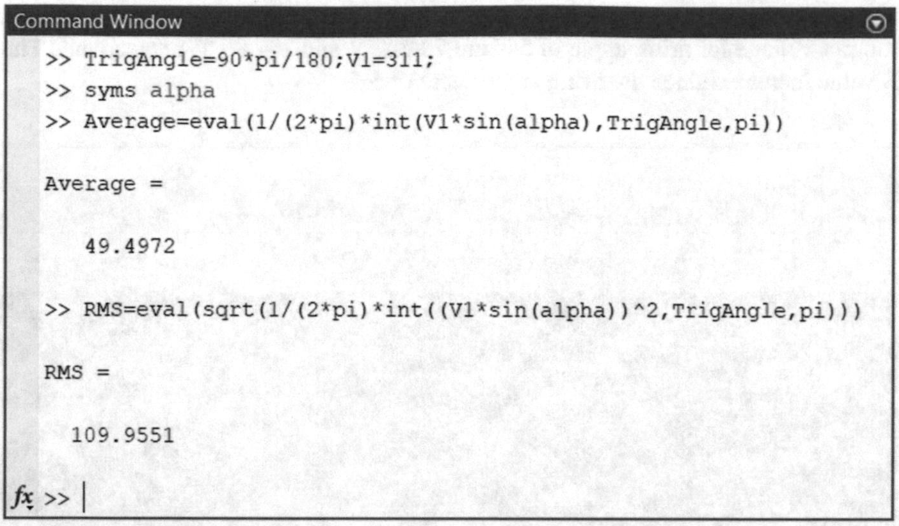

Fig. 4.136 MATLAB commands

Right click on the voltage source V2 and change its settings to what is shown in Fig. 4.137. These settings trigger the thyristor at t = 3 ms, 23 ms, 43 ms, ... The firing angle for Tdelay = 3 ms is $\frac{3\,\text{m}}{20\,\text{m}} \times 360° = 54°$.

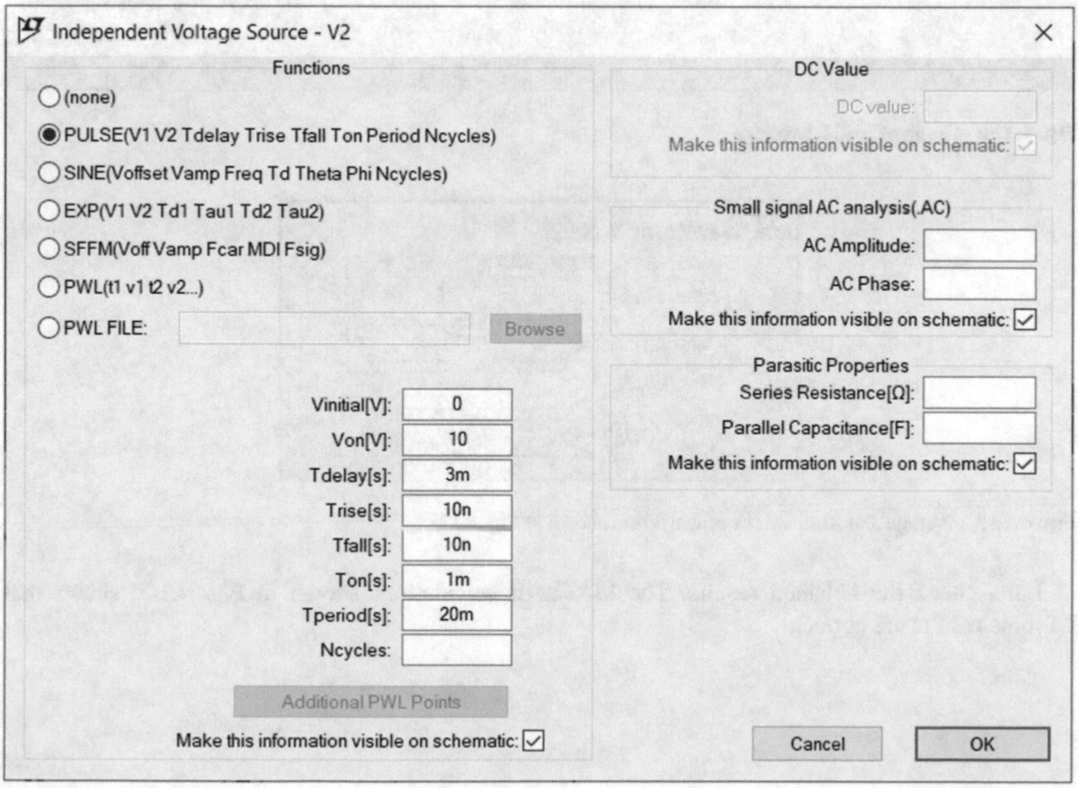

Fig. 4.137 Settings of voltage source V2

Run the simulation. The result is shown in Fig. 4.138. According to Fig. 4.139, the average and RMS of output voltage for firing angle of 54° are 78.222 V and 142.86 V, respectively. The average and RMS value increased since the firing angle decreased.

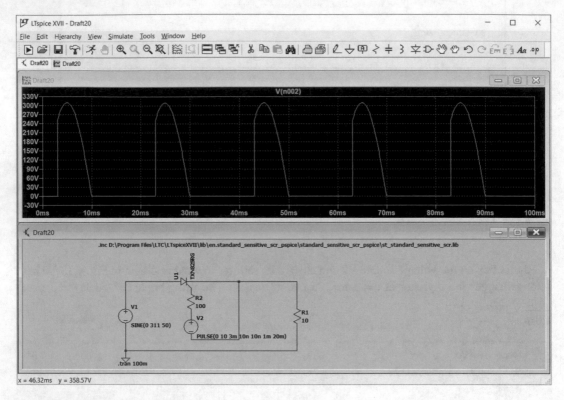

Fig. 4.138 Graph of load R1 voltage

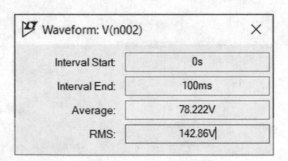

Fig. 4.139 Average and RMS values of waveform shown in Fig. 4.138

Let's check the obtained results. The MATLAB calculations shown in Fig. 4.140 shows that LTspice results are correct.

```
Command Window                                                              ⊙
  >> TrigAngle=54*pi/180;V1=311;
  >> syms alpha
  >> Average=eval(1/(2*pi)*int(V1*sin(alpha),TrigAngle,pi))

  Average =

     78.5909

  >> RMS=eval(sqrt(1/(2*pi)*int((V1*sin(alpha))^2,TrigAngle,pi)))

  RMS =

     143.4790

fx >>
```

Fig. 4.140 MATLAB commands

4.12   Example 11: Single-Phase Full Wave Controlled Rectifier (I)

In this example, we want to simulate a single-phase full wave controlled rectifier. The schematic of this example is shown in Fig. 4.141. Settings of voltage source V2 and V3 are shown in Figs. 4.142 and 4.143, respectively. Note that the phase difference between the control signal of thyristor U1 and U2 is 180° (10 ms). The firing angle is controlled with the aid of variable Td. For instance for Td = 3 ms, the firing angle is $\dfrac{\text{Td}}{\text{T}} \times 360° = \dfrac{3\,\text{m}}{20\,\text{m}} \times 360° = 30°$. T shows the period of input AC source. Since it is a 50 Hz input, the period is 20 ms.

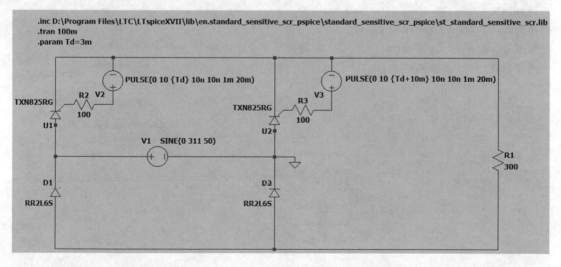

Fig. 4.141 Schematic of Example 11

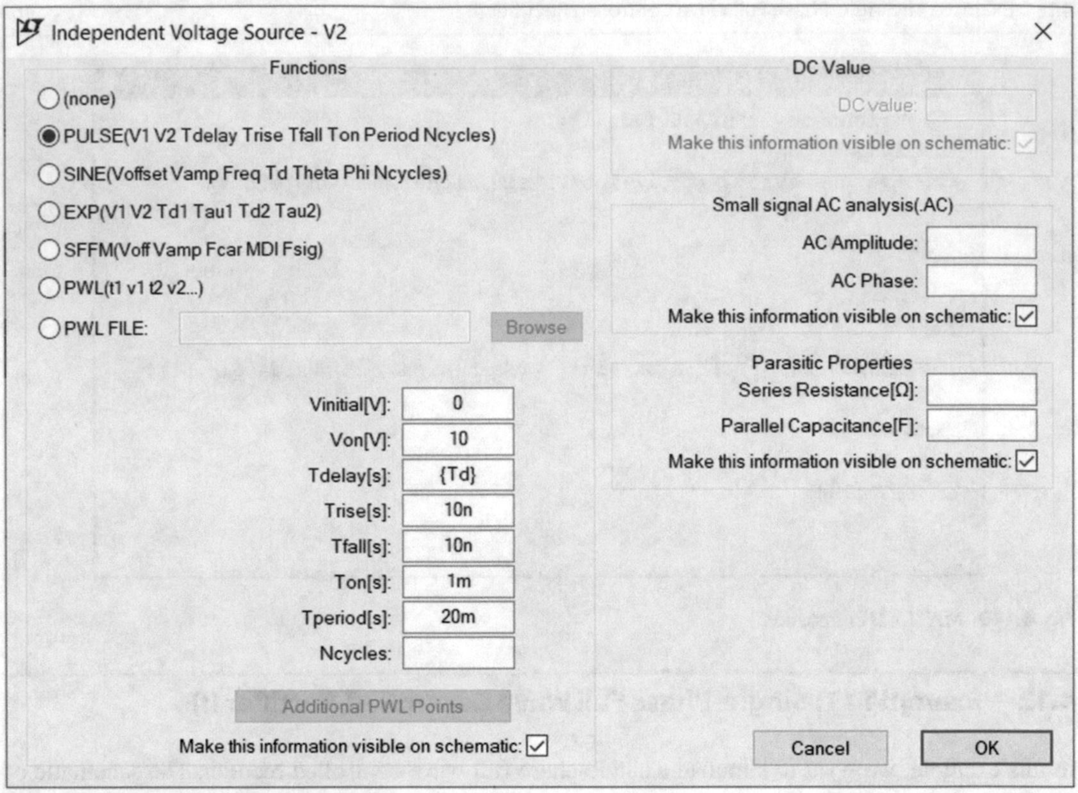

Fig. 4.142 Settings of voltage source V2

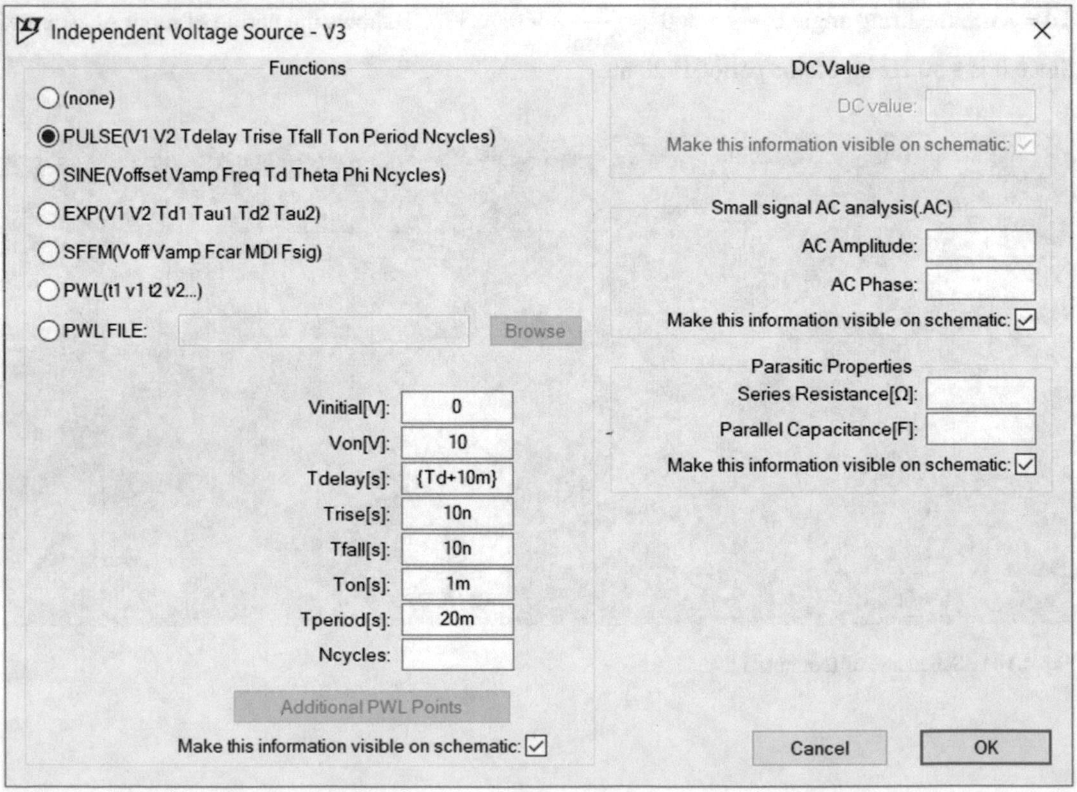

Fig. 4.143 Settings of voltage source V3

Run the simulation. The simulation result is shown in Fig. 4.144. Note that output frequency is two times bigger than the input AC source frequency. Let's measure the average and RMS value of output voltage. According to Fig. 4.145, the average and RMS of output voltage is 156.1 V and 201.69 V, respectively.

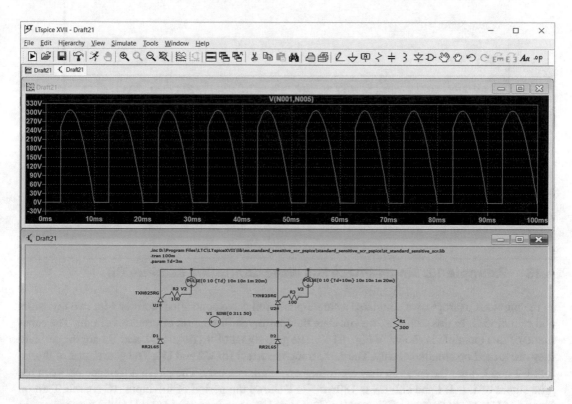

Fig. 4.144 Voltage across the load resistor R1

Fig. 4.145 Average and RMS values of waveform shown in Fig. 4.144

Let's check the obtained results. The MATLAB calculations shown in Fig. 4.146 shows that LTspice results are correct. Note that in MATLAB calculations, the voltage drop of the thyristor is ignored. So, LTspice results (which consider the voltage drop of the thyristor and other non-linear effects) are more accurate.

```
Command Window                                                              ⊙
  >> Td=3e-3;T=20e-3;TrigAngle=Td/T*2*pi;V1=311;
  >> syms alpha
  >> Average=eval(1/(pi)*int(V1*sin(alpha),TrigAngle,pi))

  Average =

    157.1818

  >> RMS=eval(sqrt(1/(pi)*int((V1*sin(alpha))^2,TrigAngle,pi)))

  RMS =

    202.9100

fx >>
```

Fig. 4.146 MATLAB commands

4.13   Example 12: Single-Phase Full Wave Controlled Rectifier (II)

In the previous example, we simulated a single-phase full wave controlled rectifier with two thyristors and two diodes. In this example, we simulate the four thyristor version of previous circuit. The schematic of this example is shown in Fig. 4.147. The gate signal of thyristors U1 and U3 are the same so they are turned on simultaneously. The same thing is correct for U2 and U4: The gate signal of thyristors U2 and U4 are the same so they are turned on simultaneously. The phase difference between the gate signal of U1–U3 and U2–U4 is 180°or $\frac{T}{2}$. T shows the period of the input AC source. Turn on instant of U1–U3 is the turn off instant for U2–U4 and Turn on instant of U2–U4 is the turn off instant for U1–U3. The voltage-dependent voltage source E1 and E2 replicates the voltage of nodes S1 and S2, respectively. So, the same gate signal is applied to U1–U3 and U2–U4. Settings of E1 and E2 are shown in Figs. 4.148 and 4.149, respectively.

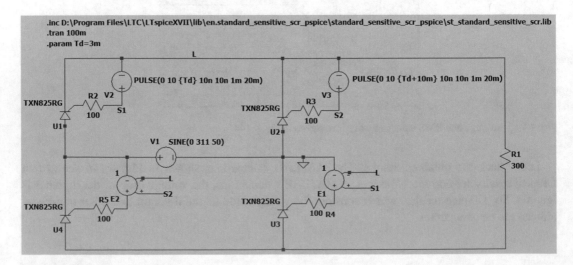

Fig. 4.147 Schematic of Example 12

Fig. 4.148 Settings of E1

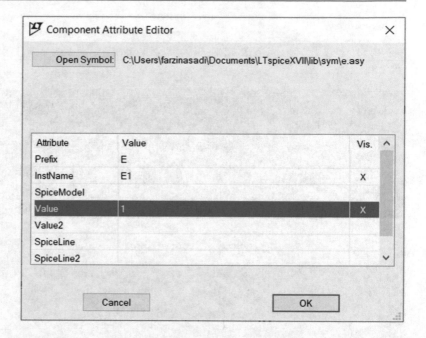

Fig. 4.149 Settings of E2

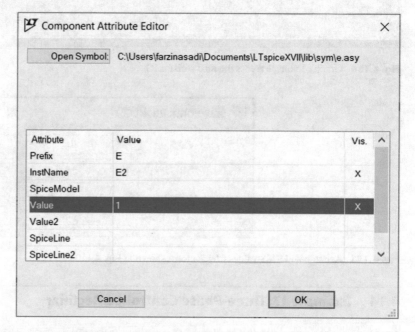

   Run the simulation. The result (for Td = 3 ms) is shown in Fig. 4.150. Let's measure the average and RMS of output voltage. According to Fig. 4.151, the average and RMS of output voltage are around 156.21 V and 201.82 V, respectively. The obtained results are quite close the results of previous example.

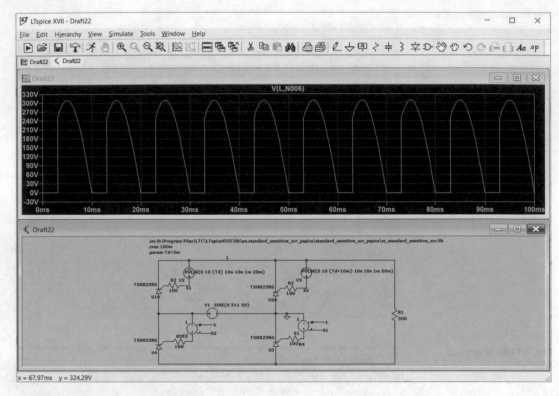

Fig. 4.150 Graph of voltage across the load resistor R1

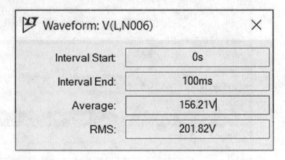

Fig. 4.151 Average and RMS values of waveform shown in Fig. 4.150

4.14   Example 13: Three-Phase Controlled Rectifier

In this example, we want to simulate a three-phase controlled rectifier. Schematic of a three-phase rectifier is shown in Fig. 4.152. The gate signal of thyristors for two firing angle (0° and 30°) is shown in Fig. 4.153. Note that T5-T6 is closed at t = 0 and firing angle (α) is measured with respect to $\omega t = \dfrac{\pi}{6} = 30°$.

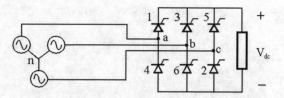

Fig. 4.152 Three-phase rectifier

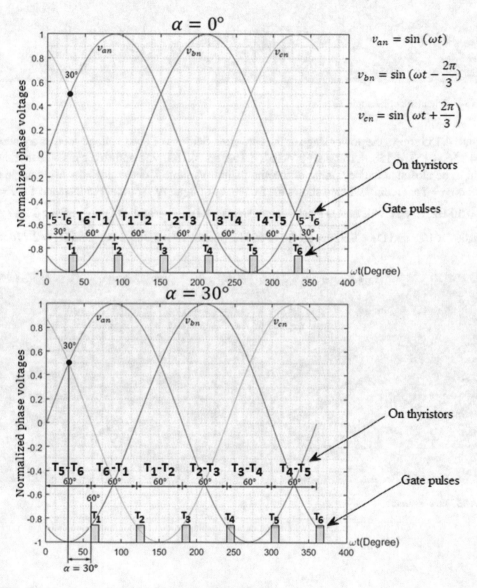

$$v_{an} = \sin(\omega t)$$

$$v_{bn} = \sin\left(\omega t - \frac{2\pi}{3}\right)$$

$$v_{cn} = \sin\left(\omega t + \frac{2\pi}{3}\right)$$

Fig. 4.153 Gate signal of thyristors for firing angle of 0° and 30°

Let's simulate the three-phase rectifier in LTspice. Draw the schematic shown in Fig. 4.154.

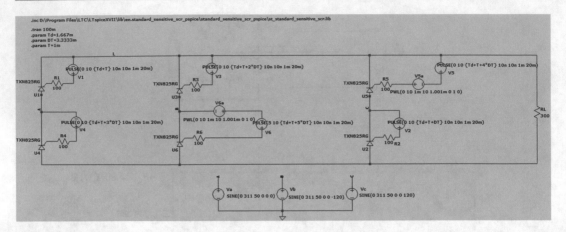

Fig. 4.154 Schematic of Example 13

Figure 4.155 shows the power stage of the converter. Settings of used voltage sources are shown in Figs. 4.156, 4.157, 4.158, 4.159, 4.160, 4.161, 4.162, and 4.163. V5a and V6a cause the thyristor U5 and U6 to be closed at the beginning of the simulation. Variable T determines the firing angle of the thyristors: $\alpha = T \times f \times 360°$ where f shows the frequency of input AC source. For instance, for $T = 1$ ms and $f = 50$ Hz, $\alpha = 18°$. Td and DT are two constants: $\mathrm{Td} = \dfrac{1}{12f}$ and $\mathrm{DT} = \dfrac{1}{6f}$. For 50 Hz system, $\mathrm{Td} = 1.667 \times 10^{-3}$ and $\mathrm{DT} = 3.333 \times 10^{-3}$. For 60 Hz system, $\mathrm{Td} = 1.389 \times 10^{-3}$ and $\mathrm{DT} = 2.778 \times 10^{-3}$.

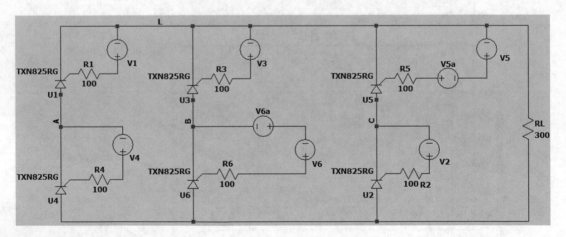

Fig. 4.155 Power circuit

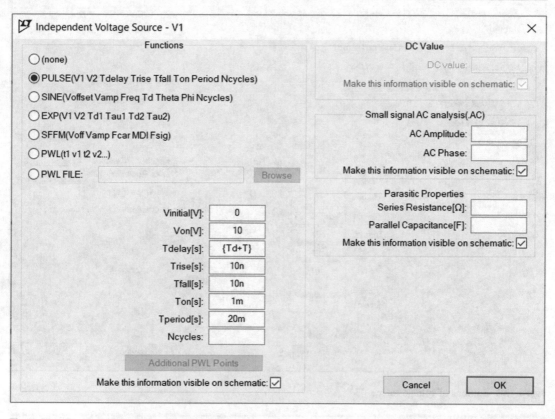

Fig. 4.156 Settings of voltage source V1

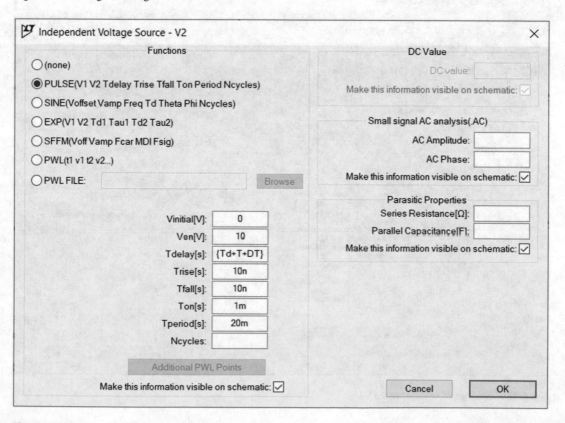

Fig. 4.157 Settings of voltage source V2

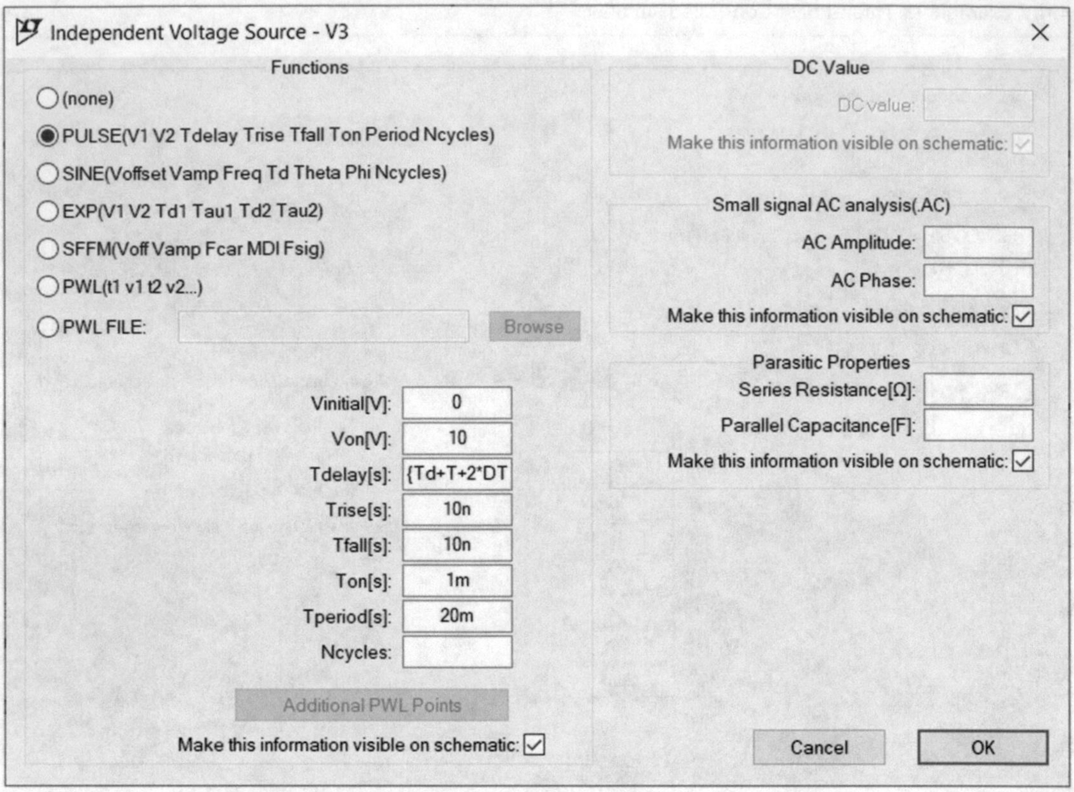

Fig. 4.158 Settings of voltage source V3

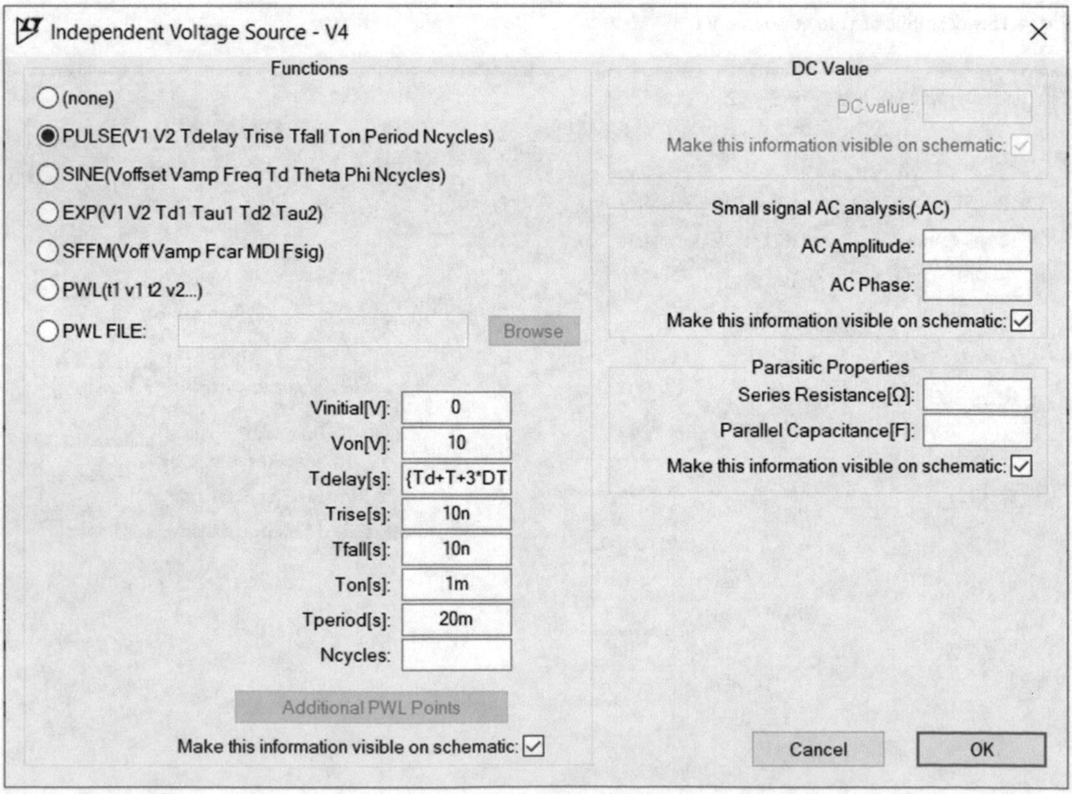

Fig. 4.159 Settings of voltage source V4

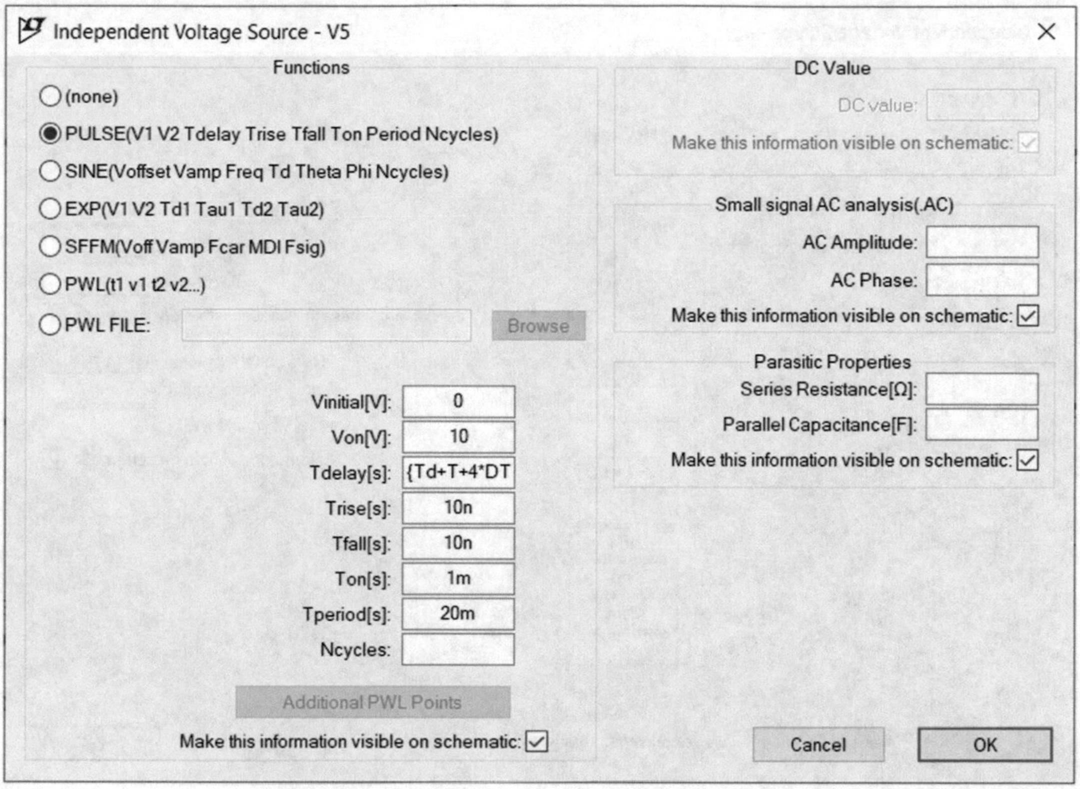

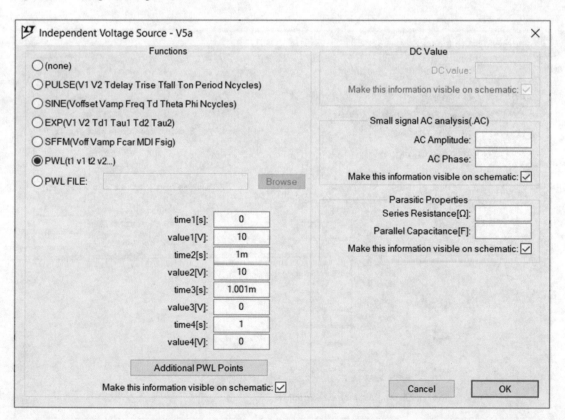

Fig. 4.160 Settings of voltage source V5

Fig. 4.161 Settings of voltage source V5a

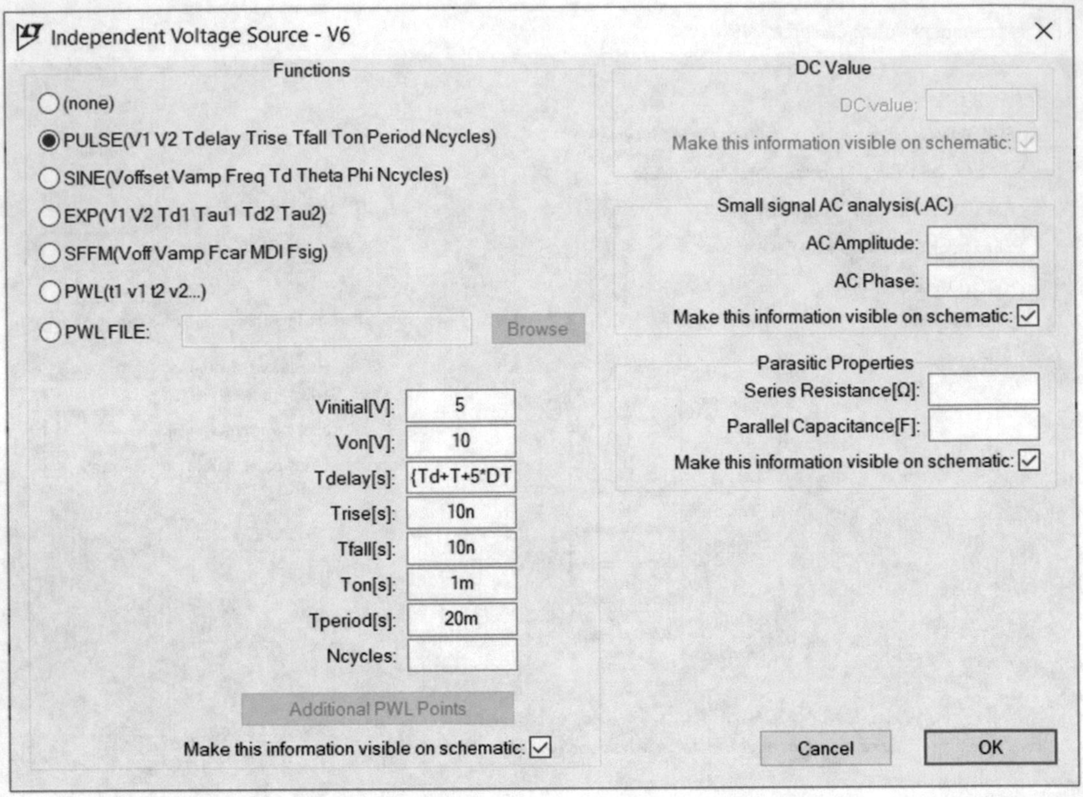

Fig. 4.162 Settings of voltage source V6

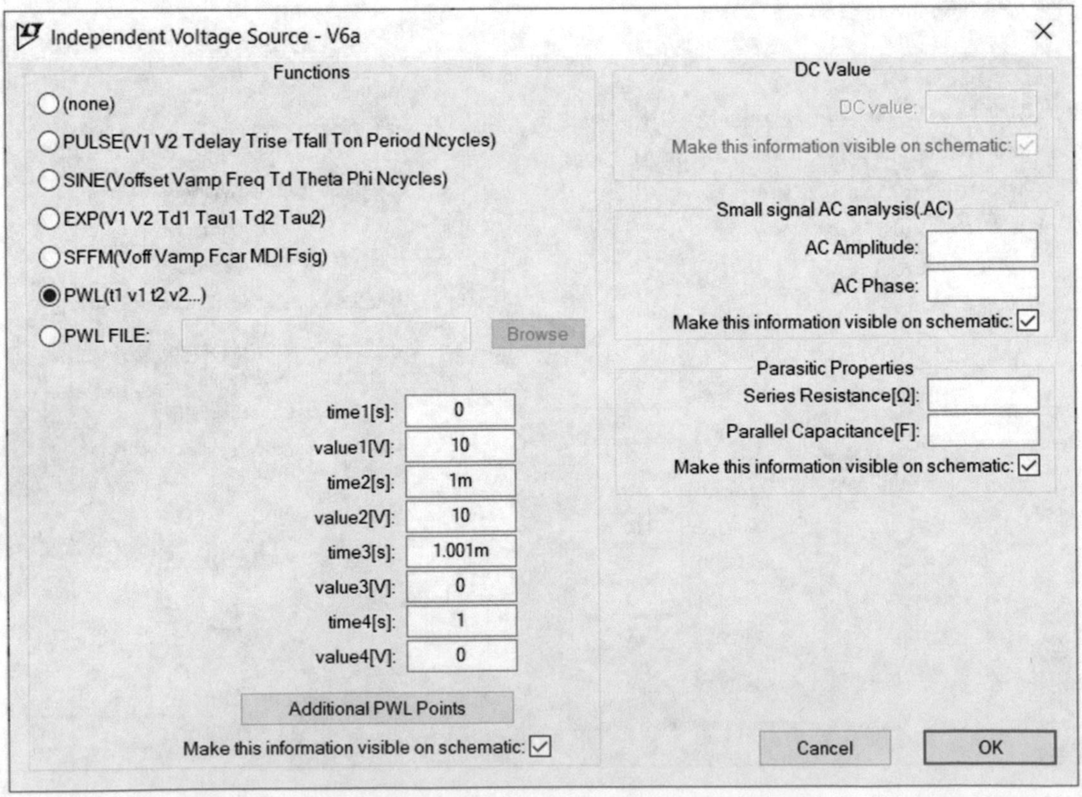

Fig. 4.163 Settings of voltage source V6a

The input three-phase AC source is shown in Fig. 4.164. Settings of the sources are shown in Figs. 4.165, 4.166, and 4.167.

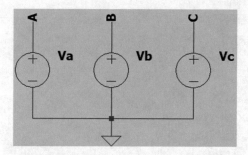

Fig. 4.164 Input three-phase source

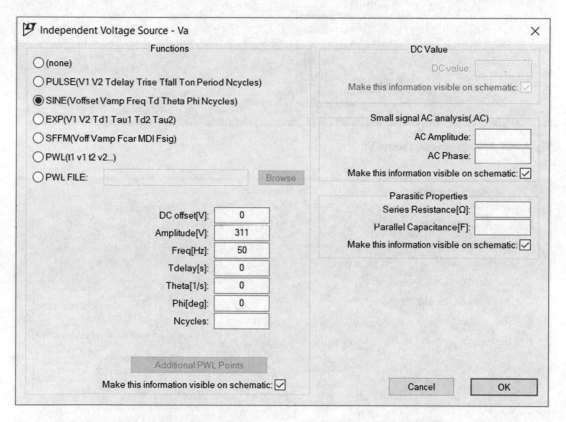

Fig. 4.165 Settings of voltage source Va

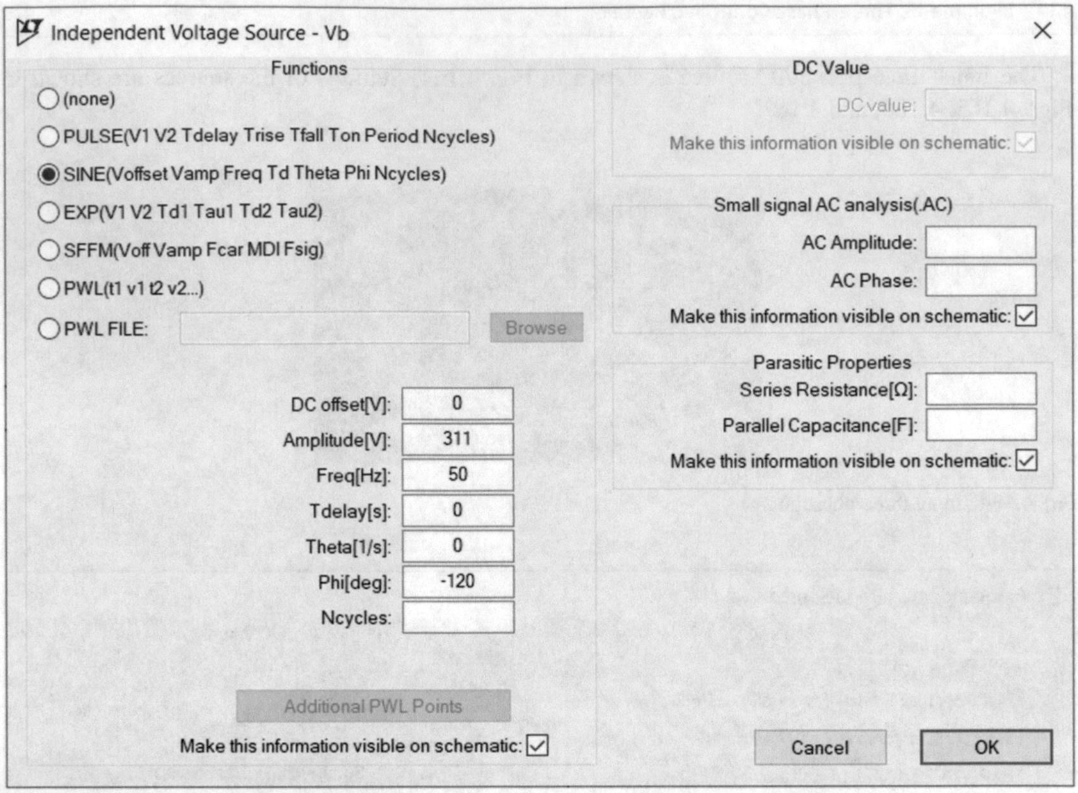

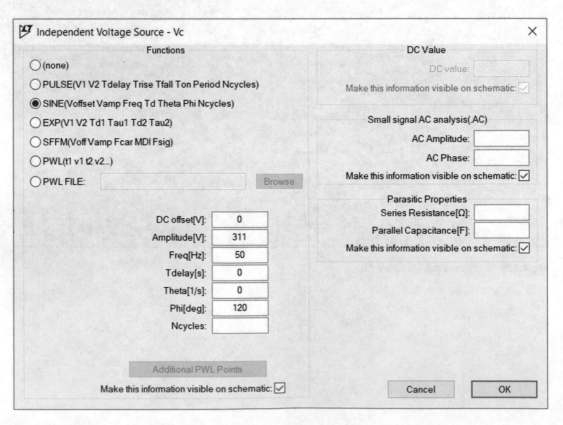

Fig. 4.166 Settings of voltage source Vb

Fig. 4.167 Settings of voltage source Vc

The commands used in the schematic 154 are shown in Fig. 4.168.

```
.inc D:\Program Files\LTC\LTspiceXVII\lib\en.standard_sensitive_scr_pspice\standard_sensitive_scr_pspice\st_standard_sensitive_scr.lib
.tran 100m
.param Td=1.667m
.param DT=3.3333m
.param T=1m
```

Fig. 4.168 SPICE commands

Run the simulation. The result is shown in Fig. 4.169. Let's measure the average and RMS value of the obtained waveform. According to Fig. 4.170, the average and RMS value are around 487.58 V and 490.46 V, respectively.

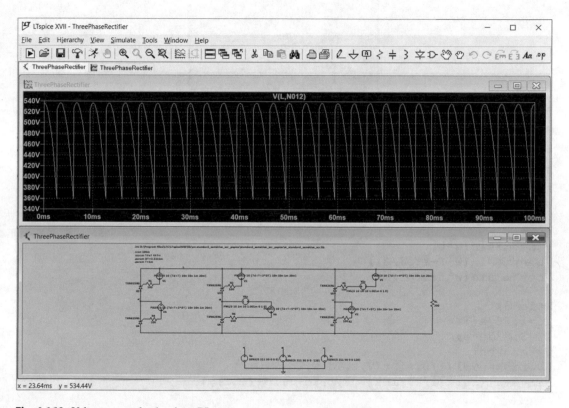

Fig. 4.169 Voltage across load resistor RL

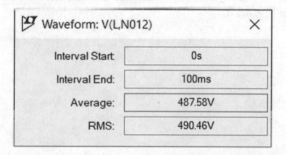

Fig. 4.170 Average and RMS values of waveform shown in Fig. 4.169

Let's check the obtained result. The average value of output voltage can be calculated with the aid of $V_o = 1.35V_{LL} \cos(\alpha)$ formula. V_o, V_{LL} and α show the average value of output voltage, RMS of line-line voltage that supplies the rectifier and triggering angle, respectively. According to the calculations shown in Fig. 4.171, the output voltage must be around 489.24 V. The LTspice result is quite close to this value. Note that the $V_o = 1.35V_{LL} \cos(\alpha)$ formula ignores the voltage drop of thyristors. So, the value calculated by this formula is a little bit bigger than the correct average value.

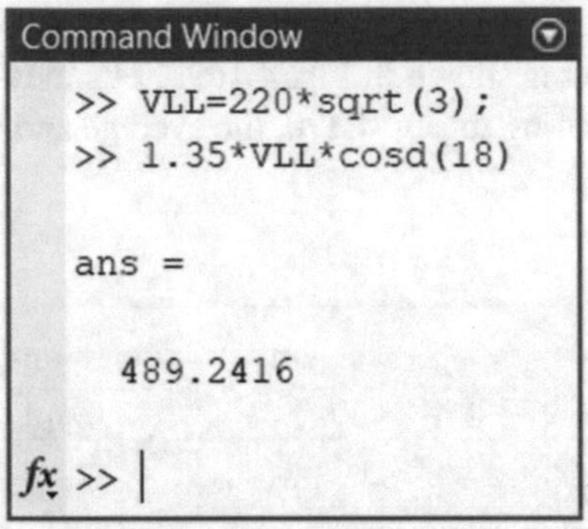

```
Command Window                    ⊙

  >> VLL=220*sqrt(3);
  >> 1.35*VLL*cosd(18)


  ans =


    489.2416

fx >> |
```

Fig. 4.171 MATLAB commands

The following MATLAB code calculates the average and RMS of output voltage of the converter.

```
f=50;
Vm=311;
w=2*pi*f;

syms t
Van=Vm*sin(w*t);
Vbn=Vm*sin(w*t-2*pi/3);
Vcn=Vm*sin(w*t+2*pi/3);

T=1e-3;
Td=1.667e-3;
DT=3.333e-3;
V=Van-Vbn;
Average=eval(1/DT*int(V,t,Td+T,Td+T+DT))
RMS=eval(sqrt(1/DT*int(V^2,t,Td+T,Td+T+DT)))
```

After running the above code, the result shown in Fig. 4.172 is obtained. Note that the above code ignored the voltage drop of the thyristors. So, the obtained results are a little bit bigger than the LTspice results shown in Fig. 4.170.

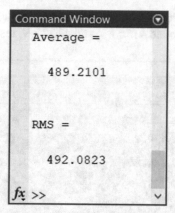

Fig. 4.172 Output of MATLAB code

Let's assume that the voltage drop across the anode-cathode of a closed thyristor is 0.75 V. So, voltage drop across two closed thyristors is 1.5 V. The following MATLAB code, considers this voltage drop.

```
f=50;
Vm=311;
w=2*pi*f;

syms t
Van=Vm*sin(w*t);
Vbn=Vm*sin(w*t-2*pi/3);
Vcn=Vm*sin(w*t+2*pi/3);

T=1e-3;
Td=1.667e-3;
DT=3.333e-3;
V=Van-Vbn-1.5; % voltage drop of 1.5 V is assumed
Average=eval(1/DT*int(V,t,Td+T,Td+T+DT))
RMS=eval(sqrt(1/DT*int(V^2,t,Td+T,Td+T+DT)))
```

After running the code, the results shown in Fig. 4.173 is obtained. Note that the obtained results are quite close to the LTspice results shown in Fig. 4.170.

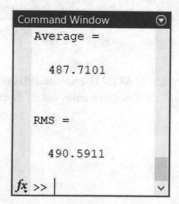

Fig. 4.173 Output of MATLAB code

Figure 4.174 shows the simulation result for firing angle of 36° (T = 2 ms). According to Fig. 4.175, the average and RMS value are around 415.57 V and 426.17 V, respectively.

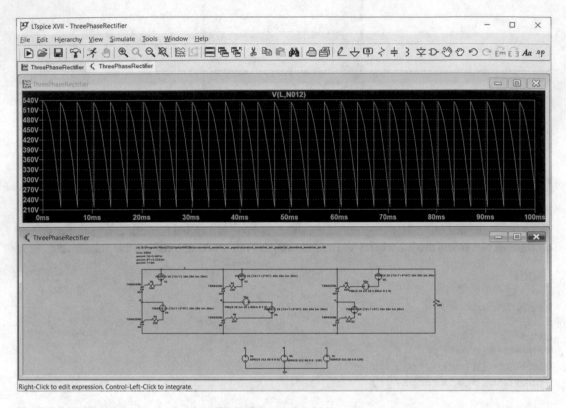

Fig. 4.174 Voltage across the load resistor RL for firing angle of 36°

🛂 Waveform: V(L,N012)	✕
Interval Start	0s
Interval End:	100ms
Average:	415.57V
RMS:	426.17V

Fig. 4.175 Average and RMS values of waveform shown in Fig. 4.174

Let's check the result. The following MATLAB code calculates the average and RMS value of output voltage with the assumption of 1.5 V voltage drop for on thyristors.

```
f=50;
Vm=311;
w=2*pi*f;

syms t
Van=Vm*sin(w*t);
Vbn=Vm*sin(w*t-2*pi/3);
Vcn=Vm*sin(w*t+2*pi/3);

T=2e-3;
Td=1.667e-3;
DT=3.333e-3;
V=Van-Vbn-1.5; % voltage drop of 1.5 V is assumed
Average=eval(1/DT*int(V,t,Td+T,Td+T+DT))
RMS=eval(sqrt(1/DT*int(V^2,t,Td+T,Td+T+DT)))
```

After running the code, the result shown in Fig. 4.176 is obtained. The obtained results are quite close to LTspice results shown in Fig. 4.175.

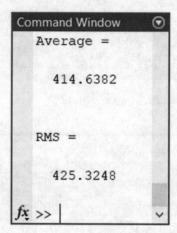

Fig. 4.176 Output of MATLAB code

Figure 4.177 shows the simulation result for firing angle of 0° (T = 0). According to Fig. 4.178, the average and RMS value are around 512.85 V and 513.3 V, respectively. In fact, firing angle of zero simulates the three-phase diode rectifier shown in Fig. 4.179.

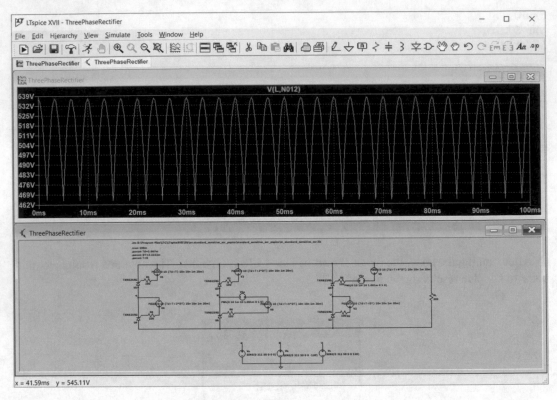

Fig. 4.177 Voltage across the load resistor RL for firing angle of 0°

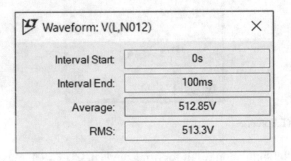

Fig. 4.178 Average and RMS values of waveform shown in Fig. 4.177

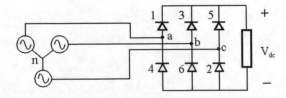

Fig. 4.179 Three-phase diode rectifier

Let's check the result. The following MATLAB code calculates the average and RMS value of output voltage with the assumption of 1.5 V voltage drop for on thyristors.

```
f=50;
Vm=311;
w=2*pi*f;
% voltage drop of 1.5 V is assumed
syms t
Van=Vm*sin(w*t);
Vbn=Vm*sin(w*t-2*pi/3);
Vcn=Vm*sin(w*t+2*pi/3);

T=0;
Td=1.667e-3;
DT=3.333e-3;
V=Van-Vbn-1.5;
Average=eval(1/DT*int(V,t,Td+T,Td+T+DT))
RMS=eval(sqrt(1/DT*int(V^2,t,Td+T,Td+T+DT)))
```

After running the code, the result shown in Fig. 4.180 is obtained. The obtained results are quite close to LTspice results shown in Fig. 4.178.

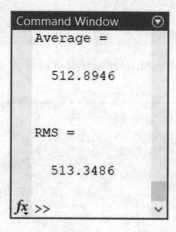

Fig. 4.180 Output of MATLAB code

Let's draw the graph of circuit currents. Studying the graph of currents helps you to select suitable thyristors for your circuit. The load current for triggering angle of 18° (T = 1 ms) is shown in Fig. 4.181. According to Fig. 4.182, the average value of load current is 1.6251 A.

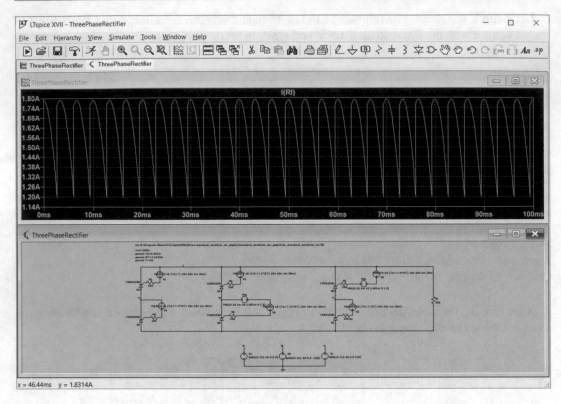

Fig. 4.181 Current through the load resistor RL for firing angle of 36°

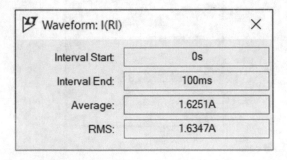

Fig. 4.182 Average and RMS values of waveform shown in Fig. 4.181

You can see the current passed from thyristors easily. In Fig. 4.183, voltage source V7 (with value of 0 V) is added in series with thyristor U1. The current of voltage source V7 is shown in Fig. 4.184. According to Fig. 4.185, the average value of current passed from thyristor U1 is 546.01 mA. Average value of current passed from other thyristors is the same as the average value of current passed from thyristor U1. Note that the average value of current passed from the thyristors is one third of the average value of current passed from the load. This is expected since each thyristor conducts the load current for one third of the cycle, i.e., 120°.

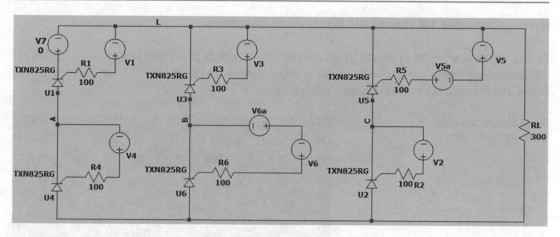

Fig. 4.183 Voltage source V7 measures the current through thyristor U1

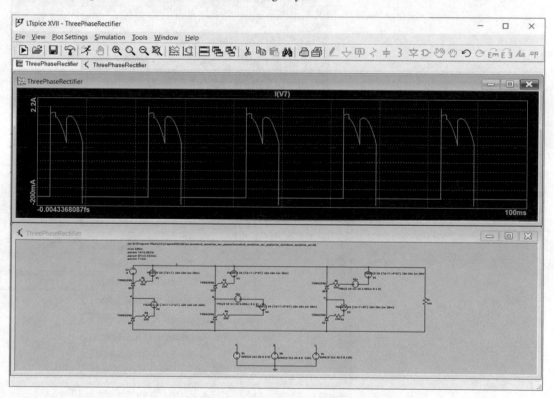

Fig. 4.184 Current entered into the + terminal of V7 (−Current through thyristor U1)

Waveform: I(V7)	×
Interval Start	0s
Interval End:	100ms
Average:	546.01mA
RMS:	952.55mA

Fig. 4.185 Average and RMS values of waveform shown in Fig. 4.184

4.15   Example 14: Harmonic Analysis of Rectifiers

In this example, we want to see the harmonic content of a single-phase half wave rectifier which has a freewheeling diode. The schematic of this example is shown in Fig. 4.186. The two .options lines in Fig. 4.186 increase the accuracy of the calculations.

Fig. 4.186 Schematic of Example 14

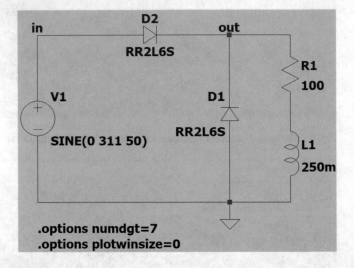

Set up a transient analysis with the settings shown in Fig. 4.187. After clicking the OK button in Fig. 4.187, the schematic changes to what is shown in Fig. 4.188.

Fig. 4.187 Simulation settings

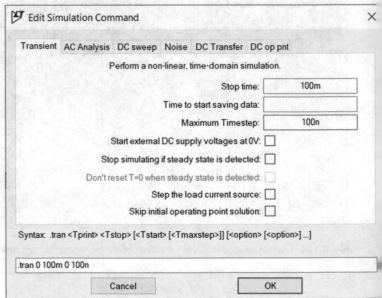

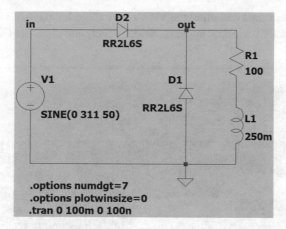

Fig. 4.188 .trans command is added to the schematic

Run the simulation and draw the voltage of node "out" (Fig. 4.189). Note that the frequency of the output voltage is 50 Hz.

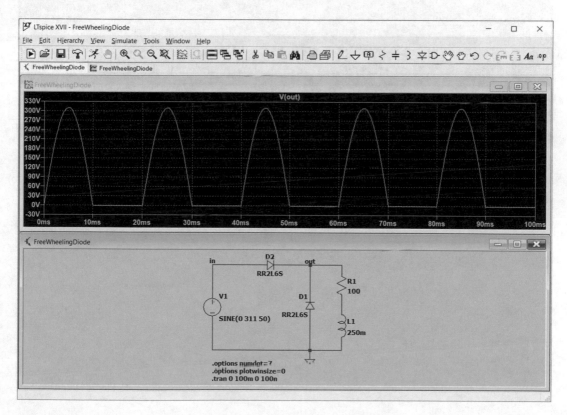

Fig. 4.189 Graph of voltage across the RL load

Right click on the black area and click the View> FFT (Fig. 4.190). After clicking the FFT, Select Waveforms to include in FFT window appears. Select V(out) and click the OK button (Fig. 4.191).

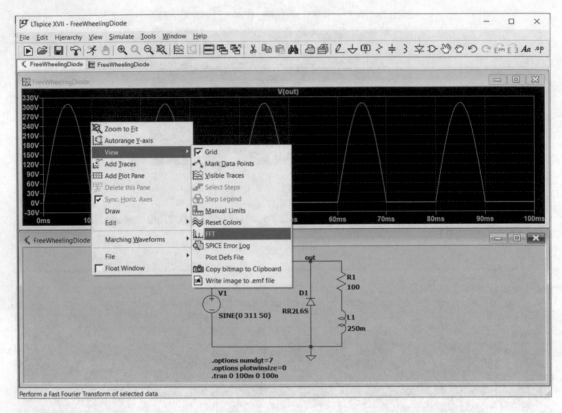

Fig. 4.190 View> FFT

Fig. 4.191 Select
Waveforms to include in
FFT window

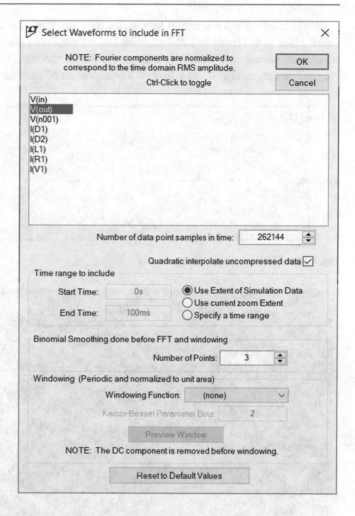

After clicking the OK button in Fig. 4.191, the graph shown in Fig. 4.192 appears on the screen.

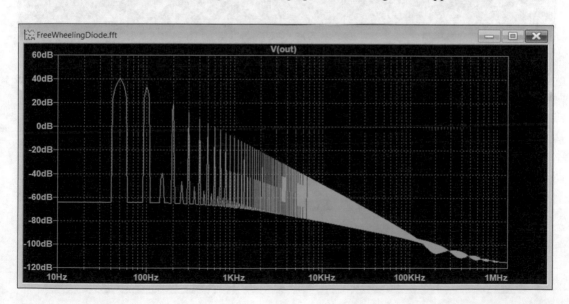

Fig. 4.192 Simulation result

Right click on the horizontal axis. After right clicking, the window shown in Fig. 4.193 appears. Enter 1kHz to the Right box and click the OK button. After clicking the OK button, the graph changes to what is shown in Fig. 4.194. This graph shows the amplitude of harmonics for [10 Hz, 1 kHz] range. You can use the cursors to read the amplitude of each harmonic.

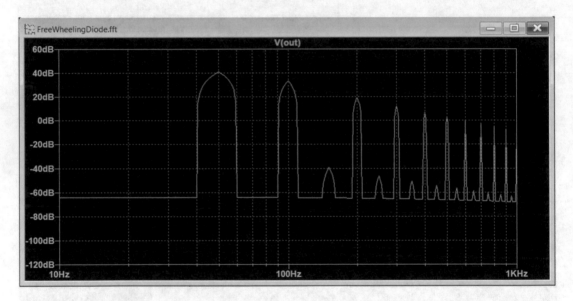

Fig. 4.193   Horizontal Axis window

Fig. 4.194   Simulation result ([10 Hz, 10 kHz] range)

Right click on the vertical axis and select the Linear (Fg. 4.195). Now, the vertical axis shows the amplitude of harmonics with unit of Volts (Fig. 4.196).

Fig. 4.195   Left Vertical Axis -- Magnitude window

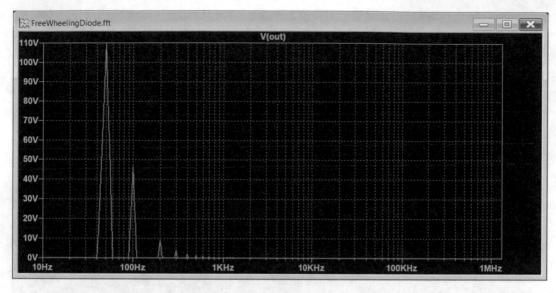

Fig. 4.196   Simulation result

Right click on the horizontal axis and enter 1kHz to the Right box (Fig. 4.197). Now the graph shows the harmonics in the [10 Hz, 1 kHz] range (Fig. 4.198).

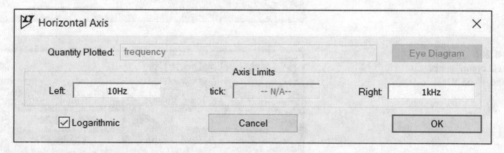

Fig. 4.197   Horizontal Axis window

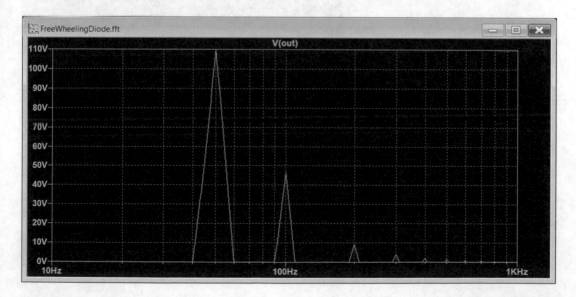

Fig. 4.198   Simulation result

You can use the cursors to read the value of harmonics. For instance, according to Fig. 4.199, the RMS of fundamental harmonic (50 Hz) is 109.896 V and according to Fig. 4.200, the RMS of second harmonic (100 Hz) is 46.65 V.

Fig. 4.199 RMS of
50 Hz component is
109.896 V

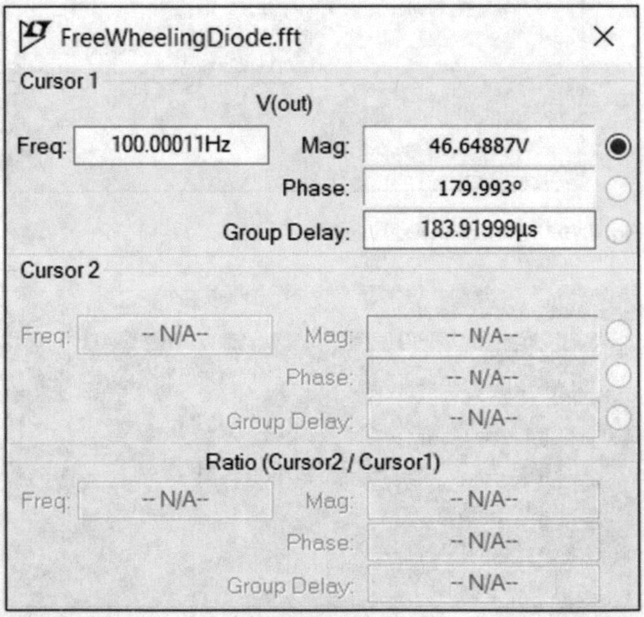

Fig. 4.200 RMS of
100 Hz component is
around 46.65 V

Let's check the obtained results. Fourier series of half wave rectified waveform is shown in Fig. 4.201.

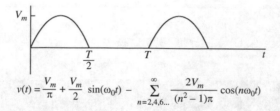

Fig. 4.201 Fourier series of half wave rectified signal

According to Fig. 4.201, the amplitude of fundamental harmonic is $\dfrac{311}{2} = 155.5\,\text{V}$. RMS of fundamental harmonic is $\dfrac{155.5}{\sqrt{2}} = 109.95\,\text{V}$. Amplitude of second harmonic is $\dfrac{2 \times 311}{\left(2^2 - 1\right)\pi} = 66\,\text{V}$. RMS of this value is $\dfrac{66}{\sqrt{2}} = 46.67\,\text{V}$. So, the LTspice results are correct.

4.16   Example 15: Measurement of Power Factor for Rectifier Circuits

In this example, we want to measure the power factor of rectifier circuit of Example 14. The power factor is nothing more than ratio of average power drawn from the input AC source to the apparent power of the input AC source. In linear circuits, the power factor can be calculated with the aid of $\cos(\varphi)$ formula. φ shows the phase difference between the voltage and current. However, in nonlinear circuits, the current is full of harmonics and you can't use the aforementioned formula. In this case, you need to calculate the power factor using the ratio of the average power to the apparent power.

Let's start. Remove the .option commands from the schematic of previous example (Fig. 4.202).

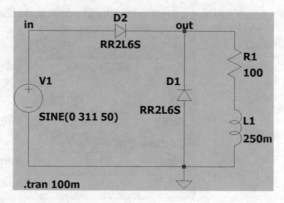

Fig. 4.202 Schematic of Example 15

Run the simulation and draw the current of diode D2 (Fig. 4.203). According to Fig. 4.204, the frequency of diode current is 50 Hz.

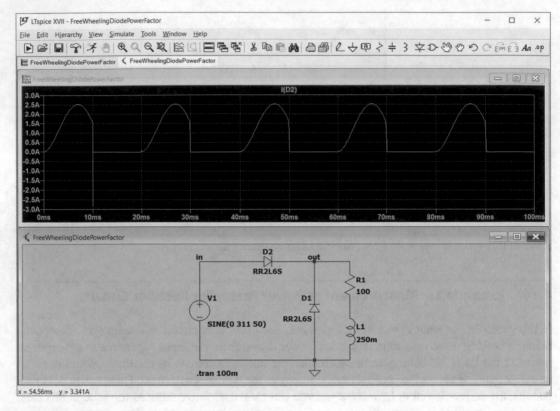

Fig. 4.203 Current through diode D2

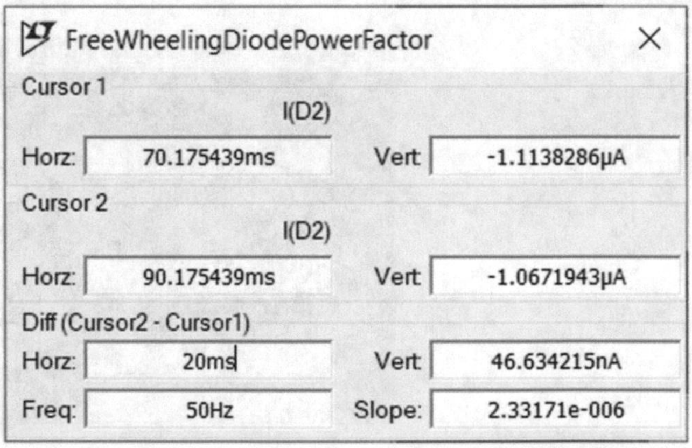

Fig. 4.204 Frequency
of waveform shown in
Fig. 4.203 is 50 Hz

Use the Zoom to rectangle icon (Fig. 4.205) and try to select one cycle of diode current (Fig. 4.206).
Select the cycle from steady-state portion of the graph.

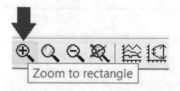

Fig. 4.205 Zoom to rectangle icon

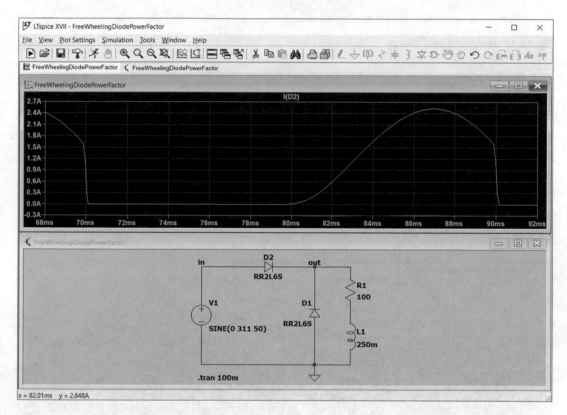

Fig. 4.206 One cycle from steady-state region of the waveform is shown

Right click on the horizontal axis (Fig. 4.207). According to Fig. 4.207, the length that we selected is 92 ms − 68 ms = 24 ms. However, according to Fig. 4.204, the length of one cycle must be 20 ms. So, enter 88 ms to the Right box and click the OK button (Fig. 4.208). After clicking the OK button, the graph changes to what is shown in Fig. 4.209.

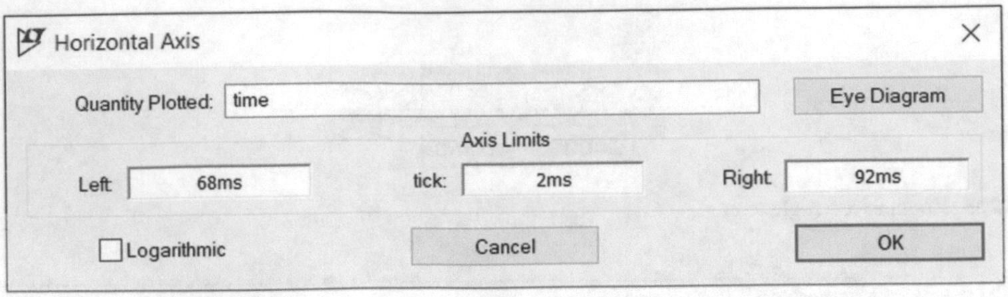

Fig. 4.207 Horizontal Axis window

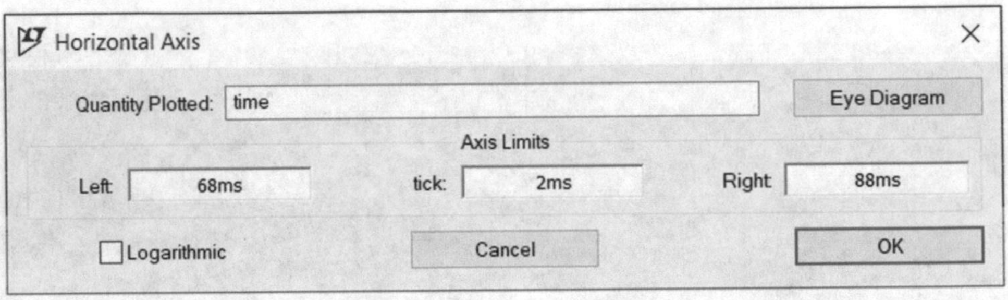

Fig. 4.208 Right box is changed to 88 ms

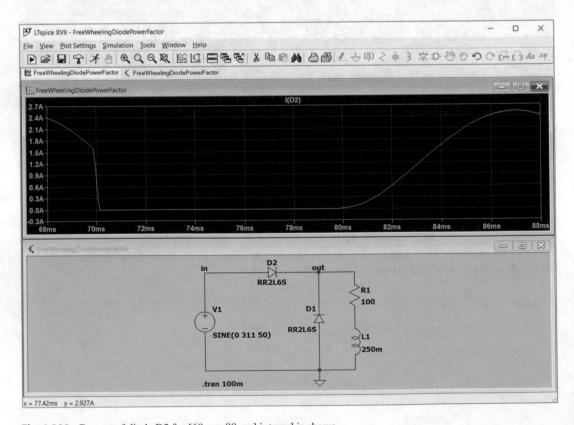

Fig. 4.209 Current of diode D2 for [68 ms, 88 ms] interval is shown

Hold down the Ctrl key and click on the I(D2) in Fig. 4.209. After clicking the I(D2), the window shown in Fig. 4.210 appears and shows the RMS of the current passed from the diode D2. The diode and the input AC source are connected in series, so, the RMS of the current drawn from the input AC source is the same as the RMS of the current drawn from the input AC source. According to Fig. 4.210, the RMS of the current is 1.2742 A. So, the apparent power is $\dfrac{311}{\sqrt{2}} \times 1.2742 = 280.2123\,\text{VA}$.

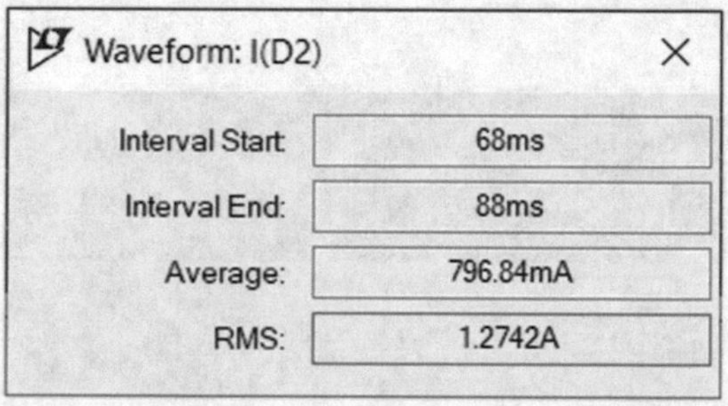

⚑ Waveform: I(D2)   ✕

Interval Start:	68ms
Interval End:	88ms
Average:	796.84mA
RMS:	1.2742A

Fig. 4.210   RMS of current waveform in Fig. 4.209 is 1.2742 A

Let's measure the average power drawn from the input AC source. Run the simulation and draw the graph of V(in)*I(V1) (Fig. 4.211). You can hold down the Alt key and click on the voltage source V1 to see its instantaneous power waveform of V1 as well.

The instantaneous power waveform of voltage source V1 is shown in Fig. 4.212. Note that V(in)*I(V1) shows the instantaneous power of the input AC source. According to Fig. 4.213, the frequency of the power waveform is 50 Hz. So, duration of one cycle of power waveform is 20 ms.

Fig. 4.211   Add Traces to Plot window

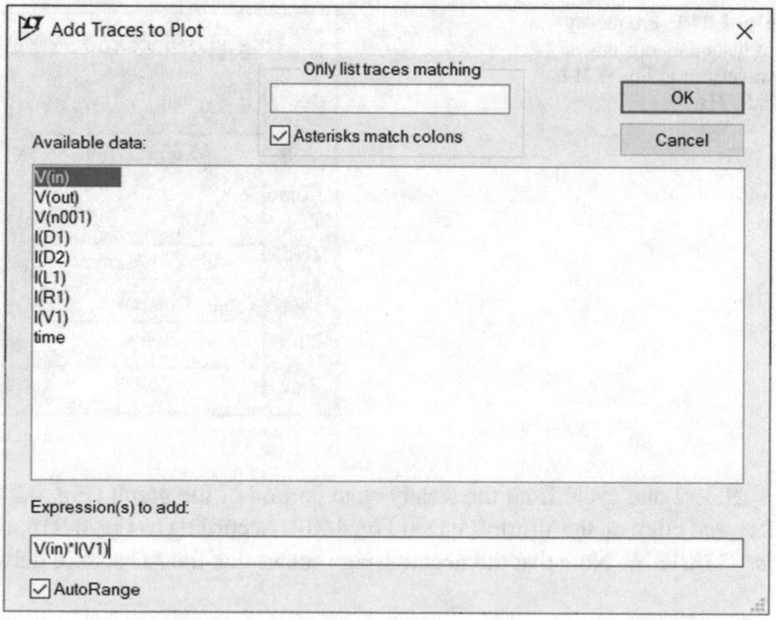

⚑ Add Traces to Plot   ✕

Only list traces matching

[   ]   [OK]

Available data: ☑ Asterisks match colons   [Cancel]

V(in)
V(out)
V(n001)
I(D1)
I(D2)
I(L1)
I(R1)
I(V1)
time

Expression(s) to add:

V(in)*I(V1)

☑ AutoRange

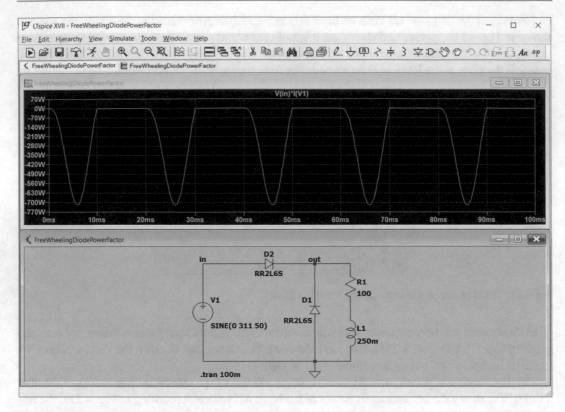

Fig. 4.212 Graph of instantaneous power drawn from the source

Fig. 4.213 Frequency of instantaneous power waveform in Fig. 4.212 is 50 Hz

Select one cycle from the steady-state portion of the graph (Fig. 4.214). Now hold down the Ctrl key and click on the V(in)*I(V1) in Fig. 4.214. According to Fig. 4.215, the average value of one cycle is −178.18 W. Note that the negative sign shows that the AC source delivered power to the circuit.

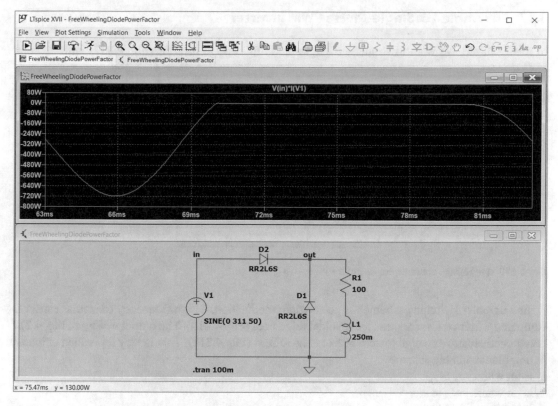

Fig. 4.214 One cycle from steady-state region is selected

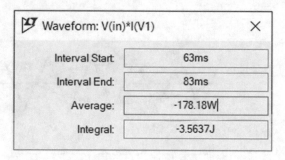

Fig. 4.215 Average power is around 178.18 W

According to the calculations shown in Fig. 4.216, the power factor of the circuit is 0.6359.

Fig. 4.216 MATLAB
command

```
Command Window
>> P=178.18;
>> Vrms=(311/sqrt(2));Irms=1.2742;S=Vrms*Irms;
>> pf=P/S

pf =

    0.6359

fx >> |
```

4.17   Example 16: Single-Phase PWM Inverter

In this example, we want to simulate a single-phase full bridge inverter (Fig. 4.217) with unipolar PWM.

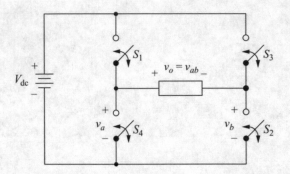

Fig. 4.217 Full-bridge converter for unipolar PWM

In a unipolar switching scheme for pulse-width modulation, a high frequency triangular carrier is compared with two low frequency sinusoidal references (Fig. 4.218). The output $v_o = v_{ab}$ in Fig. 4.217 is switched either from high to zero or from low to zero (Fig. 4.218). That is why it is called unipolar. In unipolar switching scheme,

S_1 is on when $v_{sine} > v_{tri}$.
S_2 is on when $-v_{sine} < v_{tri}$.
S_3 is on when $-v_{sine} > v_{tri}$.
S_4 is on when $v_{sine} < v_{tri}$.

Fig. 4.218 (a) Reference and carrier signals (b) bridge voltages v_a and v_b (c) Output voltage

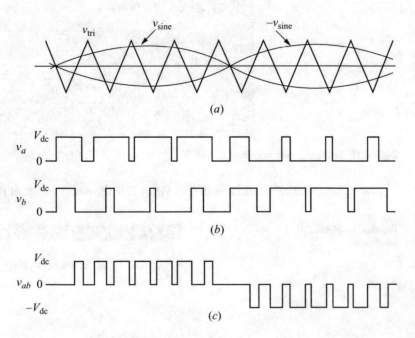

The schematic of this example is shown in Fig. 4.219. Variables "m" and "fm" determine the amplitude modulation ratio and frequency of output voltage, respectively. Amplitude modulation ratio is the ratio of amplitude of sinusoidal reference to the amplitude of triangular carrier, i.e., $m = \dfrac{V_{\text{Sinusoidal Reference}}}{V_{\text{Triangular Carrier}}}$. Frequency modulation ratio is the ratio of carrier frequency to frequency of sinusoidal reference, i.e., $m_f = \dfrac{f_{\text{Triangular Carrier}}}{f_{\text{Sinusoidal Reference}}}$.

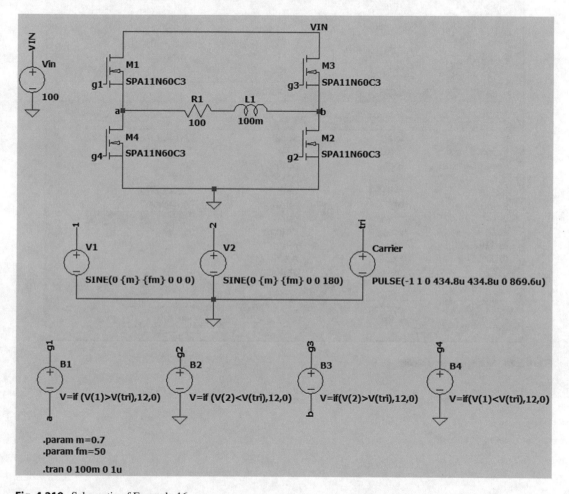

Fig. 4.219 Schematic of Example 16

Settings of V1, V2, and Carrier voltage sources (Fig. 4.220) are shown in Figs. 4.221, 4.222, and 4.223, respectively. Carrier is a triangular waveform with amplitude of 1 V. Frequency of Carrier is 1150 Hz ($m_f = 23$). Voltage source V1 and V2 play the role of reference signal and their frequency is 50 Hz.

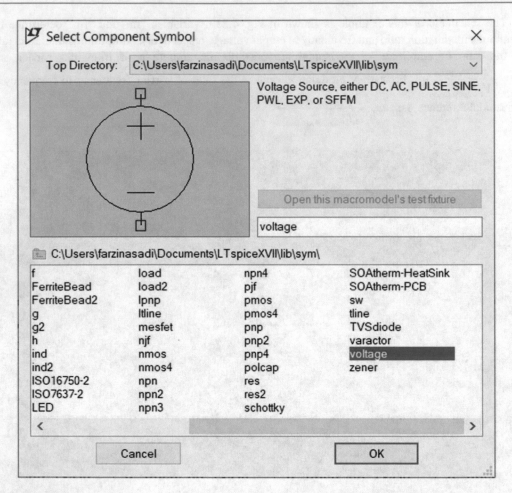

Fig. 4.220 Voltage source block

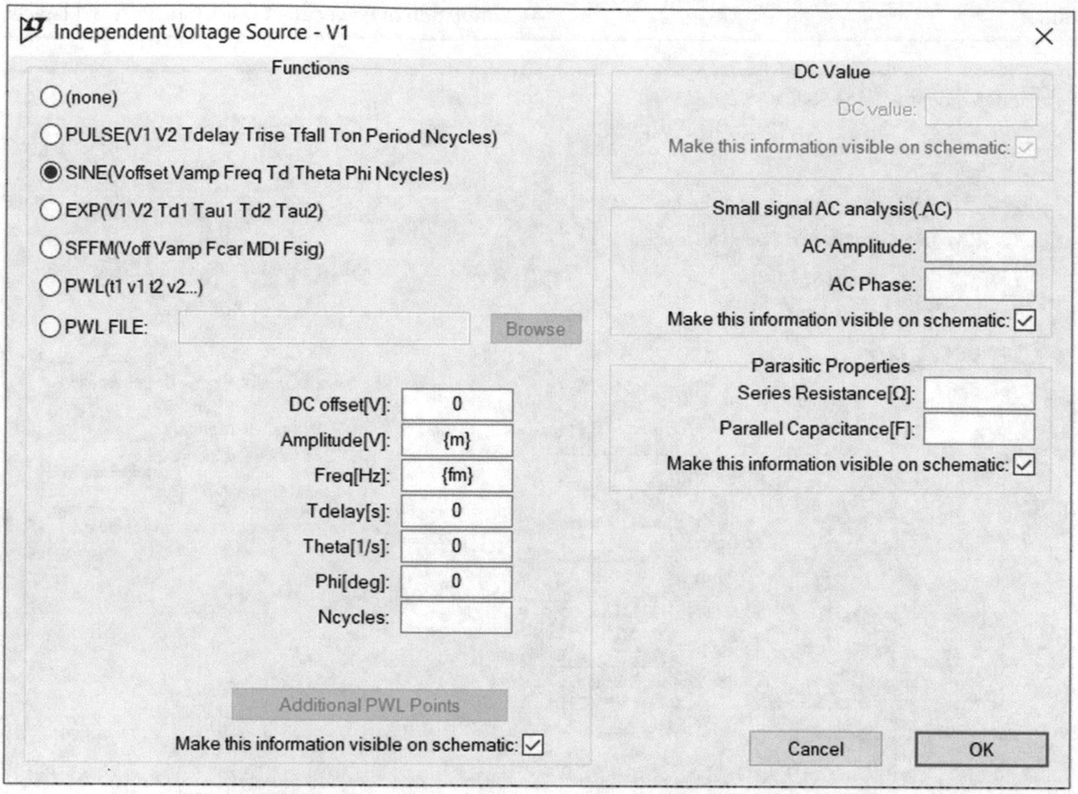

Fig. 4.221 Settings of voltage source V1

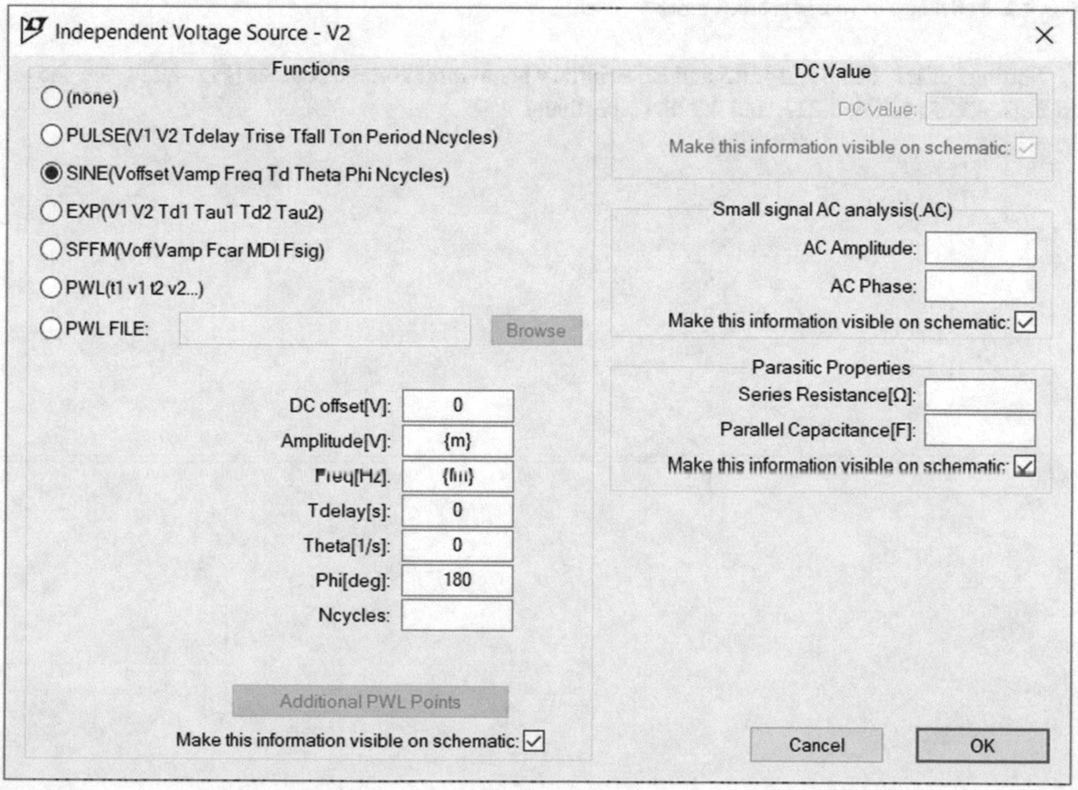

Fig. 4.222 Settings of voltage source V2

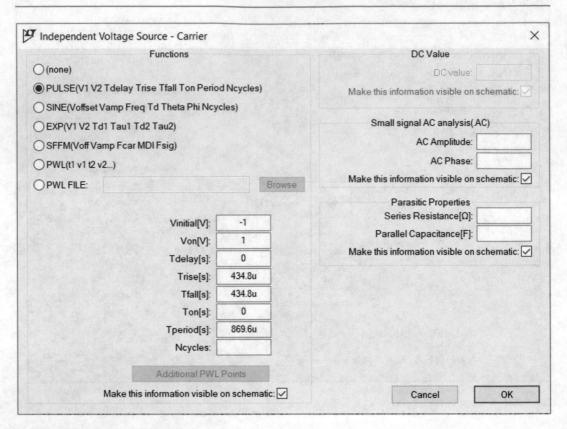

Fig. 4.223 Settings of voltage source Carrier

Settings of B1, B2, B3, B4, B5 and B6 Arbitrary behavioral voltage sources (Fig. 4.224) are shown in Figs. 4.225, 4.226, 4.227, and 4.228, respectively.

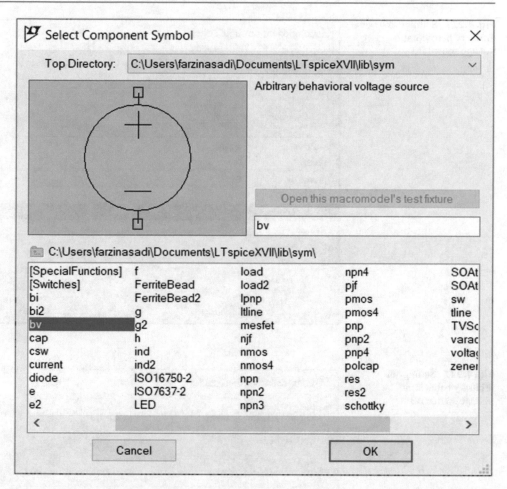

Fig. 4.224 Arbitrary behavioral voltage source block

Fig. 4.225 Settings of arbitrary behavioral voltage source B1

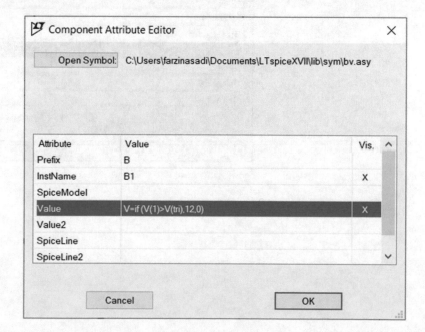

Fig. 4.226 Settings of arbitrary behavioral voltage source B2

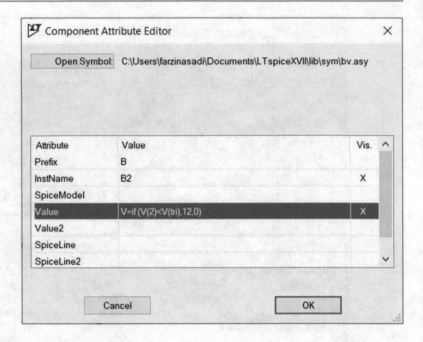

Fig. 4.227 Settings of arbitrary behavioral voltage source B3

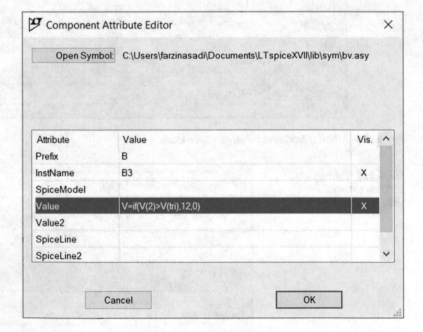

Fig. 4.228 Settings of arbitrary behavioral voltage source B4

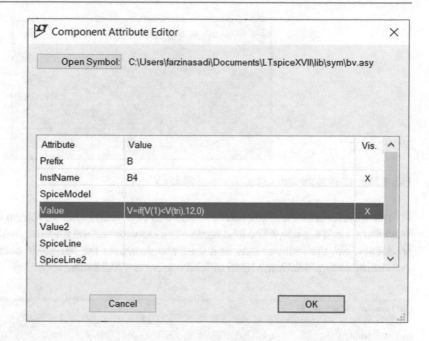

Run the simulation and draw the voltage of the load (Fig. 4.229). According to Fig. 4.230, the RMS of this waveform is 66.556 V.

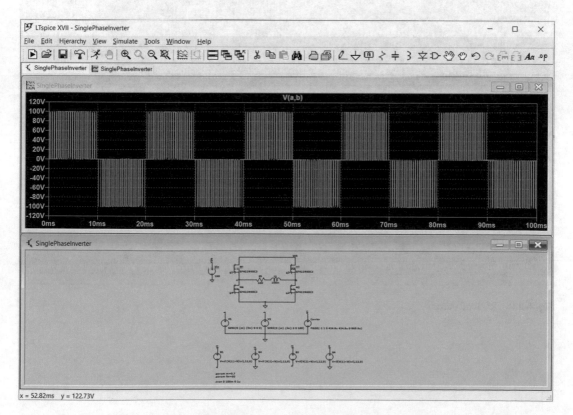

Fig. 4.229 Graph of voltage across the RL load

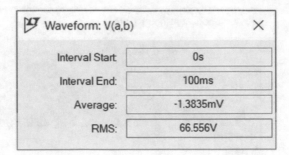

Fig. 4.230 RMS of the load voltage is around 66.556 V

The waveform of the load current is shown in Fig. 4.231. This waveform is similar to a sinusoidal waveform. Let's see why? The load contains an inductor and the impedance of inductor increases with frequency. So, the inductor acts as a filter and decreases the harmonic content of the load current. According to Fig. 4.232, the RMS of load current is 470.32 mA.

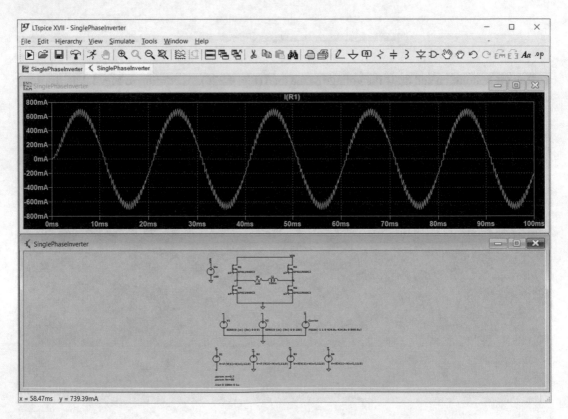

Fig. 4.231 RL load current

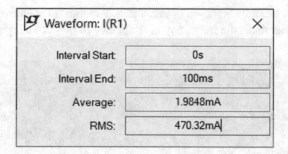

Fig. 4.232 RMS of load current is 470.32 mA

4.18 Example 17: Three-Phase PWM Inverter

In this example, we want to simulate a three-phase inverter (Fig. 4.233) which uses the Sine-PWM (SPWM) modulation technique.

Fig. 4.233 Three-phase inverter

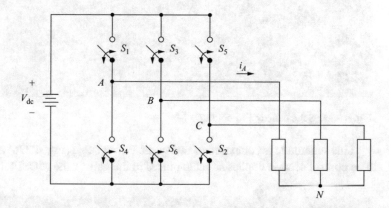

In SPWM, we need three reference signals. The three reference sinusoids are 120° apart (Fig. 4.234). Harmonics will be minimized if the carrier frequency is chosen to be an odd triple multiple of the reference frequency, that is, 3, 9, 15, 21, 27 … times the reference. In SPWM scheme,

S_1 is on when $v_a > v_{tri}$.

S_2 is on when $v_c > v_{tri}$.

S_3 is on when $v_b > v_{tri}$.

S_4 is on when $v_a < v_{tri}$.

S_5 is on when $v_c < v_{tri}$.

S_6 is on when $v_b < v_{tri}$.

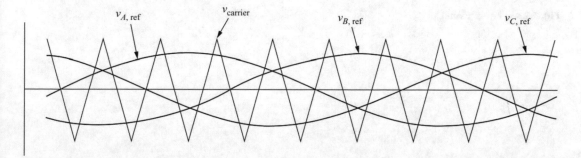

Fig. 4.234 PWM generation for three-phase inverter

The schematic of this example is shown in Fig. 4.235. Variable "m" determines the index of modulation. Variable "fm" determines the frequency of frequency of output voltage of the inverter.

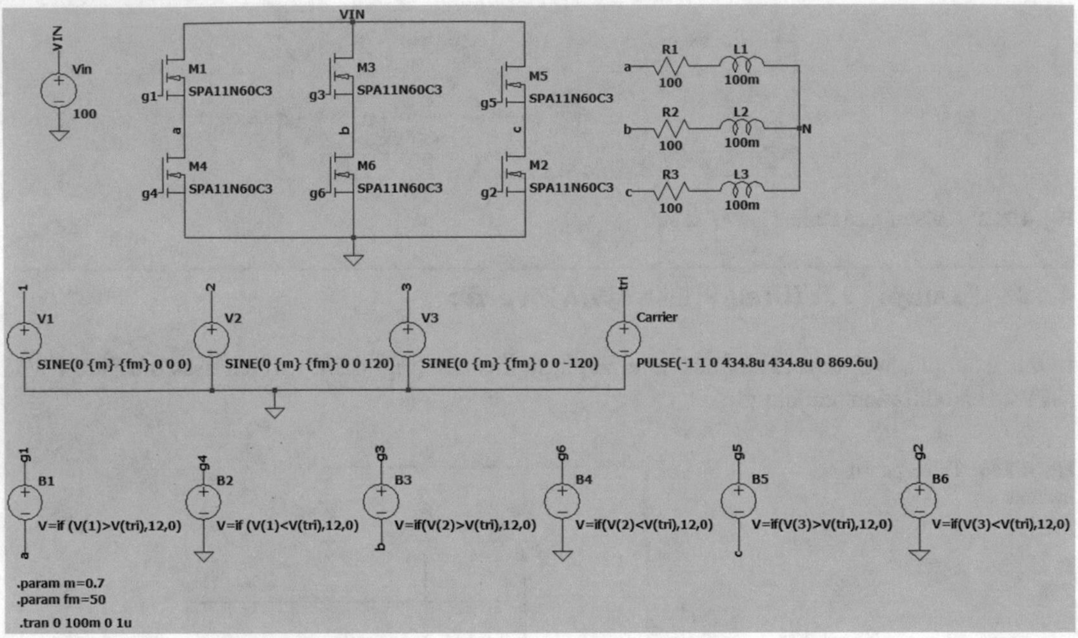

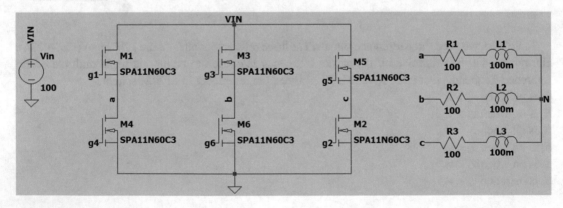

Fig. 4.235 Schematic of Example 17

This schematic is composed of two parts: Power stage (Fig. 4.236) and control section (Fig. 4.237). The control section applies suitable pulses to the gate of the MOSFETs.

Fig. 4.236 Power circuit

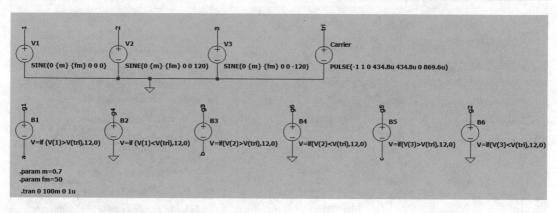

Fig. 4.237 Control circuit

Settings of V1, V2, V3, and carrier voltage sources (Fig. 4.238) are shown in Figs. 4.239, 4.240, 4.241, and 4.242.

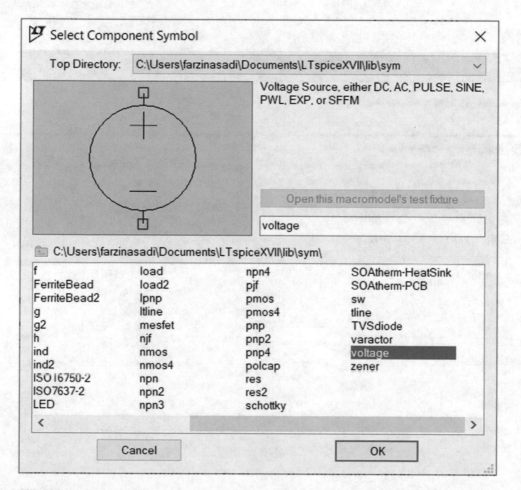

Fig. 4.238 Voltage source block

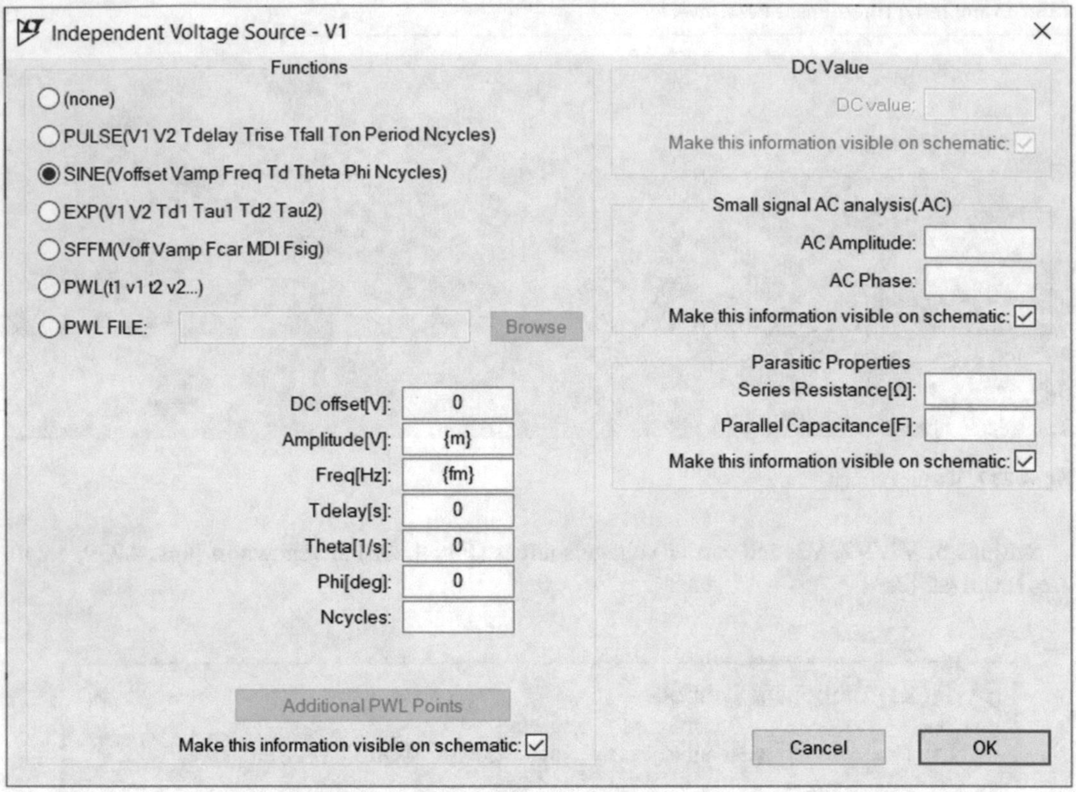

Fig. 4.239 Settings of voltage source V1

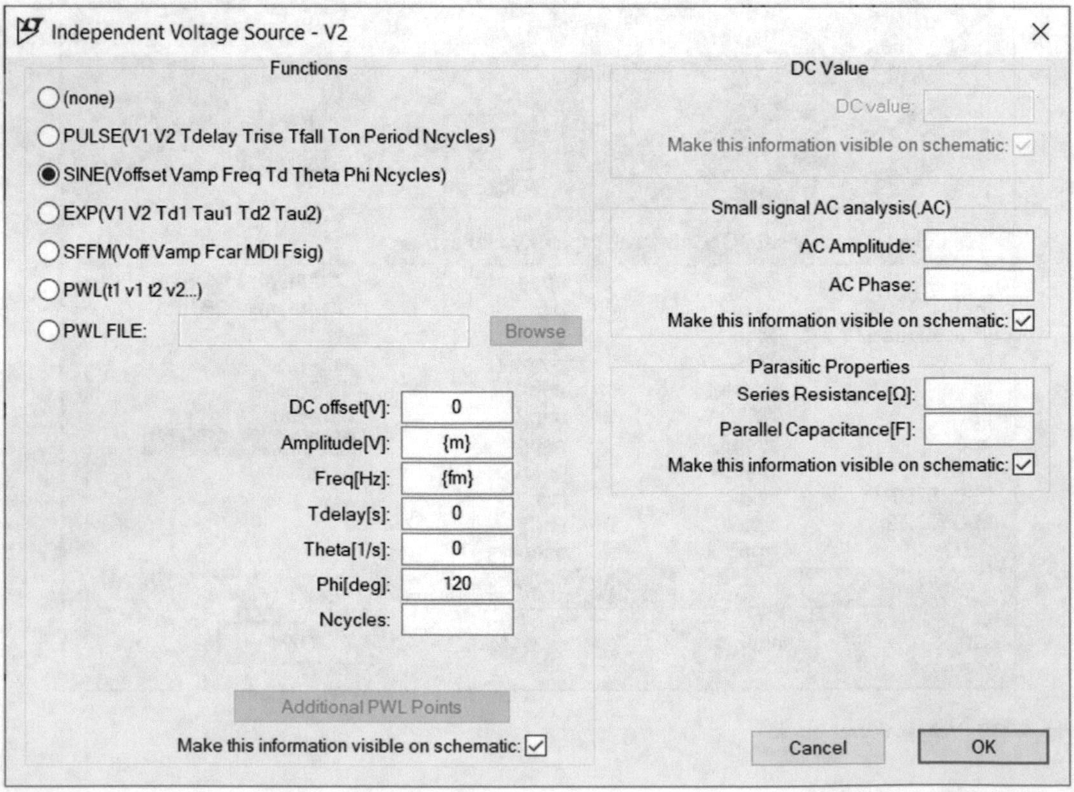

Fig. 4.240 Settings of voltage source V2

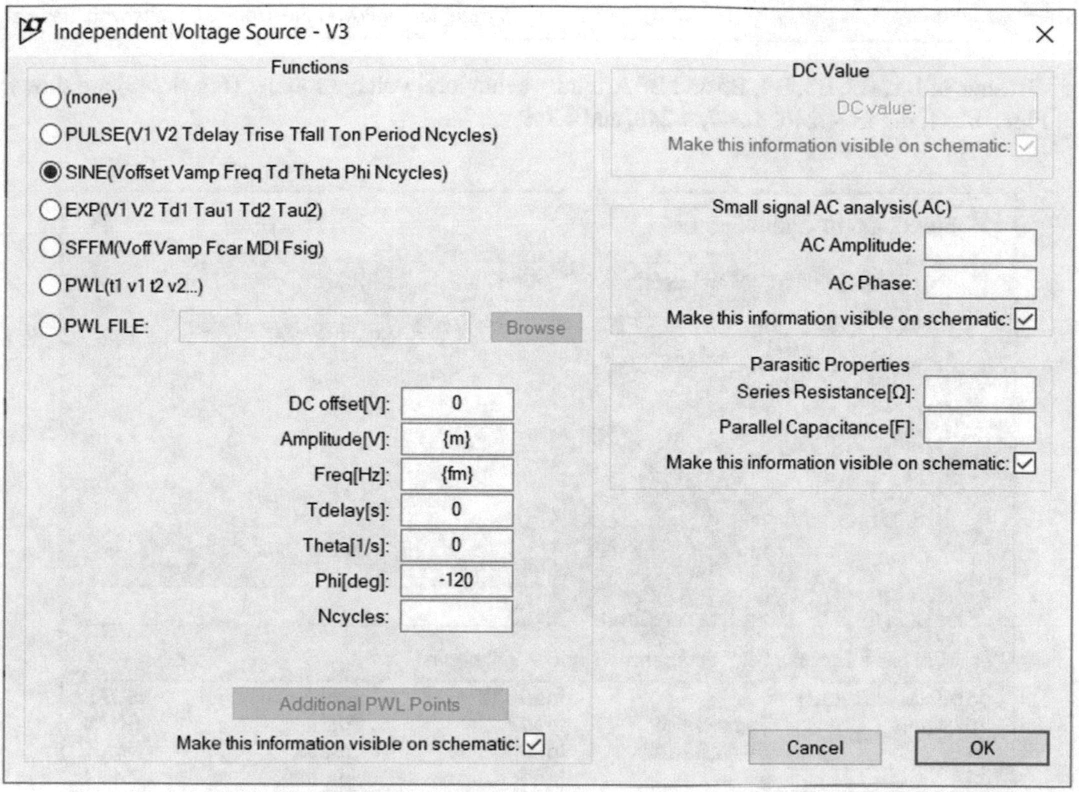

Fig. 4.241 Settings of voltage source V3

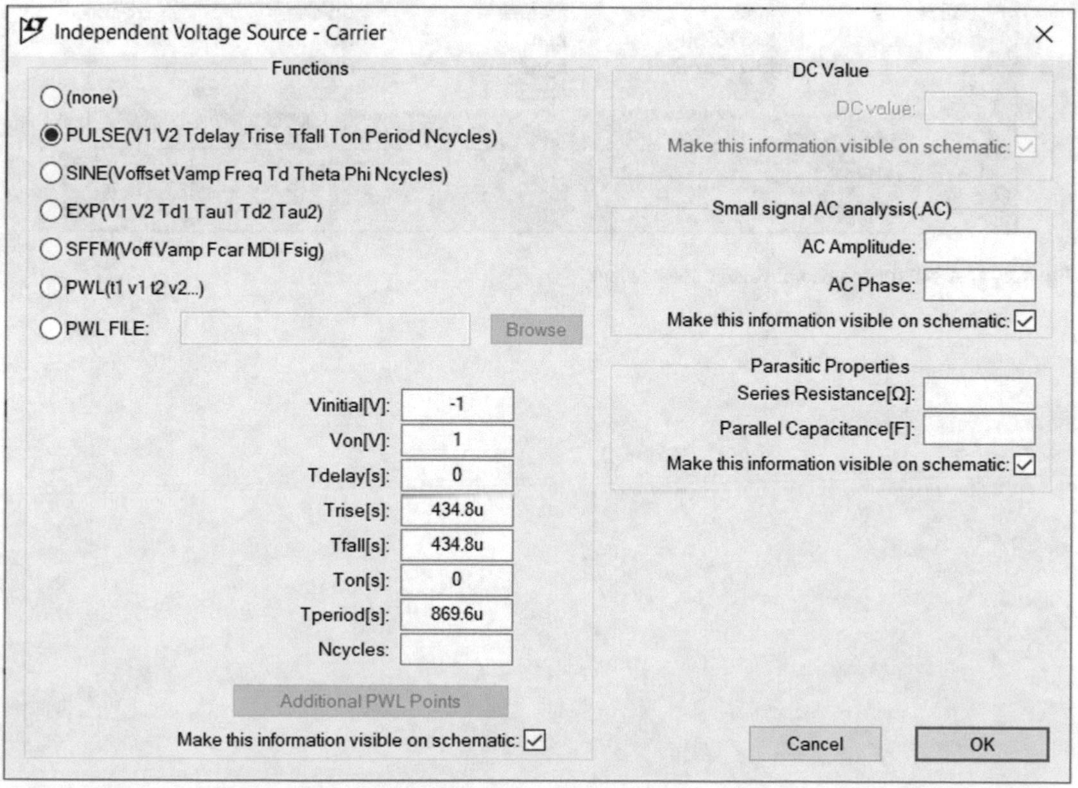

Fig. 4.242 Settings of voltage source Carrier

Settings of B1, B2, B3, B4, B5 and B6 Arbitrary behavioral voltage sources (Fig. 4.243) are shown in Figs. 4.244, 4.245, 4.246, 4.247, 4.248, and 4.249.

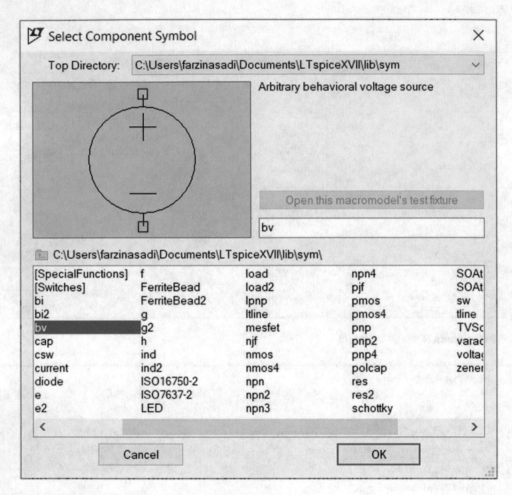

Fig. 4.243 Arbitrary behavioral voltage source block

Fig. 4.244 Arbitrary behavioral voltage source B1

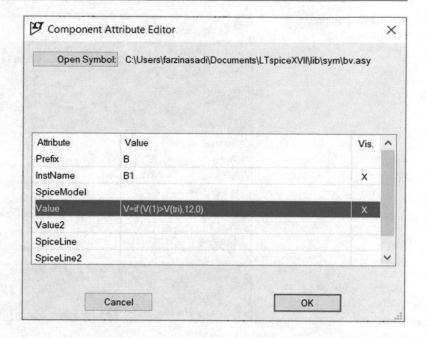

Fig. 4.245 Arbitrary behavioral voltage source B2

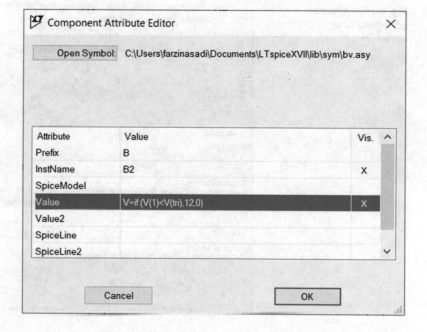

Fig. 4.246 Arbitrary
behavioral voltage
source B3

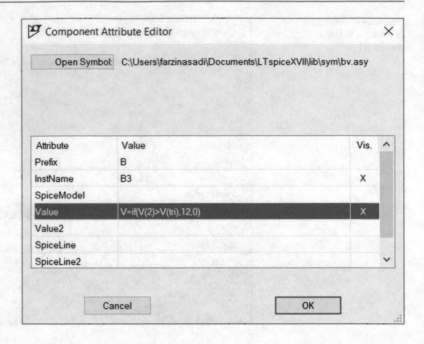

Fig. 4.247 Arbitrary
behavioral voltage
source B4

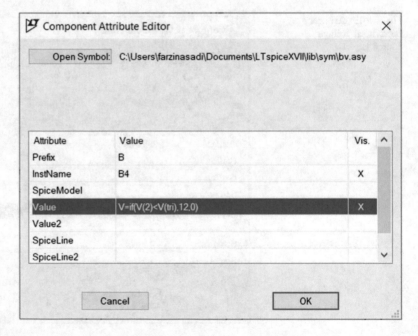

Fig. 4.248 Arbitrary behavioral voltage source B5

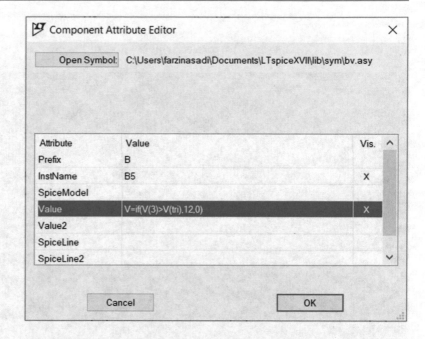

Fig. 4.249 Arbitrary behavioral voltage source B6

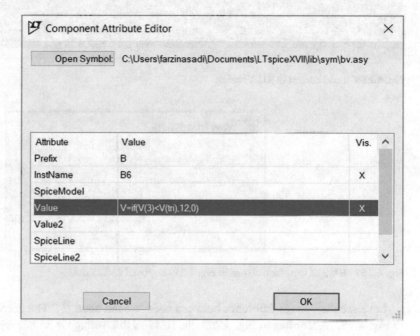

Run the simulation. The waveform of resistor R1 current is shown in Fig. 4.250. According to Fig. 4.251, the RMS of resistor R1 current is 236.25 mA.

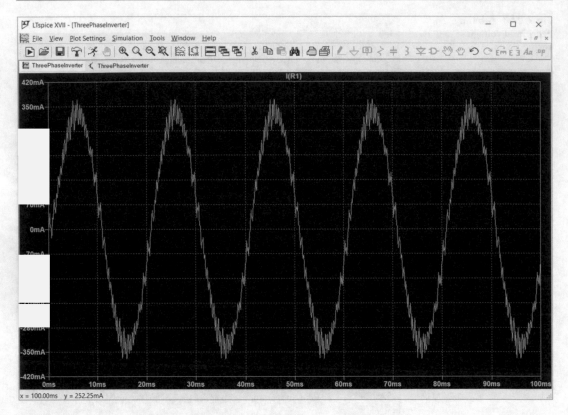

Fig. 4.250 Load current in R1L1 branch

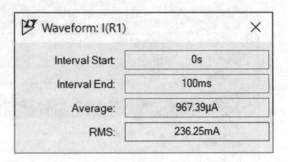

Fig. 4.251 RMS of waveform shown in Fig. 4.250 is around 236.25 mA

Let's see the voltage difference between node "a" and node "n." The graph of this voltage is shown in Fig. 4.252. According to Fig. 4.253, the RMS of this voltage is 35.821 V.

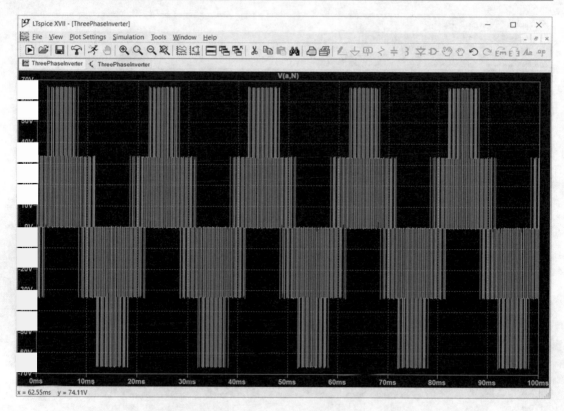

Fig. 4.252 Waveform of voltage between node a and n

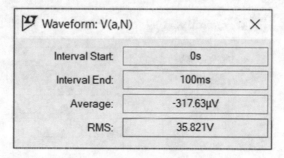

Fig. 4.253 RMS of waveform shown in Fig. 4.252 is around 35.821 V

Let's see the line-line voltage, i.e., voltage difference between node "a" and node "b." The graph of line-line voltage is shown in Fig. 4.254. According to Fig. 4.255, the RMS of line-line voltage is 62.091 V.

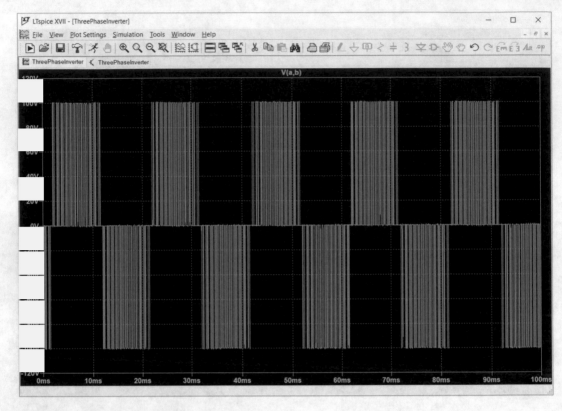

Fig. 4.254 Waveform of line-line voltage

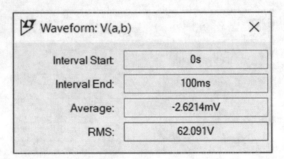

Fig. 4.255 RMS of waveform shown in Fig. 4.254 is around 62.091 V

4.19   Example 18: Harmonic Content of Three-Phase Inverter

In this example, we want to see the harmonic content of the output voltage of the inverter circuit of Example 17. Change the schematic of Example 17 to what is shown in Fig. 4.256. Added blocks are shown in Fig. 4.257. Block E1 and E2 are Voltage-dependent voltage sources (Fig. 4.258).

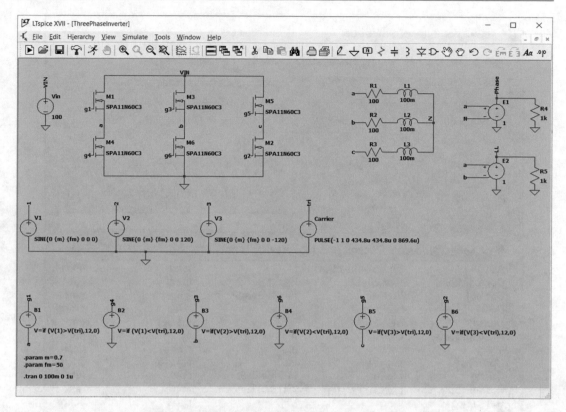

Fig. 4.256 Schematic of Example 18

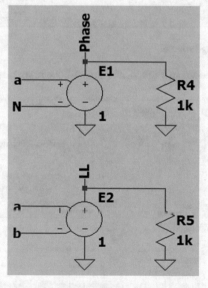

Fig. 4.257 Added blocks measure the phase and line-line voltages

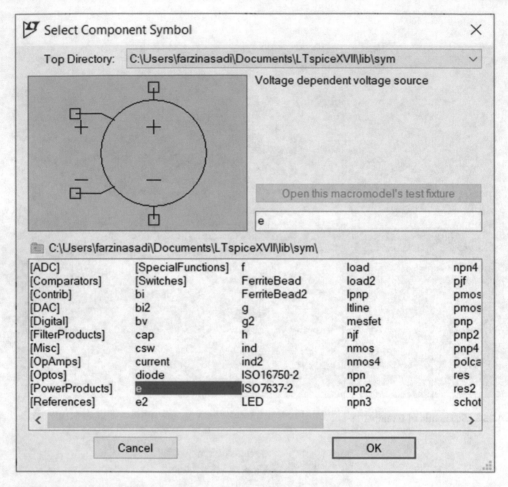

Fig. 4.258 Voltage-dependent voltage source

Settings of E1 and E2 are shown in Figs. 4.259 and 4.260, respectively.

Fig. 4.259 Settings of voltage-dependent voltage source E1

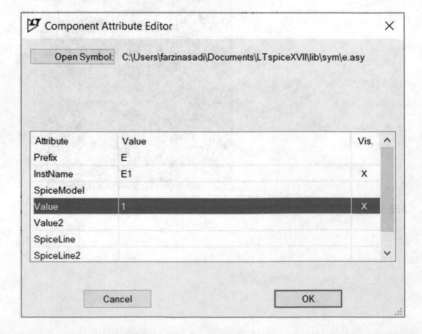

Fig. 4.260 Settings of
voltage-dependent
voltage source E2

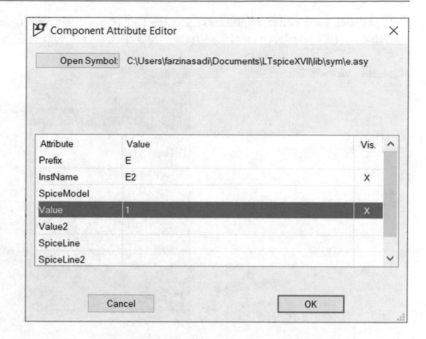

Run the simulation (Fig. 4.261).

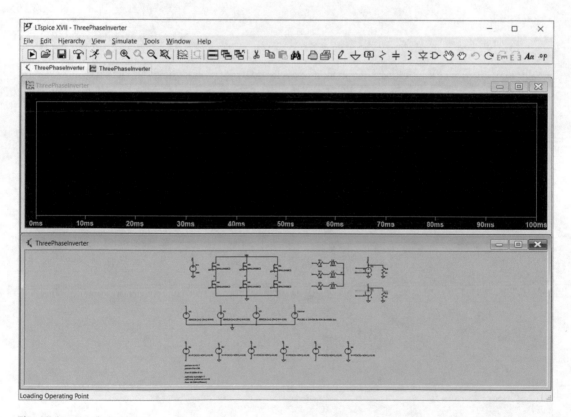

Fig. 4.261 Simulation is run

Right click on the black region and select the FFT (Fig. 4.262). This opens the Select Waveforms to include in FFT window. Select the V(phase) and click the OK button (Fig. 4.263).

Fig. 4.262 View> FFT

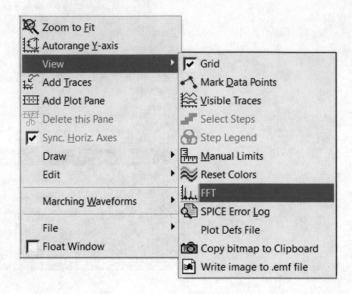

Fig. 4.263 V(phase) is
selected

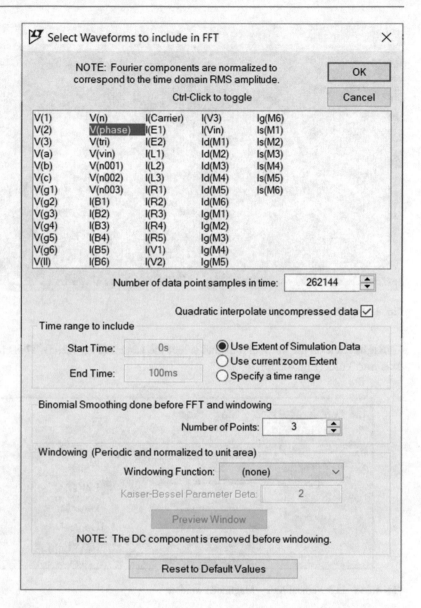

After clicking the OK button in Fig. 4.263, the result shown in Fig. 4.264 appears on the screen.

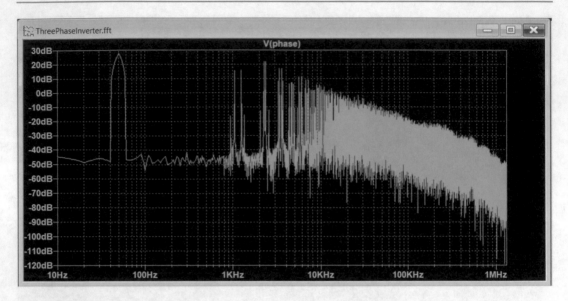

Fig. 4.264 Simulation result

Right click on the vertical axis and select the Linear (Fig. 4.265). This changes the unit of vertical axis into Volts (Fig. 4.266).

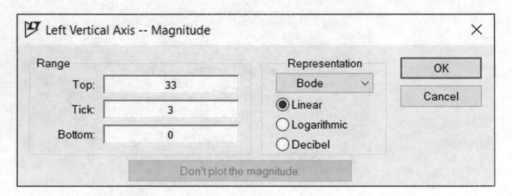

Fig. 4.265 Left Vertical Axis – Magnitude window

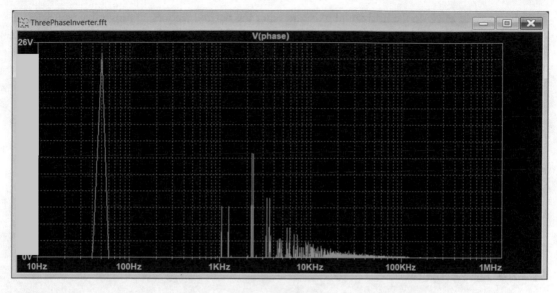

Fig. 4.266 FFT of V(phase)

You can use cursors to measure the RMS of harmonics (use the magnifier icon (Fig. 4.267) to zoom in and put the cursor at peak of harmonic that you want). According to Fig. 4.268, RMS value of the fundamental harmonic of phase voltage is 24.66 V. So, the amplitude of the fundamental harmonic of phase voltage is $24.66\sqrt{2} = 34.87\,\text{V}$.

Fig. 4.267 Magnifier icon

Fig. 4.268 RMS of
fundamental component
of phase voltage is
around 24.67 V

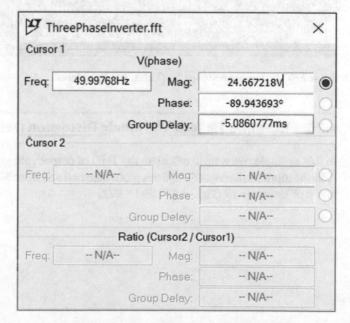

You can right click on the horizontal axis and enter the desired frequency range that you want to see. For instance, if you right click on the horizontal axis and enter 1kHz to the Left box, enter 20kHz to the Right box and click the OK button (Fig. 4.269), [1 kHz, 20 kHz] portion of the graph appears on the screen (Fig. 4.270). Use the magnifier icon shown in Fig. 4.267 to put the cursors at peak of the harmonic that you want.

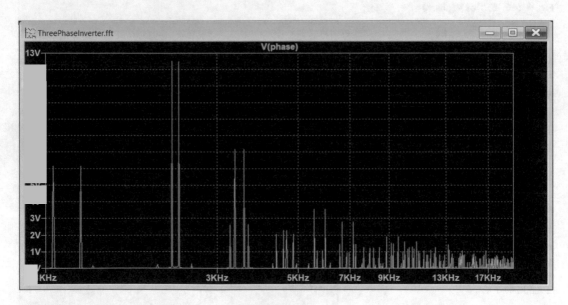

Fig. 4.269 Horizontal Axis window

Fig. 4.270 FFT of phase voltage for [1 kHz, 20 kHz] interval

4.20   Example 19: Total Harmonic Distortion (THD) of Three-Phase Inverter

In this example, we want to calculate the THD of output voltage of the inverter circuit of Example 17. Add the .options numdgt=7, .options plotwinsize=0 and .four 50 300 V(phase) commands to the schematic of Example 17 (Figs. 4.271 and 4.272).

```
.param m=0.7
.param fm=50

.tran 0 100m 0 1u

.options numdgt=7
.options plotwinsize=0
.four 50 300 V(Phase)
```

Fig. 4.271 SPICE commands used for measurement of THD

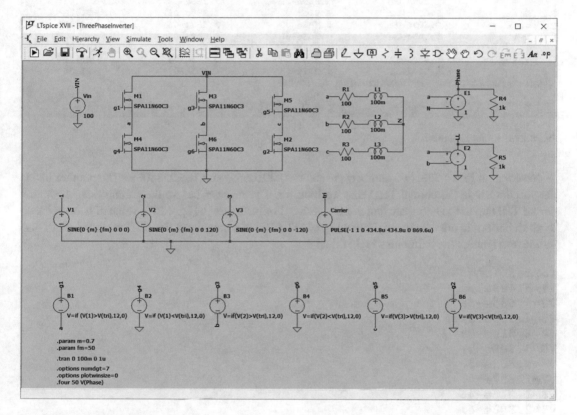

Fig. 4.272 Schematic of Example 19

Run the simulation. After simulation is done press the Ctrl+L to see the log file (Fig. 4.273). According to Fig. 4.273, the THD of V(phase) is 105.253808%.

```
SPICE Error Log: C:\Users\farzinasadi\Documents\LTspiceXVII\ThreePhaseInverter.log          ×

Heightened Def Con from 0.0739888 to 0.0739888
Heightened Def Con from 0.0922945 to 0.0922946
Heightened Def Con from 0.0966099 to 0.0966099
N-Period=1
Fourier components of V(phase)
DC component:-4.9359e-006

Harmonic            Frequency         Fourier          Normalized         Phase
 Number               [Hz]            Component         Component         [degree]
    1              5.000e+01          3.489e+01         1.000e+00           0.06°
    2              1.000e+02          6.085e-06         1.744e-07         131.57°
    3              1.500e+02          4.059e-05         1.163e-06        -129.57°
    4              2.000e+02          6.910e-05         1.980e-06         -81.36°
    5              2.500e+02          5.786e-05         1.658e-06         -83.78°
    6              3.000e+02          9.803e-05         2.809e-06        -111.55°
    7              3.500e+02          1.682e-04         4.822e-06        -100.84°
    8              4.000e+02          1.946e-04         5.576e-06         -90.69°
    9              4.500e+02          2.298e-04         6.585e-06         -95.36°
Total Harmonic Distortion: 0.001066%(105.253808%)
```

Fig. 4.273 Simulation result

Note that in Fig. 4.273, two numbers are shown 0.001066% and 105.253808%. The number inside the parenthesis is the correct THD that we look for. The number before the parenthesis, shows the partial THD up to the harmonic that you requested. For instance, in Fig. 4.273, value of harmonics up to ninth harmonic are calculated. Following MATLAB code, calculates the THD up to the ninth harmonic and ignores the harmonics higher than 9.

```
format long
V1=3.489e1;
V2=6.085e-6;
V3=4.059e-5;
V4=6.910e-5;
V5=5.786e-5;
V6=9.803e-5;
V7=1.682e-4;
V8=1.946e-4;
V9=2.298e-4;
V=[V2 V3 V4 V5 V6 V7 V8 V9];
totalHarmonicDistortin=100*sqrt(.5*V*V')/(V1/sqrt(2))
```

After running the code, the result shown in Fig. 4.274 is obtained which is the same as the number shown by LTspice in Fig. 4.273.

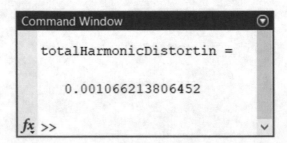

Fig. 4.274 Output of MATLAB code

Change the .four command to what is shown in Fig. 4.275. This command measures the harmonics up to 20th harmonic.

```
.param m=0.7
.param fm=50

.tran 0 100m 0 1u

.options numdgt=7
.options plotwinsize=0
.four 50 20 V(Phase)
```

Fig. 4.275 SPICE commands used for measurement of THD

Simulation result is shown in Fig. 4.276. The first number (number before the parenthesis) increased to 0.642121%.

```
SPICE Error Log: C:\Users\farzinasadi\Documents\LTspiceXVII\ThreePhaseInverter.log                 ✕

Fourier components of V(phase)
DC component:-4.9359e-006

Harmonic        Frequency       Fourier        Normalized       Phase
 Number           [Hz]          Component       Component       [degree]
    1           5.000e+01       3.489e+01       1.000e+00         0.06°
    2           1.000e+02       6.085e-06       1.744e-07       131.57°
    3           1.500e+02       4.059e-05       1.163e-06      -129.57°
    4           2.000e+02       6.910e-05       1.980e-06       -81.36°
    5           2.500e+02       5.786e-05       1.658e-06       -83.78°
    6           3.000e+02       9.803e-05       2.809e-06      -111.55°
    7           3.500e+02       1.682e-04       4.822e-06      -100.84°
    8           4.000e+02       1.946e-04       5.576e-06       -90.69°
    9           4.500e+02       2.298e-04       6.585e-06       -95.36°
   10           5.000e+02       3.235e-04       9.271e-06       -99.81°
   11           5.500e+02       4.349e-04       1.246e-05       -93.57°
   12           6.000e+02       5.042e-04       1.445e-05       -92.79°
   13           6.500e+02       6.591e-04       1.889e-05       -96.15°
   14           7.000e+02       8.729e-04       2.502e-05       -95.00°
   15           7.500e+02       1.108e-03       3.176e-05       -92.68°
   16           8.000e+02       1.437e-03       4.117e-05       -93.53°
   17           8.500e+02       1.967e-03       5.639e-05       -93.13°
   18           9.000e+02       2.870e-03       8.224e-05       -92.71°
   19           9.500e+02       2.239e-01       6.416e-03        88.55°
   20           1.000e+03       8.154e-03       2.337e-04       -91.93°
Total Harmonic Distortion: 0.642121%(105.253808%)
```

Fig 4.276 Simulation result

Following MATLAB code measures the partial THD up to 20th harmonic.

After running the code, the result shown in Fig. 4.277 is obtained which is almost the same as the number shown in Fig. 4.276.

```
format long
V1=3.489e1;
V2=6.085e-6;
V3=4.059e-5;
V4=6.910e-5;
V5=5.786e-5;
V6=9.803e-5;
V7=1.682e-4;
V8=1.946e-4;
V9=2.298e-4;
V10=3.235e-04;
V11=4.349e-04;
V12=5.042e-04;
V13=6.591e-04;
V14=8.729e-04;
V15=1.108e-03;
V16=1.437e-03;
V17=1.967e-03;
V18=2.870e-03;
V19=2.239e-01;
V20=8.154e-03;
V=[V2 V3 V4 V5 V6 V7 V8 V9 V10 V11 V12 V13 V14 V15 V16 V17
V18 V19 V20];
totalHarmonicDistortin=100*sqrt(.5*V*V')/(V1/sqrt(2))
```

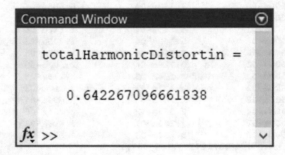

Fig. 4.277 Output of MATLAB code

The simulation results for 50, 100, and 300 harmonics are shown in Figs. 4.278, 4.279, and 4.280, respectively. As the number of harmonics increases, the first number (i.e., number outside the parenthesis) become closer to the second number (i.e., number inside the parenthesis).

SPICE Error Log: C:\Users\farzinasadi\Documents\LTspiceXVII\ThreePhaseInverter.log   ×

25	1.250e+03	8.690e+00	2.491e-01	88.53°
26	1.300e+03	1.084e-02	3.106e-04	89.13°
27	1.350e+03	2.349e-01	6.732e-03	88.57°
28	1.400e+03	5.536e-03	1.587e-04	90.46°
29	1.450e+03	4.679e-03	1.341e-04	90.86°
30	1.500e+03	4.181e-03	1.198e-04	91.63°
31	1.550e+03	3.875e-03	1.111e-04	92.44°
32	1.600e+03	3.640e-03	1.043e-04	93.91°
33	1.650e+03	3.461e-03	9.919e-05	94.58°
34	1.700e+03	3.391e-03	9.720e-05	95.66°
35	1.750e+03	3.343e-03	9.580e-05	98.27°
36	1.800e+03	3.280e-03	9.400e-05	100.70°
37	1.850e+03	3.302e-03	9.465e-05	103.26°
38	1.900e+03	3.424e-03	9.813e-05	107.98°
39	1.950e+03	9.426e-03	2.702e-04	16.64°
40	2.000e+03	3.966e-03	1.137e-04	124.14°
41	2.050e+03	3.449e-01	9.884e-03	-2.43°
42	2.100e+03	4.901e-03	1.405e-04	135.06°
43	2.150e+03	8.418e-03	2.413e-04	153.81°
44	2.200e+03	2.169e-02	6.217e-04	168.02°
45	2.250e+03	1.772e+01	5.078e-01	-2.96°
46	2.300e+03	6.552e-02	1.878e-03	0.18°
47	2.350e+03	1.768e+01	5.068e-01	177.02°
48	2.400e+03	2.163e-02	6.199e-04	166.84°
49	2.450e+03	8.666e-03	2.484e-04	149.31°
50	2.500e+03	5.496e-03	1.575e-04	127.80°

Total Harmonic Distortion: 79.930030%(105.253808%)

Fig. 4.278 Simulation result (up to 50th harmonic)

SPICE Error Log: C:\Users\farzinasadi\Documents\LTspiceXVII\ThreePhaseInverter.log   ×

76	3.800e+03	1.368e-02	3.920e-04	-93.46°
77	3.850e+03	3.279e-02	9.397e-04	-93.90°
78	3.900e+03	1.016e-02	2.913e-04	-92.15°
79	3.950e+03	9.792e-03	2.806e-04	-91.70°
80	4.000e+03	8.325e-03	2.386e-04	-89.38°
81	4.050e+03	7.570e-03	2.169e-04	-98.62°
82	4.100e+03	7.216e-03	2.068e-04	-83.66°
83	4.150e+03	7.007e-03	2.008e-04	-77.65°
84	4.200e+03	7.299e-03	2.092e-04	-65.15°
85	4.250e+03	4.160e-01	1.192e-02	174.86°
86	4.300e+03	9.220e-03	2.642e-04	-44.40°
87	4.350e+03	2.890e+00	8.281e-02	174.15°
88	4.400e+03	1.506e-02	4.317e-04	-165.39°
89	4.450e+03	1.153e-02	3.303e-04	-159.22°
90	4.500e+03	1.453e-02	4.165e-04	-165.55°
91	4.550e+03	3.220e+00	9.228e-02	-6.05°
92	4.600e+03	1.947e-02	5.580e-04	-20.50°
93	4.650e+03	3.218e+00	9.223e-02	174.13°
94	4.700e+03	1.443e-02	4.134e-04	-166.48°
95	4.750e+03	1.139e-02	3.264e-04	-161.03°
96	4.800e+03	1.508e-02	4.322e-04	-167.50°
97	4.850e+03	2.883e+00	8.263e-02	-6.06°
98	4.900e+03	8.609e-03	2.467e-04	-40.97°
99	4.950e+03	4.233e-01	1.213e-02	-6.64°
100	5.000e+03	6.207e-03	1.779e-04	-62.06°

Total Harmonic Distortion: 92.818377%(105.253808%)

Fig. 4.279 Simulation result (up to 100th harmonic)

SPICE Error Log: C:\Users\farzinasadi\Documents\LTspiceXVII\ThreePhaseInverter.log   ×

275	1.375e+04	1.451e-01	4.160e-03	-19.22°
276	1.380e+04	3.736e-03	1.071e-04	-141.23°
277	1.385e+04	1.544e-01	4.425e-03	162.30°
278	1.390e+04	7.831e-03	2.244e-04	-172.58°
279	1.395e+04	1.638e-02	4.695e-04	97.39°
280	1.400e+04	1.191e-02	3.414e-04	-179.66°
281	1.405e+04	8.590e-01	2.462e-02	-18.23°
282	1.410e+04	1.776e-02	5.089e-04	-45.26°
283	1.415e+04	1.298e+00	3.721e-02	144.98°
284	1.420e+04	2.180e-02	6.247e-04	179.27°
285	1.425e+04	1.008e+00	2.890e-02	71.64°
286	1.430e+04	2.969e-02	8.510e-04	131.11°
287	1.435e+04	1.520e+00	4.356e-02	-17.54°
288	1.440e+04	7.753e-03	2.222e-04	62.76°
289	1.445e+04	1.104e+00	3.165e-02	-9.02°
290	1.450e+04	2.404e-02	6.891e-04	34.53°
291	1.455e+04	1.124e+00	3.222e-02	-109.04°
292	1.460e+04	1.007e-02	2.887e-04	-89.23°
293	1.465e+04	9.668e-02	2.771e-03	-19.72°
294	1.470e+04	3.021e-03	8.659e-05	-35.54°
295	1.475e+04	8.377e-02	2.401e-03	58.33°
296	1.480e+04	6.937e-03	1.988e-04	61.17°
297	1.485e+04	4.941e-01	1.416e-02	-109.29°
298	1.490e+04	4.130e-03	1.184e-04	-111.88°
299	1.495e+04	2.593e-03	7.432e-05	-23.40°
300	1.500e+04	4.276e-03	1.226e-04	89.59°

Total Harmonic Distortion: 101.099454%(105.253808%)

Fig. 4.280 Simulation result (up to 300th harmonic)

The THD of line-line voltage can be calculated in the same way. The commands shown in Fig. 4.281 measure the THD of line-line voltage. According to Fig. 4.282, THD of line-line voltage is 105.401496%.

```
.param m=0.7
.param fm=50

.tran 0 100m 0 1u

.options numdgt=7
.options plotwinsize=0
.four 50 V(LL)
```

Fig. 4.281 SPICE commands used for measurement of THD

SPICE Error Log: C:\Users\farzinasadi\Documents\LTspiceXVII\ThreePhaseInverter.log   ×

```
Heightened Def Con from 0.0739888 to 0.0739888
Heightened Def Con from 0.0922945 to 0.0922946
Heightened Def Con from 0.0966099 to 0.0966099
N-Period=1
Fourier components of V(11)
DC component:-0.00120367
```

Harmonic Number	Frequency [Hz]	Fourier Component	Normalized Component	Phase [degree]
1	5.000e+01	6.044e+01	1.000e+00	-29.95°
2	1.000e+02	2.469e-03	4.085e-05	-90.79°
3	1.500e+02	2.452e-03	4.057e-05	-90.93°
4	2.000e+02	2.518e-03	4.166e-05	-90.91°
5	2.500e+02	2.578e-03	4.265e-05	-91.71°
6	3.000e+02	2.620e-03	4.335e-05	-92.16°
7	3.500e+02	2.703e-03	4.472e-05	-92.50°
8	4.000e+02	2.813e-03	4.654e-05	-93.03°
9	4.500e+02	2.891e-03	4.784e-05	-93.34°

```
Total Harmonic Distortion: 0.012330%(105.401496%)
```

Fig. 4.282 Simulation result

You can measure the THD of load current as well. In Fig. 4.283, voltage source V4 (with value of 0 V) is added in series to R1. This voltage source acts as a current sensor.

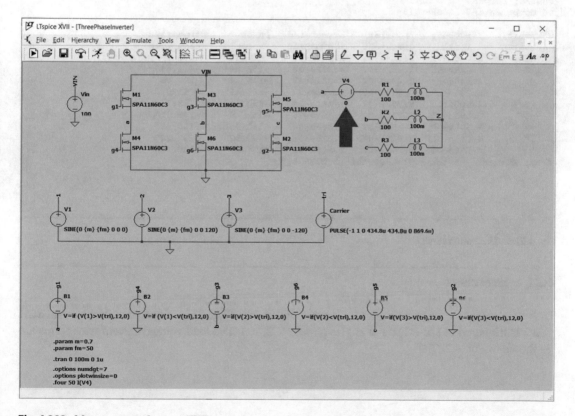

Fig. 4.283 Measurement of current THD

The commands shown in Fig. 4.284 measures the THD of the current that pass through V4. According to the result shown in Fig. 4.285, the THD of load current is 7.675878%, which is quite smaller than the THD of voltage. The inductive nature of the load acts as a filter and decrease the amplitude of higher harmonics. Reduction of higher harmonics from the load current decreases the THD of current. If you remove the inductor (i.e., purely resistive load), the THD of current and THD of voltage become equal.

```
.param m=0.7
.param fm=50

.tran 0 100m 0 1u

.options numdgt=7
.options plotwinsize=0
.four 50 I(V4)
```

Fig. 4.284 Commands for measurement of current THD

SPICE Error Log: C:\Users\farzinasadi\Documents\LTspiceXVII\ThreePhaseInverter.log   ×

```
Heightened Def Con from 0.0992417 to 0.0992417
N-Period=1
Fourier components of I(v4)
DC component:-6.33169e-008

Harmonic    Frequency    Fourier      Normalized    Phase       Normalized
Number      [Hz]         Component    Component     [degree]    Phase [deg]
   1        5.000e+01    3.329e-01    1.000e+00    -17.39°        0.00°
   2        1.000e+02    1.442e-07    4.331e-07   -132.06°     -114.67°
   3        1.500e+02    2.674e-07    8.033e-07   -145.24°     -127.86°
   4        2.000e+02    2.929e-07    8.798e-07   -158.46°     -141.07°
   5        2.500e+02    3.799e-07    1.141e-06   -150.55°     -133.17°
   6        3.000e+02    5.327e-07    1.600e-06   -162.60°     -145.21°
   7        3.500e+02    6.483e-07    1.948e-06   -167.45°     -150.07°
   8        4.000e+02    7.689e-07    2.310e-06   -169.72°     -152.34°
   9        4.500e+02    8.573e-07    2.575e-06   -166.84°     -149.46°
Total Harmonic Distortion: 0.000461%(7.675878%)
```

Fig. 4.285 Simulation result

4.21   Exercises

1. A single-phase voltage controller is shown in Fig. 4.286. SCR S1 is triggered at $2k\pi + \alpha$ and SCR S2 is triggered at $(2k + 1)\pi + \alpha$ angles ($k = 0, 1, 2, \ldots$). The load voltage/current waveforms are shown in Fig. 4.287.

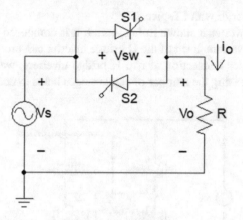

Fig. 4.286 Schematic of Exercise 1

Fig. 4.287 Waveforms
of circuit in Fig. 4.286

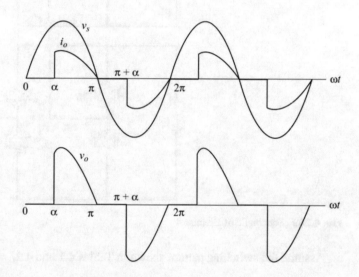

Use LTspice to simulate the circuit for $\alpha = 30°$, $60°$ and $90°$.

2. Simulate the boost converter shown in Fig. 4.288 with LTspice and measure the output voltage, output voltage ripple, output power and efficiency. Signal S1 has frequency of 200 kHz and duty cycle of 0.5. Voltage drop of diode is assumed to be 0.7 V.

Fig. 4.288 Schematic
of Exercise 2

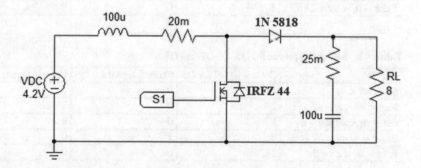

3. Simulate a fly back converter with LTspice.
4. A cascaded multilevel inverter is shown in Fig. 4.289. It is composed of series connection of two single-phase H bridge inverters. Each of the H bridge inverter can produce output voltage of $-V_{dc}$, 0 and $+V_{dc}$. So, with series connection of two H bridge inverters, we can produce $-2V_{dc}$, $-V_{dc}$, 0, $+V_{dc}$, and $+2V_{dc}$. Increasing the number of series stages helps to decrease the output THD.

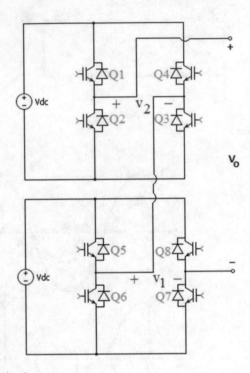

Fig. 4.289 Schematic of Exercise 4

Assume the switching pattern shown in Tables 4.1 and 4.2.

Table 4.1 Switching pattern for Q1, Q2, Q3 and Q4

Interval	Control signal for Q1	Control signal for Q2	Control signal for Q3	Control signal for Q4
A:$0 < \omega t < \alpha_2$	1	0	0	1
B:$\alpha_2 < \omega t < \pi - \alpha_2$	1	0	1	0
C:$\pi - \alpha_2 < \omega t < \pi + \alpha_2$	1	0	0	1
D:$\pi + \alpha_2 < \omega t < 2\pi - \alpha_2$	0	1	0	1
E:$2\pi - \alpha_2 < \omega t < 2\pi$	1	0	0	1

Table 4.2 Switching pattern for Q5, Q6 Q7 and Q8

Interval	Control signal for Q5	Control signal for Q6	Control signal for Q7	Control signal for Q8
a:$0 < \omega t < \alpha_1$	1	0	0	1
b:$\alpha_1 < \omega t < \pi - \alpha_1$	1	0	1	0
c:$\pi - \alpha_1 < \omega t < \pi + \alpha_1$	1	0	0	1
d:$\pi + \alpha_1 < \omega t < 2\pi - \alpha_1$	0	1	0	1
e:$2\pi - \alpha_1 < \omega t < 2\pi$	1	0	0	1

The switching patterns given in Tables 4.1 and 4.2 produce the voltage waveforms shown in Fig. 4.290.

Fig. 4.290 Waveforms generated with switching pattern shown in Tables 4.1 and 4.2

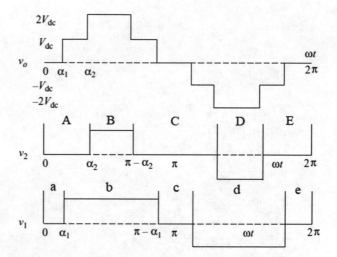

Use LTspice to simulate the circuit for $\alpha_1 = 20°$ and $\alpha_1 = 40°$ and measure the output THD.

References

1. Hart, D.W.: Power Electronics, McGraw-Hill (2010)
2. Mohan, N., Undeland, T.M., Robbins, W.P.: Power Electronics: Converters, Applications, and Design, 3rd edition, John Wiley and Sons (2007)
3. Rashid, M. H.: Power Electronics: Devices, Circuits and Applications, Pearson (2013)
4. Erikson, R.W., Maksimovic, D.: Fundamentals of Power Electronics, 3rd edition, Springer (2020)

Index

A
Amplitude modulation ratio, 505
Average value, 82

B
Band pass filter, 358
Bandwidth of the filter, 360
Boost converter, 541
Buck converter, 391

C
Center tap transformer, 188
Change the color of graph, 38
CMRR of differential pair, 323
Colpitts oscillator, 340
Common emitter amplifier, 219
Common Mode Rejection Ratio (CMRR), 317
Component icon, 12
Continuous Conduction Mode (CCM), 425
Counter block, 382
Coupled inductor, 145
Coupling coefficient, 145
Current controlled switche block, 69
Current dependent current source, 116
Current dependent voltage source, 124
Current source block, 12
Curve Fitting Toolbox, 171
Custom transistor, 268

D
DC operating point analysis, 52
DC transfer function analysis, 55
Delta connected three phase source, 129
D flip flop, 377
Diac, 446
Dimmer circuit, 446
Discontinuous Conduction Mode (DCM), 425

E
Early voltage, 277
Efficiency, 427

F
Filter block, 352
Fourier analysis, 239
Fourier series of half wave rectified waveform, 496
Freewheeling diode, 490
Frequency modulation ratio, 505
Frequency response of electric circuits, 205
Frequency response of the common emitter amplifier, 250
Full wave rectifier, 191

G
Ground icon, 19

H
Harmonic content of a single phase half wave rectifier, 490
Harmonic content of three phase inverter, 524
Harmonic contents of the common emitter amplifier, 231
High pass filter, 357

I
Impulse response, 163
Initial conditions, 60
Input impedance of differential pair, 334
Input impedance of electric circuits, 195
Installation of LTspice, 1
I-V characteristic of diode, 211
I-V characteristics of zener diode, 215

L
Label Net icon, 25
List of shortcut keys, 11

© The Editor(s) (if applicable) and The Author(s), under exclusive license to Springer Nature
Switzerland AG 2023
F. Asadi, *Essential Circuit Analysis using LTspice®*, https://doi.org/10.1007/978-3-031-09853-6

LM 741, 299
Logic circuit, 365
Low pass filter, 353, 354
LTspice help, 3
LTspice prefixes, 23

M
Maximum power transfer, 137
Midband gain, 255
MOSFET transistor, 220

N
NE 555, 347
New Schematic icon, 8

O
Op amp, 286
Op-amp clamp circuit, 361
Optocoupler, 344
Output impedance of the common emitter amplifier, 263

P
Phase difference, 78
Power factor of linear circuits, 112
Power factor of rectifier circuit, 497
Pulse width modulator, 399
PWM block, 409

Q
Quality factor (Q), 360

R
Resistor icon, 9
RMS, 82
RS flip flop, 377

S
Schmitt-Triggered buffer block, 374

Search icon, 28
Single phase full wave thyristor rectifier, 467
Single phase half wave thyristor
          rectifier, 453
Single phase inverter, 504
Single phase voltage controller, 540
SPICE Directive icon, 49
SPICE Error Log, 90
Stability of non-inverting op amp
          amplifier, 294
State diagram, 389
Step response, 154

T
Test fixture, 432
Text icon, 29
THD of three phase inverter, 532
Thevenin equivalent circuit, 116
Three phase circuit, 129
Three phase controlled rectifier, 472
Three phase inverter, 513
Total Harmonic Distortion (THD), 239
Transformer, 177
Triac, 446
Turn ratio, 178
Two-bit binary counter, 386

U
Unipolar PWM, 504

V
Voltage controlled switch block, 69
Voltage dependent current source, 127
Voltage dependent voltage source, 116
Voltage divider circuit, 7
Voltage regulation, 444
Voltage regulator, 440
Voltage source block, 12

Y
Y connected three phase source, 133

Printed in the United States
by Baker & Taylor Publisher Services